P9-DMV-928

PHYSICS, PHILOSOPHY, AND THEOLOGY: A COMMON QUEST FOR UNDERSTANDING

EDITED BY

ROBERT J. RUSSELL, WILLIAM R. STOEGER, S.J.,
AND GEORGE V. COYNE, S.J.

2000

VATICAN OBSERVATORY – VATICAN CITY STATE

SECOND EDITION

Distributed (except in Italy and Vatican City State) by:

UNIVERSITY OF NOTRE DAME PRESS
Notre Dame, Indiana 46556
USA

Distributed in Italy and Vatican City State by:

LIBRERIA EDITRICE VATICANA
V–00120 CITTÀ DEL VATICANO
VATICAN CITY STATE

ISBN 0-268-01576-7
ISBN 0-268-01577-5 (pbk.)

CONTENTS

Preface by GEORGE V. COYNE, S.J. 11
Message of His Holiness Pope John Paul II 15
List of Participants 17

I. HISTORICAL AND CONTEMPORARY RELATIONS IN SCIENCE AND RELIGION

I.G. BARBOUR: Ways of Relating Science and Theology

Introduction 21
1. Conflict 21
1.1 Scientific Materialism 22
1.2 Biblical Literalism 25
2. Independence 27
2.1 Contrasting Methods 28
2.2 Differing Languages 30
3. Dialogue 33
3.1 Boundary Questions 33
3.2 Methodological Parallels 37
4. Integration 40
4.1 Doctrinal Reformulation 40
4.2 Systematic Synthesis 42

E. MCMULLIN: Natural Science and Belief in a Creator: Historical Notes

1. Introduction 49
2. Jerusalem 49
3. Athens 53
4. Augustine 55
5. How Could a Christian Be an Aristotclian? 59
6. The Rise and Fall of Physico-Theology 63
7. Contemporary Natural Science and Belief in a Creator . . . 68
8. The Ways Divide 71

M.J. BUCKLEY, S.J.: The Newtonian Settlement and the Origins of Atheism

1. Introduction 81
2. Posing the Issues: Mechanics, Mathematics, and Theology . 82
3. From Mechanics to the First Cause 84
4. Mechanics as a Foundation for Theology 87

5. The Historical Context of the Newtonian Settlement 91
6. Beginnings of the Divorce: Physics and Theology 94
7. The Beginnings of Atheism 96
8. The Consequences for Religion 97

W.N. Clarke, S.J.: Is a Natural Theology Still Possible Today?

1. Introduction 103
2. Relation to Science 103
3. Philosophical Obstacles 105
4. Clearing the Obstacles 106
5. Constructing a Natural Theology 111
6. Postscripts on Deriving the Divine Attributes 120

O. Pedersen: Christian Belief and the Fascination of Science

1. Science Itself 125
2. Scientific Statements 126
3. The Fascination of Science. 128
4. The Role of Mind 130
5. The Fundamental Role of Mathematics 131
6. The Fundamental Scientific Experience 132
7. Finding and Giving. 133
8. Signposts in Nature? 135
9. Understanding Creation 136
10. Possible Implications 137

M. Heller: Scientific Rationality and Christian Logos

1. Introduction 141
2. Faith in Reason 142
3. From Theology to Science 144
4. Two Streams of Knowledge 145
5. What Should Be Done? 146
6. The Christian Logos 147

R.J. Clifford, S.J.: Creation in the Hebrew Bible

Introduction 151

1. The Ancient Near Eastern Background 151
   1.1 Mesopotamia. 151
   1.2 Canaanite (Ugaritic) Cosmogonies 154
2. Differences Between Creation in Ancient Near Eastern and Modern Usage 155
   2.1 Process 155
   2.2 Product or Emergent 155
   2.3 Manner of Reporting 156
   2.4 Criterion of Truth. 156

3. The Bible 157
3.1 The Psalms 157
3.2 Second Isaiah 158
3.3 Wisdom Literature 159
3.4 Wisdom Literature: Job 160
3.5 Genesis 1-11 164
4. The Distinctiveness of Biblical Creation 166
5. Conclusions 166

II. EPISTEMOLOGY AND METHODOLOGY

J. Soskice: Knowledge and Experience in Science and Religion: Can We Be Realists?

1. Introduction. 173
2. Models and Metaphors in Science and Religion 174
3. The Argument for Realism 177
4. Theological Realism 179
5. Religious Experience and Authoritative Others 180
6. Conclusions: Revelation Revisited. 181

M.B. Hesse: Physics, Philosophy, and Myth

1. Husserl's Objectivized Science 185
2. Realism and Mathematical Structure 187
3. Types of Causality 189
4. Reduction, Explanation, and Chance 192
5. "Many Worlds" Hypotheses in Quantum Physics 192
6. Anthropic Principles 194
7. Single Law Explanations 195
8. Objectivization, Myth, and Value 198

N. Lash: Observation, Revelation, and the Posterity of Noah

1. *Religio Laici*. 203
2. Interaction? 204
3. On Joining the Conversation 206
4. Words and Stories 207
5. Learning and Listening 208
6. Protocols Against Idolatry. 210

III. CONTEMPORARY PHYSICS AND COSMOLOGY IN PHILOSOPHICAL AND THEOLOGICAL PERSPECTIVE

W.R. Stoeger, S.J.: Contemporary Cosmology and Its Implications for the Science-Religion Dialogue

1. Cosmology and Its Present Directions 219
1.1 Cosmology's Principle Object 219
1.2 The Evolving Universe 221
1.3 The Origin and the End-Point of the Universe 222
1.4 Cosmology and Unification 223
1.5 Some Key Assumptions in Cosmology 224
2. Some Reflections on Cosmology and Its Conclusions . . . 225
2.1 Cosmology as a Discipline 225
2.2 The Limits of Verification in Cosmology 229
2.3 The Ontological Status of Cosmology's Principal Object 230
2.4 Linguistic Problems in Relating Cosmology to Philosophy and to Theology 231
2.5 The Gaps Between Cosmology and Philosophy 232
3. The Focus and the Experiential Grounds of a Discipline . . 232
4. The Implications for Dialogue Between Religion and Science . 238

S. McFague: Models of God for an Ecological, Evolutionary Era: God as Mother of the Universe

1. Introduction 249
2. Heuristic Theology 250
3. Metaphor 252
4. Language About God 253
5. God as Mother of the Universe 255
6. Conclusion 262

T. Peters: On Creating the Cosmos

1. Introduction 273
2. Hypothetical Consonance 274
3. Creation out of Nothing 276
4. The Bare Logic of *Creatio ex Nihilo* 280
5. Consonance with Thermodynamics and Big Bang Cosmology . 282
6. The Scientific Debate: Creation out of Nothing vs. Continuing Creation 284
7. The Theological Debate: Creation out of Nothing vs. Continuing Creation 286
8. Creation and Change 288
9. What Does *Creatio Continua* Mean?. 290
10. Conclusion 291

J. Leslie: How to Draw Conclusions from a Fine-Tuned Universe

1. Introduction 297
2. The Fine-Tuned Universe 298
3. Explanations of the Fine-Tuning 300
4. Concluding Remarks 309

F.J. Tipler: The Omega Point Theory: A Model of an Evolving God

1. Introduction . . . 313
2. The Idea of an Evolving God . . . 314
3. The Omega Point Theory . . . 315
4. Is There Only One Possible Physical Universe? . . . 323
5. The Implications of the Omega Point Theory . . . 326

J. Polkinghorne: The Quantum World

1. Introduction . . . 333
2. Issues Which Arise from the Quantum World . . . 333
3. The Non-Consequences of Quantum Theory . . . 340

R.J. Russell: Quantum Physics in Philosophical and Theological Perspective

1. Introduction . . . 343
2. Philosophical Issues in Quantum Physics . . . 343
2.1 A Short Tour of Quantum Physics . . . 343
2.2 Survey of Competing Interpretations of Quantum Physics 348
3. Quantum Nature in Theological Perspective . . . 355
3.1 Heuristic Role in Metaphorical Theology . . . 355
3.2 Constructive Role in Systematic Theology . . . 359
4. Summary and Closing Comments . . . 365

C.J. Isham: Creation of the Universe as a Quantum Process

1. Introduction . . . 375
2. Scientific Perspectives on Creation . . . 379
2.1 Causality and State Space . . . 380
2.2 The Acausality of Classical Creation . . . 383
2.3 Creation Within the Framework of Pre-Existing Space-time . . . 385
2.4 Creation of Time . . . 387
3. General Relativity . . . 388
4. Quantum Theory and Quantum Gravity . . . 392
4.1 Some Conceptual Issues in Quantum Theory . . . 392
4.2 Quantum States and Transitional Probabilities . . . 394
4.3 Quantum Gravity . . . 395
5. Quantum Creation of the Universe . . . 397
5.1 The Hartle-Hawking Proposal . . . 398
5.2 Achievements . . . 400
5.3 Assumptions . . . 401
5.4 Problems . . . 402
6. Implications for Theology . . . 404

PREFACE

PREFACE

Small beginnings are usually pleasant, especially when one senses that there is something new aborning. There is the excitement of something being created, but small enough as not to require the careful conformity with things past that big initiatives demand. Such, I believe, is what we present in this book. It may at first glance appear to be something big, a major new undertaking. After all, it contains a special message from Pope John Paul II in addition to significant contributions from major scholars. But it is still a small beginning whose full fruits lie in the future. Allow me briefly to record its history.

Three hundred years had passed since Sir Isaac Newton published his epoch-making book, *Philosophia Naturalis Principia Mathematica.* Pope John Paul II wished that the Holy See would remember that event in such a way that it would be not just a simple commemoration but rather a serious contribution to the efforts which He himself had already made to the dialogue between the culture of religious belief and the scientific culture. Therefore, in December 1986 the Vatican Observatory was asked to organize on behalf of the Holy See a conference on precisely that topic: the meeting of the two cultures.

We began with a great deal of enthusiasm, generated both by the topic itself and by the knowledge that the Holy Father himself was very interested. "We" means, besides myself, the Scientific Organizing Committee: Michael Heller, The Pontifical Academy of Theology of Cracow; Arthur R. Peacocke, Ian Ramsey Center, Oxford; Robert J. Russell (Chairperson), The Center for Theology and the Natural Sciences, Berkeley; and William R. Stoeger, S.J., Vatican Observatory.

It was decided among us from the very beginning that this would be a research Study Week, that is, one in which carefully prepared questions would be formulated, circulated to the invited participants in ample time for personal study and reflection, and addressed at the meeting through preliminary drafts of research papers, but without the requirement that these papers be *already* prepared for publication. We had decided, in fact, that, if there were to be a publication, it would reflect the result of the research and interaction carried out together during the week. Put briefly, publishable papers would be prepared after the meeting and not before. As to participation, limited in number because of the very nature of the meeting, it was decided that, in addition to the fundamental consideration of scholarly excellence, we wished the meeting to be *ecumenical* and *interdisciplinary*. The *ecumenical* dimension really posed no problem. In principle none of the three disciplines was allied to any one way of believing nor, as a matter of fact, to unbelief. In practice we knew of scholars in all three disciplines who ranged over the whole spectrum of religious belief. However, to respect the *interdisciplinary* character, which was essential to the meeting, was not as easy a task. Scholarly expertise in more than one of the disciplines was, to the best of our knowledge, not very common. Nonetheless, we required that participants would be invited on the basis that they were established scholars in at least one of the three disciplines of physics, philosophy and theology, and that in at least one of the remaining disciplines they had a serious cultivated interest and knowledge. After exploratory conversations with various scholars preliminary lists of questions to be addressed and participants to be invited were formulated.

A separate meeting to commemorate the Newton tercentenary, held in Cracow in May 1987, offered the opportunity for the Organizing Committee together with a few other scholars to meet in order to finalize plans for the Study Week. The Cracow meeting, although focused in its content more directly on Newton (the book, "Newton and the New Direction in Science," eds. G.V. Coyne, M. Heller, and J. Życiński [Libreria Editrice Vaticana: Vatican City State] is the result of that meeting), served not only as a convenience for organization but also as a stimulus for the ideas that were to surface more clearly at the Vatican Study Week.

And so from 21 to 26 September 1987 in the bucolic setting of the Papal Summer Residence at Castel Gandolfo the twenty-one scholars whose names appear in the List of Participants met to study and discuss problems associated with "Our Knowledge of God and Nature: Physics, Philosophy and Theology" — the title of the Study Week. We gave specific attention to such issues as: historical and contemporary relations between religion and science; modes of reasoning and practice in religion and science; creation as understood in modern physics, in philosophy, and in biblical and systematic theology; the status of philosophical realism in science and religion; "fine-tuning" in the early universe, the anthropic principle, and corresponding arguments for the existence of God; philosophical and theological issues arising from cosmology and quantum physics; God's action in the world; the viability of natural theology today; metaphors and models which relate theology to science; viewpoints from physical cosmology about the long-term future of life and the meaning of God. Although this list of topics is quite substantial, there are still many topics that were not treated. One cannot do everything in an instant and yet one must begin somewhere. We have begun, but it is only that, a beginning.

This book is principally the result of that Study Week, although for various reasons it cannot be said in any way to be the "proceedings" of the Week. Among these reasons are the following: (1) the nature of the meeting, as I have said, was research; these are not the proceedings of what happened there but the fruit in writing of what was undertaken there; (2) one entire paper and many of the ideas contained in the book are the outcome of a second meeting held at the Center for Theology and the Natural Sciences, Berkeley, on 15-16 January 1988, sponsored jointly by that Center and the Vatican Observatory as a continuance of the discussion held at Castel Gandolfo and concentrating on the possibilities for a natural theology today; (3) although in a solemn audience to commemorate Newton's *Principia* as a part of the Study Week and with the participants in attendance, the Holy Father did present a Discourse, the message of John Paul II published here is a later result of His study and reflections on the research completed that week; (4) several of the participants, including Arthur Peacocke, Charles Misner, and Christoph Wassermann have published or are planning to publish their papers elsewhere.

It is a small beginning, indeed, with more questions, the reader will note, than answers. It, therefore, leans into the future. It is a promise, a pledge and a challenge to continue the "Common Quest." We are, as a matter of fact, planning in a very concrete way to continue the quest by sponsoring, on the part of the Vatican Observatory, the Center for Theology and the Natural Sciences, and the Pontifical Theological Academy of Cracow, gatherings of small groups of scholars at Berkeley, Castel Gandolfo, Cracow, and elsewhere on some regular basis. For my part I know that the Holy See is as delighted to pursue these efforts as it was to have organized the Study Week which stimulated the research reported herein.

It is, I believe, only right and just, at the closing of this Preface and before you the reader enter into this book, to emphasize what has been alluded to in previous paragraphs and what will become quite obvious as one reads the book, namely, the exploratory nature of the research that is presented here. Certain circumstances of the book, namely, the sponsorship by the Vatican under the auspices of the Vatican Observatory, the publication of the Papal message, etc., might lead one to think that the contents of the book are intended as doctrinal statements or related, at least, to the developement of doctrine leading even towards dogmatic formulations. The reality is quite the contrary. Each and every part of this book, including, in my opinion, the Papal message, is exploratory. With full respect for, and hopefully an adequate knowledge of, the rich traditions in each of the three disciplines, the attempt here is to explore, and to do so in an interdisciplinary area which is, to say the least, treacherous.

History bears witness to not a few great thinkers who have met unfortunate ends in their attempt to explore in this area. Waiting there for our exploration are inviting caverns, soaring peaks, enticing rich high meadows. One can accept the invitation to explore moved by the sheer enjoyment and excitement of it all, or one can also view it as a serious venture, not thereby less enjoyable. On behalf of my fellow editors and the contributors we offer this volume in the spirit of an enjoyable exploration, undertaken seriously. We hope that we have made some small contribution, even where in exploring we may eventually be judged to have been misguided, even wrong. For us it has been enjoyable. We can only hope that it will be likewise for the reader. We wish to thank Rita Callegari and Suzanne Roth for their help in preparing this publication.

Finally I should mention in regard to the Holy See that the following were also sponsors of the Study Week: The Pontifical Theological Academy of Cracow, the Pontifical Academy of Sciences, the Pontifical Gregorian University, and the Pontifical Council for Culture.

31 July, 1988 *George V. Coyne, S.J.*

MESSAGE OF HIS HOLINESS POPE JOHN PAUL II

To the Reverend George V. Coyne, S.J.
Director of the Vatican Observatory

"Grace to you and peace from God our Father and the Lord Jesus Christ" (Eph 1:2).

As you prepare to publish the papers presented at the Study Week held at Castelgandolfo on 21–26 September 1987, I take the occasion to express my gratitude to you and through you to all who contributed to that important initiative. I am confident that the publication of these papers will ensure that the fruits of that endeavour will be further enriched.

The three hundredth anniversary of the publication of Newton's Philosophiae Naturalis Principia Mathematica provided an appropriate occasion for the Holy See to sponsor a Study Week that investigated the multiple relationships among theology, philosophy and the natural sciences. The man so honoured, Sir Isaac Newton, had himself devoted much of his life to these same issues, and his reflections upon them can be found throughout his major works, his unfinished manuscripts and his vast correspondence. The publication of your own papers from this Study Week, taking up again some of the same questions which this great genius explored, affords me the opportunity to thank you for the efforts you devoted to a subject of such paramount importance. The theme of your conference, "Our Knowledge of God and Nature: Physics, Philosophy and Theology", is assuredly a crucial one for the contemporary world. Because of its importance, I should like to address some issues which the interactions among natural science, philosophy, and theology present to the Church and to human society in general.

The Church and the Academy engage one another as two very different but major institutions within human civilization and world culture. We bear before God enormous responsibilities for the human condition because historically we have had and continue to have a major influence on the development of ideas and values and on the course of human action. We both have histories stretching back over thousands of years: the learned, academic community dating back to the origins of culture, to the city and the library and the school, and the Church with her historical roots in ancient Israel. We have come into contact often during these centuries, sometimes in mutual support, at other times in those needless conflicts which have marred both our histories. In your conference we met again, and it was altogether fitting that as we approach the close of this millennium we initiated a series of reflections together upon the world as we touch it and as it shapes and challenges our actions.

So much of our world seems to be in fragments, in disjointed pieces. So much of human life is passed in isolation or in hostility. The division between rich nations and poor nations continues to grow; the contrast between northern and southern regions of our planet becomes ever more marked and intolerable. The antagonism between races and religions splits countries into warring camps; historical animosities show no signs of abating. Even within the academic community, the separation between truth and values persists, and the isolation of their several cultures – scientific, humanistic and religious – makes common discourse difficult if not at times impossible.

But at the same time we see in large sectors of the human community a growing critical openness towards people of different cultures and backgrounds, different competencies and viewpoints. More and more frequently, people are seeking intellectual coherence and collaboration, and are discovering values and experiences they have in common even within their diversities. This openness, this dynamic interchange, is a notable feature of the international scientific communities themselves, and is based on common interests, common goals and a common enterprise, along with a deep awareness that the insights and attainments of one are often important for the progress of the other. In a similar but more subtle way this has occurred and is continuing to occur among more diverse groups – among the communities that make up the Church, and even between the scientific community and the Church herself. This drive is essentially a movement towards the kind of unity which resists homogenization and relishes diversity. Such community is determined by a common meaning and by a shared understanding that evokes a sense of mutual involvement. Two groups which may seem initially to have nothing in common can begin to enter into community with one another by discovering a common goal, and this in turn can lead to broader areas of shared understanding and concern.

As never before in her history, the Church has entered into the movement for the union of all Christians, fostering common study, prayer, and discussions that "all may be one" (Jn 17:20). She has attempted to rid herself of every vestige of anti-semitism and to emphasize her origins in and her religious debt to Judaism. In reflection and prayer, she has reached out to the great world religions, recognizing the values we all hold in common and our universal and utter dependence upon God.

Within the Church herself, there is a growing sense of "world-church", so much in evidence at the last Ecumenical Council in which bishops native to every continent - no longer predominantly of European or even Western origin - assumed for the first time their common responsibility for the entire Church. The documents from that Council and of the magisterium have reflected this new world-consciousness both in their content and in their attempt to address all people of good will. During this century, we have witnessed a dynamic tendency to reconciliation and unity that has taken many forms within the Church.

Nor should such a development be surprising. The Christian community in moving so emphatically in this direction is realizing in greater intensity the activity of Christ within her: "For God was in Christ, reconciling the world to himself" (2 Cor 5:19). We ourselves are called to be a continuation of this reconciliation of human beings, one with another and all with God. Our very nature as Church entails this commitment to unity.

Turning to the relationship between religion and science, there has been a definite, though still fragile and provisional, movement towards a new and more nuanced interchange. We have begun to talk to one another on deeper levels than before, and with greater openness towards one another's perspectives. We have begun to search together for a more thorough understanding of one another's disciplines, with their competencies and their limitations, and especially for areas of common ground. In doing so we have uncovered important questions which concern both of us, and which are vital to the larger human community we both serve. It is crucial that this common search based on critical openness and interchange should not only continue but also grow and deepen in its quality and scope.

For the impact each has, and will continue to have, on the course of civilization and on the world itself, cannot be overestimated, and there is so much that each can offer the other. There is, of course, the vision of the unity of all things and all peoples in Christ, who is active and present with us in our daily lives – in our struggles, our sufferings, our joys and in our searchings – and who is the focus of the Church's life and witness. This vision carries with it into the larger community a deep reverence for all that is, a hope and assurance that the fragile goodness, beauty and life we see in the universe is moving towards a completion and fulfilment which will not be overwhelmed by the forces of dissolution and death. This vision also provides a strong support for the values which are emerging both from our knowledge and appreciation of creation and of ourselves as the products, knowers and stewards of creation.

The scientific disciplines too, as is obvious, are endowing us with an understanding and appreciation of our universe as a whole and of the incredibly rich variety of intricately related processes and structures which constitute its animate and inanimate components. This knowledge has given us a more thorough understanding of ourselves and of our humble yet unique role within creation. Through technology it also has given us the capacity to travel, to communicate, to build, to cure, and to probe in ways which would have been almost unimaginable to our ancestors. Such knowledge and power, as we have discovered, can be used greatly to enhance and improve our lives or they can be exploited to diminish and destroy human life and the environment even on a global scale.

The unity we perceive in creation on the basis of our faith in Jesus Christ as Lord of the universe, and the correlative unity for which we strive in our human communities, seems to be reflected and even reinforced in what contemporary science is revealing to us. As we behold the incredible development of scientific research we detect an underlying movement towards the discovery of levels of law and process which unify created reality and which at the same time have given rise to the vast diversity of structures and organisms which constitute the physical and biological, and even the psychological and sociological, worlds.

Contemporary physics furnishes a striking example. The quest for the unification of all four fundamental physical forces – gravitation, electro–magnetism, the strong and weak nuclear interactions – has met with increasing success. This unification may well combine discoveries from the sub–atomic and the cosmological domains and shed light both on the origin of the universe and, eventually, on the origin of the laws and constants which govern its evolution. Physicists possess a detailed though incomplete and provisional knowledge of elementary particles and of the fundamental forces through which they interact at low and intermediate energies. They now have an acceptable theory unifying the electro–magnetic and weak nuclear forces, along with much less adequate but still promising grand unified field theories which attempt to incorporate the strong nuclear interaction as well. Further in the line of this same development, there are already several detailed suggestions for the final stage, superunification, that is, the unification of all four fundamental forces, including gravity. Is it not important for us to note that in a world of such detailed specialization as contemporary physics there exists this drive towards convergence?

In the life sciences, too, something similar has happened. Molecular biologists have probed the structure of living material, its functions and its processes of replication. They have discovered that the same underlying constituents serve in the make-up of all living organisms on earth and constitute both the genes and the proteins which these genes code. This is another impressive manifestation of the unity of nature.

By encouraging openness between the Church and the scientific communities, we are not envisioning a disciplinary unity between theology and science like that which exists within a given scientific field or within theology proper. As dialogue and common searching continue, there will be growth towards mutual understanding and a gradual uncovering of common concerns which will provide the basis for further research and discussion. Exactly what form that will take must be left to the future. What is important, as we have already stressed, is that the dialogue should continue and grow in depth and scope. In the process we must overcome every regressive tendency to a unilateral reductionism, to fear, and to self-imposed isolation. What is critically important is that each discipline should continue to enrich, nourish and challenge the other to be more fully what it can be and to contribute to our vision of who we are and who we are becoming.

We might ask whether or not we are ready for this crucial endeavour. Is the community of world religions, including the Church, ready to enter into a more thorough-going dialogue with the scientific community, a dialogue in which the integrity of both religion and science is supported and the advance of each is fostered? Is the scientific community now prepared to open itself to Christianity, and indeed to all the great world religions, working with us all to build a culture that is more humane and in that way more divine? Do we dare to risk the honesty and the courage that this task demands? We must ask ourselves whether both science and religion will contribute to the integration of human culture or to its fragmentation. It is a single choice and it confronts us all.

For a simple neutrality is no longer acceptable. If they are to grow and mature, peoples cannot continue to live in separate compartments, pursing totally divergent interests from which they evaluate and judge their world. A divided community fosters a fragmented vision of the world; a community of interchange encourages its members to expand their partial perspectives and form a new unified vision.

Yet the unity that we seek, as we have already stressed, is not identity. The Church does not propose that science should become religion or religion science. On the contrary, unity always presupposes the diversity and the integrity of its elements. Each of these members should become not less itself but more itself in a dynamic interchange, for a unity in which one of the elements is reduced to the other is destructive, false in its promises of harmony, and ruinous of the integrity of its components. We are asked to become one. We are not asked to become each other.

To be more specific, both religion and science must preserve their autonomy and their distinctiveness. Religion is not founded on science nor is science an extension of religion. Each should possess its own principles, its pattern of procedures, its diversities of interpretation and its own conclusions. Christianity possesses the source of its justification within itself and does not expect science to constitute its primary apologetic. Science must bear witness to its own worth. While each can and should support the other as distinct dimensions of a common human culture, neither ought to assume that it forms a necessary premise for the other. The unprecedented opportunity we have today is for a common interactive relationship in which each discipline retains its integrity and yet is radically open to the discoveries and insights of the other.

But why is critical openness and mutual interchange a value for both of us? Unity involves the drive of the human mind towards understanding and the desire of the human spirit for love. When human beings seek to understand the multiplicities that surround them, when they seek to make sense of experience, they do so by bringing many factors into a common vision. Understanding is achieved when many data are unified by a common structure. The one illuminates the many; it makes sense of the whole. Simple multiplicity is chaos; an insight, a single model, can give that chaos structure and draw it into intelligibility. We move towards unity as we move towards meaning in our lives. Unity is also the consequence of love. If love is genuine, it moves not towards the assimilation of the other but towards union with the other. Human community begins in desire when that union has not been achieved, and it is completed in joy when those who have been apart are now united.

In the Church's earliest documents, the realization of community, in the radical sense of that word, was seen as the promise and goal of the Gospel: "That which we have seen and heard we proclaim also to you, so that you may have fellowship with us; and our fellowship is with the Father and with his Son Jesus Christ. And we are writing this that our joy may be complete" (1 Jn 1:3–3). Later the Church reached out to the sciences and to the arts, founding great universities and building momuments of surpassing beauty so that all things might be recapitulated in Christ (cf. Eph 1:10).

What, then, does the Church encourage in this relational unity between science and religion? First and foremost that they should come to understand one another. For too long a time they have been at arm's length. Theology has been defined as an effort of faith to achieve understanding, as fides quaerens intellectum. As such, it must be in vital interchange today with science just as it always has been with philosophy and other forms of learning. Theology will have to call on the findings of science to one degree or another as it pursues its primary concern for the human person, the reaches of freedom, the possibilities of Christian community, the nature of belief and the intelligibility of nature and history. The vitality and significance of theology for humanity will in a profound way be reflected in its ability to incorporate these findings.

Now this is a point of delicate importance, and it has to be carefully qualified. Theology is not to incorporate indifferently each new philosophical or scientific theory. As these findings become part of the intellectual culture of the time, however, theologians must understand them and test their value in bringing out from Christian belief some of the possibilities which have not yet been realized. The hylomorphism of Aristotelian natural philosophy, for example, was adopted by the medieval theologians to help them explore the nature of the sacraments and the hypostatic union. This did not mean that the Church adjudicated the truth or falsity of the Aristotelian insight, since that is not her concern. It did mean that this was one of the rich insights offered by Greek culture, that it needed to be understood and taken seriously and tested for its value in illuminating various areas of theology. Theologians might well ask, with respect to contemporary science, philosophy and the other areas of human knowing, if they have accomplished this extraordinarily difficult process as well as did these medieval masters.

If the cosmologies of the ancient Near Eastern world could be purified and assimilated into the first chapters of Genesis, might contemporary cosmology have something to offer to our reflections upon creation? Does an evolutionary perspective bring any light to bear upon theological anthropology, the meaning of the human person as the imago Dei, the problem of Christology -- and even upon the development of doctrine itself? What, if any, are the eschatological implications of contemporary cosmology, especially in light of the vast future of our universe? Can theological method fruitfully appropriate insights from scientific methodology and the philosophy of science?

Questions of this kind can be suggested in abundance. Pursuing them further would require the sort of intense dialogue with contemporary science that has, on the whole, been lacking among those engaged in theological research and teaching. It would entail that some theologians, at least, should be sufficiently well-versed in the sciences to make authentic and creative use of the resources that the best-established theories may offer them. Such an expertise would prevent them from making uncritical and overhasty use for apologetic purposes of such recent theories as that of the "Big Bang" in cosmology. Yet it would equally keep them from discounting altogether the potential relevance of such theories to the deepening of understanding in traditional areas of theological inquiry.

In this process of mutual learning, those members of the Church who are themselves either active scientists or, in some special cases, both scientists and theologians could serve as a key resource. They can also provide a much-needed ministry to others struggling to integrate the worlds of science and religion in their own intellectual and spiritual lives, as well as to those who face difficult moral decisions in matters of technological research and application. Such bridging ministries must be nurtured and encouraged. The Church long ago recognized the importance of such links by establishing the Pontifical Academy of Sciences, in which some of the world's leading scientists meet together regularly to discuss their researches and to convey to the larger community where the directions of discovery are tending. But much more is needed.

The matter is urgent. Contemporary developments in science challenge theology far more deeply than did the introduction of Aristotle into Western Europe in the thirteenth century. Yet these developments also offer to theology a potentially important resource. Just as Aristotelian philosophy, through the ministry of such great scholars as St Thomas Aquinas, ultimately came to shape some of the most profound expressions of theological doctrine, so can we not hope that the sciences of today, along with all forms of human knowing, may invigorate and inform those parts of the theological enterprise that bear on the relation of nature, humanity and God?

Can science also benefit from this interchange? It would seem that it should. For science develops best when its concepts and conclusions are integrated into the broader human culture and its concerns for ultimate meaning and value. Scientists cannot, therefore, hold themselves entirely aloof from the sorts of issues dealt with by philosophers and theologians. By devoting to these issues something of the energy and care they give to their research in science, they can help others realize more fully the human potentialities of their discoveries. They can also come to appreciate for themselves that these discoveries cannot be a genuine substitute for knowledge of the truly ultimate. Science can purify religion from error and superstition; religion can purify science from idolatry and false absolutes. Each can draw the other into a wider world, a world in which both can flourish.

For the truth of the matter is that the Church and the scientific community will inevitably interact; their options do not include isolation. Christians will inevitably assimilate the prevailing ideas about the world, and today these are deeply shaped by science. The only question is whether they will do this critically or unreflectively, with depth and nuance or with a shallowness that debases the Gospel and leaves us ashamed before history. Scientists, like all human beings, will make decisions upon what ultimately gives meaning and value to their lives and to their work. This they will do well or poorly, with the reflective depth that theological wisdom can help them attain, or with an unconsidered absolutizing of their results beyond their reasonable and proper limits.

Both the Church and the scientific community are faced with such inescapable alternatives. We shall make our choices much better if we live in a collaborative interaction in which we are called continually to be more. Only a dynamic relationship between theology and science can reveal those limits which support the integrity of either discipline, so that theology does not profess a pseudo-science and science does not become an unconscious theology. Our knowledge of each other can lead us to be more authentically ourselves. No one can read the history of the past century and not realize that crisis is upon us both. The uses of science have on more than one occasion proven massively destructive, and the reflections on religion have too often been sterile. We need each other to be what we must be, what we are called to be.

And so on this occasion of the Newton Tricentennial, the Church speaking through my ministry calls upon herself and the scientific community to intensify their constructive relations of interchange through unity. You are called to learn from one another, to renew the context in which science is done and to nourish the inculturation which vital theology demands. Each of you has everything to gain from such an interaction, and the human community which we both serve has a right to demand it from us.

Upon all who participated in the Study Week sponsored by the Holy See and upon all who will read and study the papers herein published I invoke wisdom and peace in our Lord Jesus Christ and cordially impart my Apostolic Blessing.

From the Vatican, 1 June, 1988

Joannes Paulus PP. II

LIST OF PARTICIPANTS

IAN G. BARBOUR, Department of Religion, Carleton College Northfield, Minnesota 55057, USA.

MICHAEL J. BUCKLEY, S.J., National Council of Catholic Bishops, 1312 Massachusetts Ave., N.W., Washington, D.C. 20005, USA.

RICHARD CLIFFORD, S.J., Weston School of Theology, 3 Phillips Place, Cambridge, Massachusetts 02138, USA.

GEORGE V. COYNE, S.J., Specola Vaticana, V-00120, Città del Vaticano.

MICHAEL HELLER, Pontifical Academy of Theology, Faculty of Philosophy, ul. Podzamcze 8, 31-003 Cracow, Poland; and Specola Vaticana, V-00120 Città del Vaticano.

MARY B. HESSE, Department of History and Philosophy of Science, University of Cambridge, Free School Lane, Cambridge CB2 3RH, UK.

CHRIS J. ISHAM, The Blackett Laboratory, Imperial College, Prince Consort Road, London SW7 2BZ, UK.

NICHOLAS LASH, Faculty of Divinity, University of Cambridge, St. John's Street, Cambridge CB2 1TW, UK.

JOHN LESLIE, Department of Philosophy, University of Guelph, Guelph, Ontario N1G 2W1, Canada.

ERNAN MCMULLIN, Department of Philosophy, University of Notre Dame, Notre Dame, Indiana 46556, USA.

SALLY MCFAGUE, Department of Theology, Vanderbilt University, Nashville, Tennessee 37240, USA.

CHARLES MISNER, Department of Physics and Astronomy, University of Maryland, College Park, Maryland 20742, USA.

ARTHUR R. PEACOCKE, Ian Ramsey Center, St. Cross College, Oxford OX1 3LZ, UK.

OLAF PEDERSEN, History of Science Department, University of Aarhus, Ny Munkegade, DK-8000 Aarhus C, Denmark.

TED PETERS, Pacific Lutheran Theological Seminary, 2770 Marin Ave., Berkeley, California 94708, USA.

JOHN POLKINGHORNE, Trinity Hall, University of Cambridge, Trinity Lane, Cambridge, CB2 1TJ, UK.

ROBERT J. RUSSELL, The Center for Theology and the Natural Sciences, The Graduate Theological Union, 2400 Ridge Road, Berkeley, California 94709, USA.

JANET M. SOSKICE, Ripon College, Cuddeson, Oxford OX9 9EX, UK.

WILLIAM R. STOEGER, S.J., Specola Vaticana, V-00120, Città del Vaticano.

FRANK J. TIPLER, Department of Mathematics and Department of Physics, Tulane University, New Orleans, Louisiana 70118, USA.

CHRISTOPH WASSERMANN, Director of Research, Faculty of Theology, University of Geneva, Geneva, Switzerland.

I.

HISTORICAL AND CONTEMPORARY RELATIONS IN SCIENCE AND RELIGION

WAYS OF RELATING SCIENCE AND THEOLOGY

IAN G. BARBOUR, Department of Religion, Carleton College

Introduction

There is great diversity in contemporary views of the relationship between science and theology. To give an overview of some of the main options, I have grouped them under four headings: *Conflict, Independence, Dialogue,* and *Integration*. Particular authors may not fall neatly under any one heading; a person may agree with adherents of a given position on some issues but not on others. The *Dialogue* viewpoint, in particular, may be combined with either *Independence* or *Integration* themes. After surveying these four broad patterns, I will suggest reasons for supporting *Dialogue* and, with some qualifications, certain versions of *Integration*.

Any view of the relationship of science and theology reflects philosophical assumptions. Our discussion must therefore draw from three disciplines: science (the empirical study of the order of nature), theology (critical reflection on the life and thought of the religious community), and philosophy, especially epistemology (analysis of the characteristics of inquiry and knowledge) and metaphysics (analysis of the most general characteristics of reality). Theology deals primarily with religious beliefs, which must always be seen against the wider background of a religious tradition that includes formative scriptures, communal rituals, individual experiences, and ethical norms. I will be particularly concerned with the epistemological assumptions of recent Western authors writing about the relationship between science and religious beliefs.

1. *Conflict*

Scientific materialism is at the opposite end of the theological spectrum from biblical literalism, but they share several characteristics which lead me to discuss them together. Both believe that there are serious conflicts between contemporary science and traditional religious beliefs. Both seek knowledge with a sure foundation — that of logic and sense-data, in the one case, that of infallible scripture, in the other. They both claim that science and theology make rival literal statements about the same domain, the history of nature, so that one must choose between them.

I will suggest that each represents a misuse of science. In both cases there is a failure to observe the proper boundaries of science. The scientific materialist starts from science but ends by making broad philosophical claims; the biblical literalist moves from theology to make claims about scientific matters. In both schools of thought, the differences between the two disciplines are not adequately respected.

1.1 *Scientific Materialism*

Scientific materialism makes two assertions: (1) the scientific method is the only reliable path to knowledge; (2) matter (or matter and energy) is the fundamental reality in the universe.

The first is an epistemological or methodological assertion about the characteristics of inquiry and knowledge. The second is a metaphysical or ontological assertion about the characteristics of reality and the world. The two assertions are linked by the assumption that only the entities and causes with which science deals are real; only science can progressively disclose the nature of the real.

In addition, many forms of materialism express *reductionism*. Epistemological reductionism claims that the laws and theories of all the sciences are in principle reducible to the laws of physics and chemistry. Metaphysical reductionism claims that the component parts of any system constitute its most fundamental reality. The materialist believes that all phenomena will eventually be explained in terms of the actions of material components, which are the only effective causes in the world. Analysis of the parts of any system has, of course, been immensely useful in science, but it need not preclude the study of higher organizational levels in larger wholes. Evolutionary naturalism sometimes avoids reductionism and holds that distinctive phenomena have emerged at higher levels of organization, but it shares the conviction that the scientific method is the only acceptable mode of inquiry.

Let us consider the assertion that the scientific method is the only reliable form of understanding. Science starts from reproducible public data. Theories are formulated and their implications are tested against experimental observations. Additional criteria of coherence, comprehensiveness and fruitfulness influence choice among theories. Religious beliefs are not acceptable, in this view, because religion lacks such public data, such experimental testing, and such criteria of evaluation. Science alone is objective, open-minded, universal, cumulative and progressive. Religious traditions, by contrast, are said to be subjective, closed-minded, parochial, uncritical and resistant to change. We will see that recent writing in the history and philosophy of science has questioned this idealized portrayal of science, but it is accepted by many scientists who think it undermines the credibility of religious beliefs.

Among philosophers, *logical positivism from* the 1920s to the 1940s asserted that scientific discourse provides the norm for all meaningful language. The only meaningful statements (apart from abstract logical relations) are empirical propositions verifiable by sense-data. Statements in ethics, metaphysics, and religion were said to be neither true nor false, but meaningless pseudo-statements, expressions of emotion or preference devoid of cognitive significance. Whole areas of human language and experience were thus eliminated from serious discussion because they were not subject to the verification which science was said to provide. But critics replied that sense-data do not provide an indubitable starting point in science, for they are already conceptually organized and theory-laden. The interaction of observation and theory is more complex than the positivists had assumed. Moreover, the positivists had dismissed metaphysical questions, but had often assumed a materialist metaphysics. Since Wittgenstein's later writings, the linguistic analysts argued that science cannot be the norm for all meaningful discourse because language has many differing uses and functions.

Among scientists, the success of molecular biology in accounting for many of the basic mechanisms of genetics and biological activity has often been taken as a vindicaton of the reductionist approach. Thus Francis Crick, co-discoverer of the structure of DNA, wrote: "The ultimate aim of the modern movement in biology is in fact to explain all biology in terms of physics and chemistry."[1] Other findings of science suggest that humanity is alone in an immense and impersonal universe. Physicist Steven Weinberg holds that scientific activity itself is the only source of consolation in a meaningless world. The earth is "just a tiny part of an overwhelmingly hostile universe."

> The more the universe seems comprehensible, the more it also seems pointless. But if there is no solace in the fruits of research, there is at least some consolation in the research itself.... The effort to understand the universe is one of the very few things that lifts human life a little above the level of farce, and gives it some of the grace of tragedy.[2]

Most of Carl Sagan's TV series and book, *Cosmos,* is devoted to a fascinating presentation of the discoveries of modern astronomy, but at intervals he interjects his own philosophical commentary. "The Cosmos is all that is or ever was or ever will be."[3] He says that the universe is eternal, or else its source is simply unknowable. Sagan attacks Christian ideas of God at a number of points, and argues that mystical and authoritarian claims threaten the ultimacy of the scientific method, which he says is "universally applicable". Nature (which he capitalizes) replaces God as the object of reverence. He expresses great awe at the beauty, vastness, and interrelatedness of the cosmos. Sitting at the console from which he shows us the wonders of the universe, he is a new kind of high priest, not only revealing the mysteries to us but telling us how we should live. We can indeed admire Sagan's great ethical sensitivity, and his deep concern for nuclear survival and environmental preservation. But perhaps we should question his unlimited confidence in the scientific method, on which he says we should rely to bring in the age of peace and justice.

Jacques Monod's *Chance and Necessity* gives a lucid account of molecular biology, interspersed with a defense of scientific materialism. He claims that biology has proved that there is no purpose in nature. "Man knows at last that he is alone in the universe's unfeeling immensity, out of which he emerged only by chance."[4] "Chance alone is the source of all novelty, all creation, in the biosphere." Chance is "blind" and "absolute" because random mutations are unrelated to the needs of the organism; the causes of individual variations are completely independent of the environmental forces of natural selection. Monod espouses a thoroughgoing reductionism: "Anything can be reduced to simple, obvious mechanical interactions. The cell is a machine. The animal is a machine. Man is a machine."[5] Consciousness is an epiphenomenon which will eventually be explained biochemically.

Monod asserts that human behavior is genetically determined; he says little about the role of language, thought, or culture in human life. Value judgments are completely subjective and arbitrary. Humanity alone is the creator of values; the assumption of almost all previous philosophies that values are grounded in the nature of reality is undermined by science. But Monod urges us to make the free axiomatic choice that knowledge itself will be our supreme value. He advocates "an ethics of knowledge", but he does not show what this might entail apart from the support of science.

Monod's reductionism is inadequate as an account of purposive behavior and consciousness in animals and human beings. There are alternative interpretations in which the interaction of chance and law is seen to be more complex than Monod's portrayal, and not incompatible with some forms of theism. Arthur Peacocke, for example, gives chance a positive role in the explorations of potentialities inherent in the created order, which would be consistent with the idea of divine purpose (though not with the idea of a precise predetermined plan).[6] At the moment, however, we are interested in Monod's attempt to rely exclusively on the methods of science (plus an arbitrary choice of ethical axioms). He says that science proves that there is no purpose in the cosmos. Surely it would be more accurate to say that science does not deal with divine purpose; it is not a fruitful concept in the development of scientific theories.

As a last example, consider the explicit defense of scientific materialism by the sociobiologist E.O. Wilson. His writings trace the genetic and evolutionary origins of social behavior in insects, animals, and humans. He asks how self-sacrificial behavior could arise and persist among social insects, such as ants, if their reproductive ability is thereby sacrificed. Wilson shows that such "altruistic" behavior enhances the survival of close relatives with similar genes (in an ant colony, for example); selective pressures would encourage such self-sacrifice. He believes that all human behaviour can be reduced to and explained by its biological origins and present genetic structure. "It may not be too much to say that sociology and the other social sciences, as well as the humanities, are the last branches of biology to be included in the Modern Synthesis."[7] The mind will be explained as "an epiphenomenon of the neural machinery of the brain."

Wilson holds that religious practices were a useful survival mechanism in humanity's earlier history because they contributed to group cohesion. But he says that the power of religion will be gone forever when religion is explained as a product of evolution; it will be replaced by a philosophy of "scientific materialism".[8] (He doesn't tell us why the power of science won't also be undermined when science is likewise understood as a product of evolution. Do evolutionary origins really have anything to do with the legitimacy of either field?) He maintains that morality is the result of deep impulses encoded in the genes, and that "the only demonstrable function of morality is to keep the genes intact."

Wilson's writing has received criticism from several quarters. For example, anthropologists have replied that most systems of human kinship are not organized in accord with coefficients of genetic similarity, and that Wilson does not even consider cultural explanations for human behavior.[9] In the present context, I would prefer to say that he has described an important area of biology which suggests some of the constraints within which human behavior occurs, but he has over-generalized and extended it as an all-encompassing explanation, leaving no room for the causal efficacy of other facets of human life and experience.

Each of these authors seems to have assumed that there is only one acceptable type of explanation, so that explanation in terms of astronomical origins, or biochemical mechanisms, or evolutionary development, excludes any other kind of explanation. Particular scientific concepts have been extended and extrapolated beyond their scientific use; they have been inflated into comprehensive naturalistic philosophies. Scientific concepts and theories have been taken to provide an exhaustive description of reality, and the abstractive and selective character of science has been ignored. Whitehead calls this "the fallacy of misplaced concreteness." It can also be described as "making a metaphysics out of a method." But because scientific materialism starts from scientific ideas, it carries considerable influence in an age that respects science.

1.2 *Biblical Literalism*

There has been a variety of views of scripture and its relation to science in the history of Christian thought. Augustine held that when there appears to be a conflict between demonstrated scientific knowledge and a literal reading of the Bible, the latter should be interpreted metaphorically, as in the case of the first chapter of Genesis. Scripture is not concerned about "the form and shape of the heavens"; the Holy Spirit "did not wish to teach men things of no relevance to their salvation."[10] Medieval writers acknowledged a variety of literary forms and levels of truth in scripture, and they gave figurative and allegorical interpretations to many problematic passages. Luther and the Anglicans continued this tradition, though some later Lutherans and Calvinists were more literalistic.

Biblical interpretation did play a part in the condemnation of Galileo. He held that God is revealed in both "the book of Nature" and "the book of Scripture"; the two books could not conflict, he said, since they both came from God. He maintained that writers of the Bible were only interested in matters essential to our salvation, and in their writing they had to "accommodate themselves to the capacity of the common people" and the mode of speech of the times. But Galileo's theories did conflict with a literal interpretation of some scriptual passages, and they called into question the Aristotelian system which the church had adopted in the Thomistic synthesis. At the 350th anniversary of the publication of the *Dialogues,* Pope John Paul II said that since then there has been "a more accurate appreciation of the methods proper to the different orders of knowledge." The church, he said, "is made up of individuals who are limited and who are closely bound up with the culture of the time in which they live.... It is only through humble and assiduous study that she learns to dissociate the essentials of faith from the scientific systems of a given age, especially when a culturally influenced reading of the Bible seemed to be linked to an obligatory cosmology."[11] In 1984 a Vatican commission acknowledged that "church officials had erred in condemning Galileo."[12]

In Darwin's day evolution was mainly taken as a challenge to design in nature, and as a challenge to human dignity (assuming that no sharp line separated human and animal forms), but it was also taken by some groups as a challenge to scripture. There were some who defended biblical inerrancy and totally rejected evolution. Yet most traditionalist theologians reluctantly accepted the idea of evolution, though sometimes only after making an exception for humanity, arguing that the soul is inaccessible to scientific investigation. Liberal theologians had already accepted the historical analysis of biblical texts ("higher criticism") which traced the influence of historical contexts and cultural assumptions on biblical writings. They saw evolution as consistent with their optimistic view of historical progress, and they spoke of evolution as God's way of creating.

In the 20th century the Roman Catholic church and most of the main-line Protestant denominations have held that scripture is the human witness to the primary revelation which occurred in the lives of the prophets and the life and person of Christ. Many traditionalists and evangelicals insist on the centrality of Christ without insisting on the infallibility of a literal interpretation of the Bible. But smaller fundamentalist groups and a large portion of some major denominations in the U.S., such as the Southern Baptists, have maintained that scripture is inerrant throughout. The 1970s and 1980s have seen a growth of fundamentalist membership and political power. For many members of "the New Right" and "the Moral Majority," the Bible provides not only certainty in a time of rapid change, but a basis for the defense of traditional values in a time of moral disintegration (sexual permissiveness, drug use, increasing crime rates, etc.).

In the Scopes trial in 1925 it was argued that the teaching of evolution in public schools should be forbidden because it is contrary to scripture. More recently, a new argument called "scientific creationism" or "creation science" has asserted that there is scientific evidence for the creation of the world within the last few thousand years. The law which was passed by the Arkansas legislature in 1981 required that "creationist theory" be given equal time with evolutionary theory in high school biology texts and classes. It was to be presented purely as a scientific theory, with no reference to God or the Bible.

In 1982 the U.S. District Court overturned the Arkansas law, primarily because it favored a particular religious view, violating the constitutional separation of church and state. Although the bill itself made no explicit reference to the Bible, it used many phrases and ideas taken from Genesis. The writings of the leaders of the creationist movement had made clear their religious purposes.[13] Many of the witnesses against the bill were theologians or church leaders who objected to its theological assumptions.[14]

The court also ruled that "creation science" is not legitimate science. It concluded that the scientific community, not the legislature or the courts, should decide the status of scientific theories. It was shown that proponents of "creation science" had not even submitted papers to scientific journals, much less had them published. At the trial, scientific witnesses showed that a long evolutionary history is central in almost all fields of science, including astronomy, geology, paleontology, and biochemistry, as well as most branches of biology. They also replied to the purported scientific evidence cited by creationists. Claims of geological evidence for a universal flood, and for the absence of fossils of transitional forms between species, were shown to be dubious.[15] In 1987 the U.S. Supreme Court struck down a Louisiana creationism law; it said that the law would have restricted academic freedom and supported a particular religious viewpoint.[16]

"Creation science" is thus a threat to both religious and scientific freedom. It is understandable that the search for certainty in a time of moral confusion and rapid cultural change has encouraged the growth of biblical literalism. But when absolutist positions lead to intolerance and attempts to impose particular religious views on others in a pluralistic society, we must object in the name of religious freedom. Some of the same forces have contributed to the revival of Islamic fundamentalism and the enforcement of orthodoxy in Iran and elsewhere.

We can also see the danger to science when proponents of ideological positions try to use the power of the state to reshape science, whether it be in Nazi Germany, Stalinist Russia, Khomeini's Iran, or creationists in the U.S. To be sure, scientists are inescapably influenced by cultural assumptions and metaphysical presuppositions — as well as by economic forces which in large measure determine the direction of scientific development. The scientific community is never completely autonomous or isolated from its social context, yet it must be protected from political pressures which would dictate scientific conclusions. Science teachers must be free to draw from this larger scientific community in their teaching.

Creationists have raised valid objections when evolutionary naturalists have promoted atheistic philosophies as if they were part of science. But the creationists err in assuming that evolutionary theory is inherently atheistic, and they thereby perpetuate the false dilemma of having to choose between science and religion. The whole controversy reflects the shortcomings of fragmented and specialized higher education. The training of scientists seldom includes any exposure to the history and philosophy of science, or any reflection on the relation of science to society, to ethics, or to religious thought. On the other hand, the clergy has little familiarity with science and is hesitant to discuss controversial subjects in the pulpit.

2. *Independence*

Conflicts between science and religion can be avoided if the two enterprises are understood to be independent and autonomous. Each has its own distinctive domain and its characteristic methods which can be justified on its own terms. Proponents of this view say that there are two jurisdictions and each party must keep off the other's turf. Each must tend to its own business and not meddle in the affairs of the other. Each mode of inquiry is selective and has its limitations. This separation into watertight compartments is motivated not simply by the desire to avoid unnecessary conflicts, but by the desire to be faithful to the distinctive character of each area of life and thought. We will look first at contrasting methods and domains in science and religion. Then we will consider their differing languages and functions.

2.1 *Contrasting Methods*

There have, of course, been many writers in the history of Western thought who have elaborated contrasts between religious and scientific knowledge. In the Middle Ages, the contrast was between revealed truth and human discovery. God can be fully known only as revealed through scripture and tradition. The structures of nature, on the other hand, can be known by unaided human reason and observation. There was, however, some middle ground in "natural theology"; it was held that the existence (though not all the attributes) of God can be demonstrated by rational arguments, including the argument from the evidence of design in nature.

This epistemological dichotomy was supported by the metaphysical dualism of spirit and matter, or soul and body. But this dualism was mitigated insofar as the spiritual realm permeated the material realm. While God's transcendence was emphasized, there was considerable reference to divine immanence, and the Holy Spirit was said to work in nature as well as in human life and history. St. Thomas held that God intervenes miraculously at particular times and also continually sustains the natural order. God as primary cause works through the secondary causes which science studies, but these two kinds of cause are on completely different levels.

In the 20th century, Protestant *neo-orthodoxy* sought to recover the Reformation emphasis on the centrality of Christ and the primacy of revelation, while fully accepting the results of modern biblical scholarship and scientific research. According to Karl Barth and his followers, God can be known only as revealed in Christ and acknowledged in faith. God is the transcendent, the wholly other, unknowable except as self-disclosed. Natural theology is suspect because it relies on human reason. Religious faith depends entirely on divine initiative, not on human discovery of the kind which occurs in science. The sphere of God's action is history, not nature. Scientists are free to carry out their work without interference from theology, and vice versa, since their methods and their subject matter are totally dissimilar. Here, then, is a clear contrast. Science is based on human observation and reason, while theology is based on divine revelation.[17]

In this view, the Bible must be taken seriously but not literally. Scripture is not itself revelation; it is a fallible human record witnessing to revelatory events. The locus of divine activity was not the dictation of a text, but the lives of persons and communities: Israel, the prophets, the person of Christ, and those in the early Church who responded to Him. The biblical writings reflect diverse interpretations of these events; we must acknowledge the human limitations of their authors and the cultural influences on their thought. Their opinions concerning scientific questions reflect the prescientific speculations of ancient times. We should read the opening chapters of Genesis as a symbolic protrayal of the basic relation of humanity and the world to God, a message about human creatureliness and the goodness of the natural order. These religious meanings can be separated from the ancient cosmology in which they were expressed.

Another movement which advocates a sharp separation of the spheres of science and religion is *existentialism*. Here the contrast is between the realm of personal selfhood and the realm of impersonal objects. The former is known only through subjective involvement; the latter is known in the objective detachment typical of the scientist. Common to all existentialists, whether atheistic or theistic, is the conviction that we can only know authentic human existence by being personally involved as unique individuals making free decisions. The meaning of life is found only in commitment and action, never in the spectatorial, rationalistic attitude of the scientist searching for abstract general concepts and universal laws.

Religious existentialists say that God can only be encountered in the immediacy and personal participation of an I-Thou relationship, not in the detached analysis and manipulative control which characterize the I-It relationships of science. Rudolf Bultmann acknowledges that the Bible often uses objective language in speaking of God's acts, but he proposes that we can retain the original experiential meaning of such passages by translating them into the language of human self-understanding, the language of hopes and fears, choices and decisions, and new possibilities for our lives. Theological formulations must be statements about the transformation of human life by a new understanding of personal existence. Such affirmations have no connection with scientific theories about external events in the impersonal order of a law-abiding world.[18]

Langdon Gilkey, in his earlier writing and in his testimony at the Arkansas trial, expresses many of these themes. He makes the following distinctions:(1) Science seeks to explain objective, public, repeatable data. Religion asks about order and beauty in the world and experiences in our inner life (such as guilt, anxiety and meaninglessness, on the one hand, and forgiveness, trust and wholeness, on the other); (2) Science asks objective how-questions. Religion asks personal why-questions about meaning and purpose and about our ultimate origin and destiny; (3) The basis of authority in science is logical coherence and experimental adequacy. The final authority in religion is God and revelation, understood through persons to whom enlightenment and insight were given, and validated in our own experience; (4) Science makes quantitative predictions which can be tested experimentally. Religion must use symbolic and analogical language because God is transcendent.[19]

In the context of the trial, it was an effective strategy to insist that science and religion ask very different questions and use very different methods. It provided methodological grounds for criticizing the attempt of biblical literalists to derive scientific conclusions from scripture. More specifically, Gilkey argued that the doctrine of creation is not a literal statement about the history of nature, but a symbolic assertion that the world is good and orderly and dependent on God in every moment of time, a religious assertion essentially independent of both prescientific biblical cosmology and modern scientific cosmology.

In some of his other writings, Gilkey has developed themes which we will consider under the heading of *Dialogue*. He says that there is a "dimension of ultimacy" in the scientist's passion to know, a commitment to the search for truth, and faith in the rationality and uniformity of nature. For the scientist, these constitute what Tillich called an "ultimate concern." But Gilkey states that there are dangers when science is extended to a total naturalistic philosophy, or when science and technology are ascribed a redemptive and saving power, as occurs in the liberal myth of progress through science. Both science and religion can be demonic when they are used in the service of particular ideologies, and when the ambiguity of human nature is ignored.[20]

Thomas Torrance has developed further some of the distinctions in neo-orthodoxy. Theology is unique, he says, because its subject-matter is God. Theology is "a dogmatic or positive and independent science operating in accordance with the inner law of its own being, developing its distinctive modes of inquiry and its essential forms of thought under the determination of its given subject-matter."[21] God infinitely transcends all creaturely reality and "can be known only as he has revealed himself," especially in the person of Christ. We can only respond in fidelity to what has been given to us, allowing our thinking to be determined by the given. In science, reason and experiment can disclose the structure of the real but contingent world. Torrance is particularly appreciative of Einstein's realist interpretation of quantum physics, and he defends a realist epistemology in both science and theology.

2.2 *Differing Languages*

An even more effective way of separating science and religion is to interpret them as languages which are unrelated because their functions are totally different. The logical positivists had taken scientific statements as the norm for all discourse, and had dismissed as meaningless any statement not subject to empirical verification. The later linguistic analysts, in response, insisted that differing types of language serve differing functions not reducible to each other. Each "language game" (as Wittgenstein and his successors called it) is distinguished by the way it is used in a social context. Science and religion do totally different jobs, and neither should be judged by the standards of the other. *Scientific language* is used primarily for prediction and control. A theory is a useful tool for summarizing data, correlating regularities in observable phenomena, and producing technological applications. Science asks carefully delimited questions about natural phenomena. We must not expect it to do jobs for which it was not intended, such as providing an over-all world-view, a philosophy of life, or a set of ethical norms. The scientist is no wiser than anyone else when he steps out of his laboratory and speculates beyond his strictly scientific work.[22]

The distinctive function of *religious language*, according to the linguistic analysts, is to recommend a way of life, to elicit a set of attitudes, and to encourage allegiance to particular moral principles. Much of religious language is connected with ritual and practice in the worshipping community. It may also express and lead to personal religious experience. One of the great strengths of the linguistic movement is that it does not concentrate on religious beliefs as abstract systems of thought, but looks at the way religious language is actually used in the lives of individuals and communities. Linguistic analysts draw on empirical studies of religion by sociologists, anthropologists, and psychologists, as well as the literature produced within religious traditions.

Some scholars have studied diverse cultures and concluded that religious traditions are ways of life which are primarly practical and normative. Stories, rituals and religious practices bind individuals in communities of shared memories, assumptions, and strategies for living. Other scholars claim that the primary aim of religion is the transformation of the person. Religious literature speaks extensively of experiences of liberation from guilt in forgiveness, overcoming anxiety in trust, or the transition from brokenness to wholeness. Eastern traditions talk about liberation from bondage to suffering and self-centeredness in the experience of peace, unity and enlightenment.[23] These are obviously activities and experiences which have little to do with science.

George Lindbeck compares the linguistic view with two other views of religious doctrines:

1) In the *propositional* view, doctrines are truth claims about objective realities. "Christianity, as traditionally interpreted, claims to be true, universally valid, and supernaturally revealed."[24] If doctrines are true or false, and rival doctrines are mutually exclusive, there can be only one true faith. (Neo-orthodoxy holds that doctrines are derived from the human interpretation of revelatory events, but it, too, understands doctrines as true or false propositions.)

2) In the *expressive* view, doctrines are symbols of inner experiences. Liberal theology has held that the experience of the holy is found in all religions. If there can be diverse symbolizations of the same core experience, adherents of different traditions can learn from each other. This view tends to stress the private and individual side of religion, with less emphasis on communal aspects. If doctrines are interpretations of religious experience, they are not likely to conflict with scientific theories about nature.

3) In the *linguistic* view, which Lindbeck himself advocates, doctrines are rules of discourse correlated with individual and communal forms of life. Religions are guides to living; they are "ways of life which are learned by practicing them." Lindbeck argues that individual experience cannot be

our starting point because it is already shaped by prevaling conceptual and linguistic frameworks. Religious stories and rituals are formative of our self-understanding. This approach allows us to accept the particularity of each religious tradition, without making exclusive or universal claims for it. There is no assumption of a universal truth or an underlying universal experience, and each cultural system is self-contained. By minimizing the role of beliefs and truth claims, the linguistic view avoids conflicts between science and theology which can occur in the propositional view, yet it escapes the individualism and subjectivity of the expressive view.

In sum, the three movements we have been considering — neo-orthodoxy, existentialism and linguistic analysis — all understand religion and science to be independent and autonomous forms of life and thought. Each field is selective and has its limitations. Every discipline abstracts from the totality of experience those features in which it is interested. The astronomer, Arthur Eddington, once told a delightful parable about a man studying deep-sea life, using a net with a three-inch mesh. After bringing up repeated samples, the man concluded that there are no deep sea fish less than three inches in length. Our methods of fishing, Eddington suggests, determine what we can catch. If science is selective, it cannot claim that its picture of reality is complete.[25]

The independence of science and religion represents a good first approximation. It preserves the distinctive character of each enterprise, and it is a useful strategy for responding to both types of conflict mentioned earlier. Religion does indeed have its characteristic methods, questions, attitudes, functions and experiences, which are different from those of science. But there are serious difficulties in each of these schools of thought.

As I see it, *neo-orthodoxy* rightly stresses the centrality of Christ and the prominence of scripture in the Christian tradition. It is more modest in its claims than biblical literalism, since it acknowledges the role of human interpretation in scripture and doctrine. But in most versions it, too, holds that revelation and salvation occur only through Christ, which seems to me problematical in a pluralistic world. Most neo-orthodox authors emphasize divine transcendence, and give short shrift to immanence. The gulf between God and the world is decisively bridged only in the Incarnation. While Barth and his followers do indeed elaborate a doctrine of creation, their main concern is with the doctrine of redemption. Nature tends to be treated as the unredeemed setting for human redemption, though it may participate in the eschatological fulfilment at the end of time.

Existentialism rightly puts personal commitment at the center of religious faith, but it ends by privatizing and interiorizing religion to the neglect of its communal aspects. If God acts exclusively in the realm of selfhood, not in the realm of nature, the natural order is devoid of religious significance — except as the impersonal stage for the drama of personal existence. This anthropocentric framework, concentrating on humanity alone, offers little protection against the modern exploitation of nature as a collection of impersonal objects. If religion deals with God and the self, and science deals with nature, who can say anything about the relationship between God and nature, or between the self and nature? To be sure, religion is concerned with the meaning of personal life, but this cannot be divorced from belief in a meaningful cosmos. I will also suggest that existentialism exaggerates the contrast between an impersonal, objective stance in science and the personal involvement which is essential in religion. Personal judgment does enter the work of the scientist, and rational reflection is an important part of religious inquiry.

Finally, *linguistic analysis* has helped us to see the diversity of functions of religious language. Religion is indeed a way of life and not simply a set of ideas and beliefs. But the religious practice of a community, including worship and ethics, presupposes distinctive beliefs. Against instrumentalism, which sees both scientific theories and religious beliefs as human constructs useful for specific human purposes, I advocate a criticial realism which holds that both communities make cognitive claims about realities beyond the human world. We cannot remain content with a plurality of unrelated languages if they are languages about the same world. If we seek a coherent interpretation of all experience, we cannot avoid the search for a unified world-view.

If science and religion are totally independent, the possibility of conflict is avoided, but the possibility of constructive dialogue and mutual enrichment is also ruled out. We do not experience life as neatly divided into separate compartments; we experience it in wholeness and interconnectedness before we develop particular disciplines to study different aspects of it. There are also biblical grounds for the conviction that God is Lord of our total lives and of nature, rather than of a separate "religious" sphere. Finally, there is a critical task in our age, the articulation of a theology of nature that will encourage a strong environmental concern. I will suggest that none of the options considered above are adequate to that task.

3. *Dialogue*

In moving beyond the *Independence* thesis, this section outlines some indirect interactions between science and theology involving boundary questions and methods of the two fields. The fourth section, entitled *Integration*, will be devoted to more direct relationships when scientific theories influence religious beliefs, or when they both contribute to the formulation of a coherent world-view or a systematic metaphysics.

3.1 *Boundary Questions*

Boundary questions refer to general presuppositions of the whole scientific enterprise. Historians have wondered why modern science arose in the Judaeo-Christian West among all world cultures. A good case can be made that the doctrine of creation helped to set the stage for scientific activity. Both Greek and biblical thought had asserted that the world is orderly and intelligible. But the Greeks held that this order is necessary, and therefore one can deduce its structure from first principles. It is not surprising that they were stronger in mathematics and logic than in experimental science. Only biblical thought held that the world's order is contingent. If God created both form and matter, the world didn't have to be as it is, so one has to observe it to discover the details of its order. Moreover, while nature is real and good, it is not itself divine, as many ancient cultures held. It is therefore permissible to experiment on nature.[26] Looking back, we can observe that the "desacralization" of nature, which enouraged scientific study, was not an unmixed blessing, for it also allowed the subsequent exploitation of nature, though there also were many other economic and cultural forces which contributed to environmental destruction.

We must be careful not to over-state the case for the role of Christian thought in the rise of science. Arab science made significant advances in the Middle Ages, while science in the West was often hampered by an other-worldly emphasis (although important practical technologies were developed, especially in some of the monastic orders). When modern science did develop in Europe, it was aided by the humanistic interests of the Renaissance, the growth of crafts, trade and commerce, and new patterns of leisure and education. Yet it does appear that the idea of creation gave a religious legitimacy to scientific inquiry. Newton and many of his contemporaries believed that in their work they were "thinking God's thoughts after him." Moreover, the Calvinist "Protestant ethic" seems to have particularly supported science. In the Royal Society the earliest institution for the advancement of science, 7 out of 10 members were Puritans, and many were clergy.

I believe the case for the historical contribution of Christianity to the rise of science is convincing. But once science was well established its own success was sufficient justification for many scientists, without the need for religious legitimation. Theistic beliefs are clearly not explicit presuppositions of science, since many atheistic or agnostic scientists do first-rate work without them. One can simply accept the contingency and intelligibility of nature as givens, and devote one's efforts to investigating the detailed structure of its order. Yet if one does raise wider questions, one is perhaps more open to religious answers. For many scientists exposure to the order of the universe, as well as to its beauty and complexity, is an occasion of wonder and reverence.

On the contemporary scene we have seen that Torrance maintains the characteristic neo-orthodox distinction between human discovery and divine revelation. But in recent writings he says that at its boundaries science raises religious questions which it cannot answer. In pressing back to the very earliest history of the cosmos astronomy forces us to ask why those particular initial conditions were present. Science shows us an order which is both rational and contingent (that is, its laws and initial conditions were not necessary). It is the combination of contingency and intelligibility which prompts us to search for new and unexpected forms of rational order. The theologian can reply that God is the creative ground and reason for the contingent but rational unitary order of the universe.

"Correlation with that rationality in God goes far to account for the mysterious and baffing nature of the intelligibility inherent in the universe, and explains the profound sense of religious awe it calls forth from us and which, as Einstein insisted, is the mainspring of science."[27]

Wolfhart Pannenberg has explored methodological issues in some detail. He accepts Karl Popper's contention that the scientist proposes testable hypotheses and then attempts to refute them experimentally. Pannenberg claims that the theologian can also use universal rational criteria in critically examining religious beliefs. However, the parallels eventually break down, he says, because theology is the study of reality as a whole; reality is an unfinished process whose future we can only anticipate, since it does not yet exist. Moreover, theology is interested in unique and unpredictable historical events. Here the theologian tries to answer another kind of limit-question with which the scientific method cannot deal, a limit not of initial conditions or ontological foundations but of openness toward the future.[28]

Three Roman Catholic authors, Ernan McMullin, Karl Rahner, and David Tracy, seem to me to be advocates of *Dialogue*, though with differing emphases. McMullin starts with a sharp distinction between religious and scientific statements which resembles the *Independence* position, but he ends with a concern for compatibility, consonance, and coherence. God as primary cause acts through the secondary causes studied by science, but these are on radically different levels within different orders of explanation. On its own level the scientific account is complete and without gaps. McMullin is critical of all attempts to derive arguments for God from phenomena unexplained by science; he is dubious about arguments from design or from the directionality of evolution. Gaps in the scientific account are usually closed by the advance of science, and in any case they would only point to a cosmic force and not to the transcendent biblical God. God sustains the whole natural sequence and "is responsible equally and uniformly for all events." The theologian has no stake in particular scientific theories, including astrophysical theories about the early cosmos.[29]

Some theologians have taken the accumulating evidence for the "Big Bang"theory as corroboration of the biblical view that the universe had a beginning in time, which would be a welcome change after the "conflicts" of the past. McMullin however, maintains that the doctrine of creation is not an explanation of cosmological beginnings at all, but an assertion of the world's absolute dependence on God in every moment. The intent of Genesis was not the specification that there was a first moment in time. Moreover, the Big Bang theory does not prove that there was a beginning in time, since the current expansion could be one phase of an oscillating or cyclic universe. He concludes: "What one cannot say is, first, that the Christian doctrine of creation 'supports' the Big Bang model, or, second, that the Big Bang model 'supports' the Christian doctrine of creation."[30] But he says that for God to choose the initial conditions and laws of the universe would not involve any gaps or violations of the sequence of natural causes. McMullin denies that there is any direct logical connection between scientific and religious assertions, but he does endorse the search for a looser kind of compatibility. The aim should be "consonance but not direct implication," which implies that in the end the two sets of assertions are not totally independent:

> The Christian cannot separate his science from his theology as though they were in principle incapable of interrelation. On the other hand, he has learned to distrust the simpler pathways from one to the other. He has to aim for some sort of coherence of world-view, a coherence to which science and theology, and indeed many others sorts of human construction like history, politics, and literature, must contribute. He may, indeed *must,* strive to make his theology and his cosmology consonant in the contributions they make to this world-view. But this consonance (as history shows) is a tentative relation, constantly under scrutiny, in constant slight shift.[31]

For Karl Rahner, the methods and the content of science and theology are independent, but there are important points of contact and correlations to be explored. God is known primarily through scripture and tradition, but he is dimly and implicity known by all persons as the infinite horizon within which every finite object is apprehended. Rahner extends Kant's transcendental method by analyzing the conditions which make knowledge possible in a neo-Thomist framework. Man knows by abstracting form from matter; in the mind's pure desire to know there is a drive beyond every limited object toward the Absolute. Authentic human experiences of love and honesty are experiences of grace; Rahner affirms the implicit faith of the "anonymous Christian" who does not explicitly acknowledge God or Christ but is commited to the true and the good.[32]

Rahner holds that the traditional doctrines of human nature and of Christology fit well with an evolutionary viewpoint. The human being is a unity of matter and spirit, which are distinct but can only be understood in relation to each other. Science studies matter and gives only part of the whole picture, for we know ourselves to be free self-conscious agents. Evolution from matter to life, mind, and spirit is God's creative action through natural causes which reach their goal in humanity and the Incarnation. Matter develops out of its inner being in the direction of spirit, empowered to achieve an active self-transcendence in higher levels of being. The Incarnation is at the same time the climax of the world's development and the climax of God's self-expression. Rahner insists that Creation and Incarnation are parts of a single process of God's self-communication. Christ as truly human is a moment in biological evolution which has been oriented toward its fulfillment in Him.[33]

David Tracy also sees a religious dimension in science. He holds that religious questions arise at the horizons or limit-situations of human experience. In everyday life, these limits are encountered in experiences of anxiety and confrontation with death, as well as in joy and basic trust. He describes two kinds of limit-situation in science: ethical issues in the uses of science, and presuppositions or conditions for the possibility of scientific inquiry. Tracy maintains that the intelligibility of the world requires an ultimate rational ground. For the Christian, the sources for understanding that ground are the classic religious texts and the structures of human experience. All our theological formulations, however, are limited and historically conditioned. Tracy is open to the reformulation of traditional doctrines in contemporary philosophical categories; he is sympathetic to many aspects of process philosophy and recent work in language and hermeneutics.[34]

How much room is there for the reformulation of traditional theological doctrines in the light of the findings of science? If the points of contact between science and theology refer only to basic presuppositions and boundary questions, no reformulation will be called for. But if there are some points of contact between particular doctrines and particular scientific theories (such as the doctrine of creation in relation to evolution or astronomy), and if it is acknowledged that all doctrines are historically conditioned, there is in principle the possibility of some doctrinal development and reformulation, not just correlation or consonance. What is the nature and extent of the authority of tradition in theology? The Thomistic synthesis of biblical and Aristotelian thought has had a dominant position in the Catholic tradition in the past, but with the help of recent biblical, patristic, and liturgical scholarship, there have been significant efforts to delineate the central biblical message with less dependence on scholastic interpretive categories (see Sec. 4 below).

3.2 *Methodological Parallels*

The positivists, along with most neo-orthodox and existentialist authors, had portrayed science as *objective*, meaning that its theories are validated by clearcut criteria and are tested by agreement with indisputable, theory-free data. Both the criteria and the data of science were held to be independent of the individual subject and unaffected by cultural influences. By contrast, religion seemed *subjective*. We have seen that existentialists made much of the contrast between objective detachment in science and personal involvement in religion.

Since the 1960's, these sharp contrasts were increasingly called into question. Science, it appeared, was not as objective, nor religion as subjective, as had been claimed. There may be differences of emphasis between the fields, but the distinctions are not as abolute as had been asserted. Scientific data are theory-laden, not theory-free. Theoretical assumptions enter the selection, reporting and interpretation of what are taken to be data. Moreover, theories do not arise from logical analysis of data, but from acts of creative imagination in which analogies and models often play a role. Conceptual models help us to imagine what is not directly observable.

subjectivity of science

Many of these same characteristics are present in religion. If the data of religion include religious experience, rituals and scriptural texts, such data are even more heavily laden with conceptual interpretations. In religious language, too, metaphors and models are prominent, as discussed in my writing and that of Sallie McFague and Janet Soskice.[35] Clearly religious beliefs are not amenable to strict empirical testing, but they can be approached with some of the same spirit of inquiry found in science. The scientific criteria of coherence, comprehensiveness and fruitfulness have their parallels in religious thought.

Thomas Kuhn's influential book, *The Structure of Scientific Revolutions,* maintains that both theories and data in science are dependent on the prevailing paradigms of the scientific community. He defined a paradigm as a cluster of conceptual, metaphysical, and methodological presuppositions embodied in a tradition of scientific work. With a new paradigm, the old data are reinterpreted and seen in new ways, and new kinds of data are sought. A paradigm shift is, in Kuhn's words, "a radical transformation of the scientific imagination," a "scientific revolution" which is not the product of experiment alone. In the choice between paradigms there are no rules for applying scientific criteria or for judging their relative importance. Their evaluation is an act of judgment by the scientific community. A paradigm defines a community which works together within a set of shared assumptions. An established paradigm is resistant to falsification, since discrepancies between theory and data can be set aside as anomalies or reconciled by introducing *ad hoc* hypotheses (though an accumulation of anomalies and *ad hoc* hypotheses may eventually lead to a paradigm shift).[36]

Now religious traditions can also be looked on as communities which share a common paradigm. The paradigm is based on shared data, such as religious experience and a memory of key stories and events, but the interpretation of the data is even more paradigm-dependent than in the case of science. There is a greater use of *ad hoc* assumptions to reconcile apparent anomalies, such as the existence of evil, so religious paradigms are even more resistant to falsification. But paradigm shifts in religion do occur, historically in movements such as the Thomisitic synthesis and the Protestant Reformation, and in the life of individuals who join another paradigm community.[37]

The status of the observer in science has also been reconsidered. The earlier accounts had identified objectivity with the separability of the observer from the object of observation. But in quantum physics the influence of the process of observation on the system observed is crucial. In relativity, the most basic measurements, such as the mass, velocity and length of an object, depend on the frame of reference of the observer. Stephen Toulmin traces the change from the assumption of a detached spectator to the recognition of the participation of the observer; he cites examples from quantum physics, ecology, and the social sciences. Every experiment is an action in which we are agents, not just observers. The observer as subject is a participant inseparable from the object of observation.[38] Fritjof Capra and other adherents of Eastern religions have seen parallels here with the mystical traditions which affirm the union of the knower and the known, deriving ultimately from the participation of the individual in the Absolute.[39]

Michael Polanyi envisions a harmony of method over the whole range of knowledge, and says that this overcomes the bifurcation of reason and faith. The unifying idea is the personal participation of the knower in all knowledge. In science, the heart of discovery is creative imagination, which is a very personal act. Science requires skills which, like those in riding a bicycle, cannot be formally specified but only learned by example and practice. In all knowledge we have to see patterns in wholes. In recognizing a friend's face, or in making a medical diagnosis, we use many clues, but we cannot identify all the particulars on which our judgment of a *gestalt* relies.

Polanyi holds that the assessment of evidence is always an act of discretionary personal judgment. There are no rules which specify whether an unexplained discrepancy between theory and experiment should be set aside as an anomaly, or taken to invalidate the theory. It is commitment to rationality and universality, not impersonal detachment, which protects such decisions from arbitrariness. Scientific activity is thus personal but not subjective. Participation in a community of inquiry is another safeguard against subjectivity, though it never removes the burden of individual responsiblity.

Polanyi holds that all these characteristics are even more important in religion. Here personal involvement is greater, but not to the exclusion of rationality and universal intent. Participation in the historical tradition and present experience of a religious community is essential. If theology is the elucidiation of the implications of worship, then surrender and commitment are preconditions of understanding. Replying to reductionism, Polanyi describes ascending levels of reality in evolutionary history and the world today:

> Admittedly, religious conversion commits our whole person and changes our whole being in a way that an expansion of natural knowledge does not do. But once the dynamics of knowing are recognized as the dominant principle of knowledge, the difference appears only as one of degree. . . . It established a continuous ascent from our less personal knowing of inanimate matter to our convivial knowing of living beings and beyond this to knowing our responsible fellow men. Such I believe is the true transition from the sciences to the humanities and also from our knowing the laws of nature to our knowing the person of God.[40]

Polanyi good!

Several authors have recently invoked similar methodological parallels. John Polkinghorne gives examples of personal judgment and theory-laden data in both fields, and he defends critical realism in both cases. The data for a religious community are its scriptural records and its history of religious experience. There are similarities between the fields in that "each is corrigible, having to relate theory to experience, and each is essentially concerned with entities whose unpicturable reality is more subtle than that of naive objectivity."[41] Holmes Rolston holds that religious beliefs interpret and correlate experience, much as scientific theories interpret and correlate experimental data. Beliefs can be tested by the criteria of consistency and congruence with experience. But Rolston acknowledges that personal involvement is more total in the case of religion, since the primary goal is the reformation of the person. Moreover there are other significiant differences: science is interested in causes, while religion is interested in personal meanings.[42]

Such methodological comparisons seem to me illuminating for both fields. There are, however, several dangers in the use of this approach: (1) In the attempt to legitimate religion in an age of science, it is tempting to dwell on similarities and pass over differences. Although science is indeed a more human enterprise than the positivists had recognized, it is clearly more objective than religion in each of the senses which have been mentioned. The kinds of data from which religion draws are radically different from those in science, and the possibility of testing religious beliefs are more limited; (2) In reacting to the absolute distinctions presented by adherents of the *Independence* thesis, it would be easy to ignore the distinctive features of religion that do exist. In particular, by treating religion as an intellectual system, and talking only about religious beliefs, one may distort the diverse characteristics of religion as a way of life, which the linguistic analysts have so well described. Religious belief must always be seen in the context of the life of the religious community and in relation to the goal of personal transformation; (3) Consideration of methodology is an important but preliminary task in the dialogue of science and theology. The issues tend to be somewhat abstract, and therefore of more interest to philosophers of science, and to philosophers of religion, than to scientists or theologians and religious believers. Yet methodological issues have rightly come under new scrutiny in both communities. Furthermore, if we acknowledge methodological similarities we are more likely to encourage attention to substantive issues. If theology at its best is a reflective enterprise which can develop and grow, it can be open to new insights, including those derived from the theories of science.

4. *Integration*

The final group of authors sees religious significance in the content of specific scientific theories and discoveries. Can science and theology be integrated without risking the kinds of conflict from which we started? There are two versions of such integration. First, scientific theories may contribute to the reformulation of theological doctrines whose main sources lie outside science. Second both science and religion may contribute jointly to the formulation of a systematic synthesis: a coherent world-view with an inclusive metaphysics.

4.1 *Doctrinal Reformulation*

A minimal use of science would be the employment of scientific concepts as analogies for communicating traditional beliefs. For example, the paradoxical character of language about the electron as both wave and particle is said to be reminiscent of paradoxical language about Christ as both human and divine.[43] Again, one might translate inherited concepts into contemporary terms to render them more intelligible, without intending any change in their essential meaning. This would represent an apologetic use of science but not a significant integration of ideas.

At the other extreme are forms of natural theology in which religious claims are derived directly from science. The most recent rendition of the argument from design is the Anthropic Principle. Astrophysicists have found that the values of many of the physical constants in the early universe were very critical; if they had been even slightly different it would have been impossible for life to emerge in the universe.[44] Hugh Montefiore has used this principle and other examples of design in the universe, including the directionality of evolution, to argue for an intelligent Designer. Some of the theories he cites, for example the Big Bang, are widely accepted. Others, such as Lovelock's "Gaia Hypothesis" and Sheldrake's "morphogenetic field," are much more controversial and have little support in the scientific community. Montefiore does not claim that these arguments prove the existence of God, but only that the latter is more probable than other explanations.[45]

Intermediate between an apologetic use and a natural theology is the integration which occurs when traditional doctrines are reformulated to take scientific theories into account. Here science and religion are considered to be relatively independent sources of ideas, but with some areas of overlap in their concerns. In particular, the doctrines of creation, providence, and human nature may be affected by the findings of science. The theologian will want to draw mainly from broad features of science which are widely accepted, rather than risk adapting to limited or speculative theories which are more likely to be abandoned in the future. Here the goal would be a theology of nature, based primarily on sources outside of science, rather than a natural theology which is more heavily dependent on science. But if religious beliefs are to be in harmony with scientific knowledge, some adjustments or modifications may be called for.

In particular, our understanding of the general characteristics of nature will affect our models of God's relation to nature. Nature is today understood to be a dynamic evolutionary process with a long history of emergent novelty, characterized throughout by chance and law. The natural order is ecological, interdependent and multi-leveled. These characteristics will modify our representation of the relation of both God and humanity to nonhuman nature. This will in turn affect our attitudes toward nature and will have practical implications for environmental ethics. The problem of evil (theodicy) will also be viewed differently in an evolutionary rather than static world.

For Arthur Peacocke, the starting point of theological reflection is past and present religious experience, together with a continuous interpretive tradition. Religious beliefs are tested by community consensus and by criteria of coherence, comprehensiveness, and fruitfulness. But Peacocke is willing to reformulate traditional beliefs in response to current science. He discusses at length how chance and law work together in cosmology, quantum physics, nonequilibrium thermodynamics and biological evolution. He describes the emergence of distinctive forms of activity at higher levels of complexity in the multi-layered hierarchy of organic life and mind. Peacocke gives chance a positive role in the exploration and expression of potentialities at all levels. God creates through the whole process of law and chance, not by intervening in gaps in the process. "The natural causal creative nexus of events is itself God's creative action."[46] God creates "in and through" the processes of the natural world which science unveils.

Peacocke provides some rich images for talking about God's action in a world of chance and law. He speaks of chance as God's radar sweeping through the range of possibilities and evoking the diverse potentialities of natural systems. In other images artistic creativity is used as an analogy in which purposefulness and open-endedness are continuously present. Peacocke identifies his position as *panentheism* (not pantheism). God is in the world, but the world is also in God, in the sense that God is more than the world. In some passages Peacocke suggests that the world is God's body, and God is the world's mind or soul.

I am sympathetic with Peacocke's position at most points. He gives us vivid images for talking about God's relation to a natural order whose characteristics science has disclosed. But I believe that, in addition to images which provide a suggestive link between scientific and religious reflection, we need philosophical categories to help us unify scientific and theological assertions in a more systematic way.

There are also theological issues which require clarification. Is some reformulation of the classical idea of God's omnipotence called for? Theologians have of course wrestled for centuries with the problem of reconciling omnipotence and omniscience with human freedom and the existence of evil and suffering. But there is a new problem raised by the role of chance in diverse fields of science. Do we defend the traditional idea of divine sovereignty and hold that, in what appears to the scientist to be chance, all events are really providentially controlled by God? Or do both human freedom and chance in nature represent a self-limitation on God's foreknowledge and power which presumably is required by the creation of this sort of world?

How do we represent God's action in the world? The traditional distinction of primary and secondary causes preserves the integrity of the secondary causal chains which science studies. God does not interfere but acts through secondary causes which at their own level provide a complete explanation of all events. This tends toward Deism if God has planned all things from the beginning so they would unfold by their own structures (deterministic and probabilistic) to achieve the goals intended. Is the biblical picture of the particularity of divine action then replaced by the uniformity of divine concurrence with natural causes? Should we then speak only of God's one action, the whole of cosmic history?

4.2 *Systematic Synthesis*

A more systematic integration can occur if both science and religion contribute to a coherent world-view elaborated in a comprehensive metaphysics. Metaphysics is the search for a set of general categories in terms of which diverse types of experience can be interpreted. An inclusive conceptual scheme is sought which can represent the fundamental characteristics of all events. Metaphysics as such is the province of the philosopher rather than of either the scientist or the theologian, but it can serve as an arena of common reflection. The Thomistic framework provided such a metaphysics, but one in which, I would argue, the dualisms of spirit/matter, mind/body, humanity/nature, and eternity/time are only partially overcome.

Process philosophy is a promising candidate for a mediating role today because it was itself formulated under the influence of both scientific and religious thought, even as it responded to persistent problems in the history of western philosophy (such as the mind-body problem). Alfred North Whitehead has been the most important exponent of process categories, though theological implications have been more fully investigated by Charles Hartshorne, John Cobb, and others. The influence of biology and physics is evident in the process view of reality as a dynamic web of interconnected events. Nature is characterized by change, chance, and novelty as well as order. It is incomplete and still coming into being. Process thinkers are critical of reductionism; they defend organismic categories applicable to activities at higher levels of organization. There is continuity as well as distintiveness among levels of reality; the characteristics of each level have rudimentary forerunners at earlier and lower levels. Against a dualism of matter and mind, or a materialism that has no place for mind, process thought envisages two aspects of all events as seen from within and from without. Because humanity is continuous with the rest of nature (despite the uniqueness of reflective self-consciusness), human experience can be taken as a clue to interpreting the experience of other beings. Genuinely new phenomena emerge in evolutionary history, but the basic metaphysical categories apply to all events.

On the religious side, God is understood to be the source of novelty and order. Creation is a long and incomplete process. God elicits the self-creation of individual entities, thereby allowing for freedom and novelty as well as order and structure. God is not the unrelated Absolute, the unmoved Mover, but instead interacts reciprocally with the world, an influence on all events though never the sole cause of any event. Process metaphysics understands every new event to be jointly the product of the entity's past, its own action, and the action of God. Here God transcends the world but is immanent in the world in a very specific way in the structure of each event. We don't have a succession of purely natural events, interrupted by gaps in which God alone operates. Process thinkers reject the idea of divine omnipotence; they believe in a God of persuasion rather than compulsion, and they have provided distinctive analyses of the place of chance, human freedom, evil and suffering in the world. Christian process theologians point out that the power of love, as exemplified in the cross, is precisely its ability to evoke a response while respecting the integrity of the other. They also hold that divine immutability is not a characteristic of the biblical God who is intimately involved with history. Hartshorne elaborates a dipolar concept of God, unchanging in purpose and character but changing in experience and relationship.[47]

Process philosophy as mediating

The writings of Teilhard de Chardin use process categories which parallel Whitehead's at many points. Some interpreters take *The Phenomenon of Man* to be a form of natural theology, an argument from evolution to the existence of God. I have suggested that it can more appropriately be viewed as a synthesis of scientific ideas with religious ideas derived from Christian tradition and experience. Teilhard's other writings make clear how deeply he was molded by his religious heritage and his own spirituality. His concept of God was modified by evolutionary ideas, even if it was not derived from an analysis of evolution. Teilhard speaks of continuing creation and a God immanent in an incomplete world. His concept of "the within of things" has close parallels in Whitehead's thought, though Teilhard's writing makes more use of evocative imagery than of philosophical analysis. His vision of the final convergence to an "Omega Point" is both a speculative extrapolation of evolutionary directionality and a distinctive interpretation of Christian eschatology which differs from the views of most process thinkers.[48]

In *The Liberation of Life* Charles Birch and John Cobb have brought together ideas from biology, process philosophy, and Christian thought. Early chapters develop an ecological or organismic model in which every being is constituted by its interaction with a wider environment, and all beings are subjects of experience, which runs the gamut from rudimentary responsiveness to reflective consciousness. Evolutionary history shows continuity but also the emergence of novelty. Humanity is continuous with and part of the natural order. Birch and Cobb develop an ethics which avoids anthropocentrism. The goal of enhancing the richness of experience in any form encourages concern for nonhuman life, without treating all forms of life as equally valuable. The volume presents a powerful vision of a just and sustainable society in an interdependent community of life.[49]

Birch and Cobb give less attention to religious ideas. They identify God with the principle of life, a cosmic power immanent in nature. At one point it is stated that God loves and redeems us, but the basis of the statement is not clarified. But earlier writings by both these authors indicate their commitment to the Christian tradition and their attempt to reformulate it in the categories of process thought. Writing with David Griffin, for example, Cobb seeks "a truly contemporary vision that is at the same time truly Christian." [50] God is understood both as "source of novelty and order" and as "creative-responsive love." Jesus' vision of the love of God opens us to creative transformation. They also show that Christian process theology can provide a sound basis for an environmental ethics.

I am in basic agreement with the *Doctrinal Reformulation* position, coupled with a cautious use of process philosophy. Still, too much reliance on science (in natural theology) or on science and process philosophy (as in Birch and Cobb) can lead to the neglect of the areas of experience which I consider most important religiously. As I see it, the center of the Christian life is the experience of redemption, the healing of our brokenness in new wholeness, and the expression of a new relationship to God and to the neighbor. Existentialists and linguistic analysts rightly point to the primacy of personal and social life in religion, and neo-orthodoxy rightly says that for the Christian community it is in response to the person of Christ that our lives can be changed. But the centrality of redemption need not lead us to belittle creation, for our personal and social lives are intimately bound to the rest of the created order. We are redeemed in and with the world, not from the world. Part of our task, then, is the articulation of a theology of nature, for which we will have to draw from both religious and scientific sources.

In that task, a systematic metaphysics can help us toward a coherent vision. But Christianity should never be equated with any metaphysical system. There are dangers if either scientific or religious ideas are distorted to fit a preconceived synthesis that claims to encompass all reality. We must always keep in mind the rich diversity of our experience. We distort it if we cut it up into separate realms or watertight compartments, but we also distort it if we force it into a neat intellectual system. A coherent vision of reality can still allow for the distinctiveness of differing types of experience, of which we can be grateful that the advocates of *Dialogue* will remind us.

NOTES

[1] Francis Crick, *Of Molecules and Men* (Seattle: University of Washington Press, 1966) 10.

[2] Steven Weinberg, *The First Three Minutes* (New York: Bantam Books, 1977) 144.

[3] Carl Sagan, *Cosmos* (New York: Random House, 1980) 4. See also Thomas M. Ross, "The Implicit Theology of Carl Sagan," *Pacific Theological Review*, **18** (Spring, 1985) 24-32.

[4] Jacques Monod, *Chance and Necessity* (New York: Vintage Books, 1972) 180.

[5] Monod, BBC lecture, quoted in John Lewis, ed., *Beyond Chance and Necessity* (London: Garnstone Press, 1974) ix. This book includes a number of interesting critiques of Monod.

[6] Arthur Peacocke, *Creation and the World of Science* (Oxford: Clarendon Press, 1979) Chap. 3.

[7] E. O. Wilson, *Sociobiology: The New Synthesis* (Cambridge, MA: Harvard University Press, 1975) 4.

[8] E. O. Wilson, *On Human Nature* (Cambridge, MA: Harvard University Press, 1978) Chaps. 8, 9.

[9] See the essays by Marshall Sahlins, Ruth Mattern, Richard Burian and others in Arthur Caplan, ed., *The Sociobiology Debate* (New York: Harper and Row, 1978).

[10] Cited by Ernan McMullin, "How Should Cosmology Relate to Theology?" In Arthur Peacocke, ed., *The Sciences and Theology in the Twentieth Century* (Notre Dame, IN: University of Notre Dame Press, 1981) 21.

[11] *Origins: NC Documentary Service*, **13** (1983) 50-51.

[12] *Origins: NC Documentary Service*, **16** (1986) 122. See Cardinal Paul Poupard, ed. *Galileo Galilei: Toward a Resolution of 350 Years of Debate. 1633-1983* (Pittsburgh: Duquesne University Press, 1987).

[13] Henry Morris, *A History of Modern Creationism* (San Diego: Master Books, 1984). The text of the ruling, *McLean v. Arkansas*, together with articles by several of the participants in the trial, is printed in *Science, Technology, and Human Values* **7** (Summer, 1982).

[14] See Langdon Gilkey, *Creationism on Trial* (Minneapolis: Winston Press, 1985); Roland Frye, ed., *Is God a Creationist: The Religious Case Against Creation-Science* (New York: Charles Scribner's Sons, 1983).

[15] In addition to the reports on the trial mentioned above, see Philip Kitcher, *Abusing Science: The Case Against Creationism* (Cambridge: MIT Press, 1982); Michael Ruse, *Darwinism Defended* (Reading: Addison-Wesley, 1982).

[16] *Washington Post*, June 20, 1987, p. A I.

[17] A good introduction is Karl Barth, *Dogmatics in Outline* (New York: Harper and Row, 1949). See also W. A. Whitehouse, *Christian Faith and the Scientific Attitude* (New York: Philosophical Library, 1952).

[18] Rudolf Bultmann, *Jesus Christ and Mythology* (New York: Charles Scribner's Sons, 1958).

[19] Langdon Gilkey, *Creationism on Trial*, pp. 108-116. See also his *Maker of Heaven and Earth* (Garden City, NY: Doubleday and Co., 1959).

[20] Langdon Gilkey, *Religion and the Scientific Future* (New York: Harper and Row, 1970) Chap. 2. Also his *Creationism on Trial*, Chap. 7.

[21] Thomas Torrance, *Theological Science* (Oxford: Oxford University Press, 1969) 281.

[22] Useful summaries are given in Frederick Ferré, *Language, Logic and God* (New York: Harper and Brothers, 1961) and William Austin, *The Relevance of Natural Science to Theology* (London: Macmillan, 1976). See also Stephen Toulmin, *The Return to Cosmology* (Berkeley: University of California Press, 1982) Part I.

[23] Frederick Streng, *Understanding Religious Life,* 3rd ed. (Belmont: Wadsworth Publishing Co., 1985).

[24] George Lindbeck, *The Nature of Doctrine: Religion and Theology in a Postliberal Age* (Philadelphia: Westminster Press, 1984) 22.

[25] Arthur Eddington, *The Nature of the Physical World* (Cambridge: Cambridge University Press, 1928) 16.

[26] Alfred North Whitehead, *Science and the Modern World* (New York: The Macmillan Company, 1925), Chap. 1; Stanley Jaki, *The Road of Science and the Ways to God* (Chicago: University of Chicago Press, 1978).

[27] Thomas Torrance, "God and Contingent Order," *Zygon,* **14** (1979) 347. See also his *God and Contingent Order* (Oxford: Oxford University Press, 1981). Torrance also defends contingency within the created order (that is, the unpredictability of particular events), as evident in the uncertainties of quantum physics. Here the invocation of Einstein seems more dubious, since he adhered to a determinist as well as a realist view of physics. Einstein was confident that quantum uncertainties would be removed when we found the underlying deterministic laws, which he believed that a rational universe must have.

[28] Wolfhart Pannenberg, *Theology and the Philosophy of Science* (Philadelphia: Westminster Press, 1976).

[29] Ernan McMullin, "Natural Science and Christian Theology," in David Byers, ed., *Religion Science, and the Search for Wisdom* (Washington, D.C.: National Conference of Catholic Bishops, 1987). See also his "Introduction: Evolution and Creation" in Ernan McMullin, ed., *Evolution and Creation* (Notre Dame: University of Notre Dame Press, 1985).

[30] Ernan McMullin, "How Should Cosmology Relate to Theology?" in *The Sciences and Theology in the Twentieth Century*, ed. Arthur Peacocke (Notre Dame: University of Notre Dame Press, 1981) 39.

[31] *Ibid.,*52.

[32] Karl Rahner, *Foundations of Christian Faith* (New York: Seabury, 1978); Gerald McCool, ed., *A Rahner Reader* (New York: Seabury Press, 1975); Leo O'Donovan, ed., *A World of Grace: An Introduction to the Themes and Foundations of Karl Rahner's Theology* (New York: Seabury, 1980).

[33] Karl Rahner, "Christology within an Evolutionary View of the World," *Theological Investigations* Vol. 5 (Baltimore: Helicon Press, 1966); also *Hominization: The Evolutionary Origin of Man as a Theological Problem* (New York: Herder and Herder, 1965).

[34] David Tracy, *Blessed Rage for Order* (New York: Seabury, 1975); also *Plurality and Ambiguity* (San Francisco: Harper and Row, 1987).

[35] Ian G. Barbour, *Myths. Models and Metaphors* (New York: Harper and Row, 1974); Sallie McFague, *Metaphorical Theology: Models of God in Religious Language* (Philadephia: Fortress Press, 1982); Janet Soskice, *Metaphor and Religious Language* (Oxford: Clarendon Press, 1985).

[36] Thomas Kuhn, *The Structure of Scientific Revolutions,* 2nd ed. (Chicago: University of Chicago Press, 1970).

[37] See, for example, James W. Jones, *The Texture of Knowledge* (Washington, DC: University Press of America, 1981); Hans Küng, "Paradigm Change in

Theology," in Hans Küng and David Tracy, eds., *Moving Toward a New Theology* (Edinburgh: T. and T. Clark, 1988).

[38] Stephen Toulmin, *The Return to Cosmology,* part III.

[39] Fritjof Capra, *The Tao of Physics* (New York: Bantam Books, 1977).

[40] Michael Polanyi, "Faith and Reason," *Journal of Religion,* **41** (1961) 244. See also his *Personal Knowledge* (Chicago: University of Chicago Press, 1958).

[41] John Polkinghorne, *One World: The Interaction of Science and Theology* (Princeton: Princeton University Press, 1987) 64.

[42] Holmes Rolston, III, *Science and Religion: A Critical Survey* (New York: Random House, 1987).

[43] John Polkinghorne, *op. cit.*, 84; Ian G. Barbour, 1974, *op.cit.*, ch. 5. See also the essay by Robert John Russell in the present volume.

[44] John Barrow and Frank Tipler, *The Anthropic Cosmological Principle* (Oxford: Clarendon Press, 1986).

[45] Hugh Montefiore, *The Probability of God* (London: SCM Press, 1985).

[46] Arthur Peacocke, *Intimations of Reality* (Notre Dame: University of Notre Dame Press, 1984) 63; see also *Creation and the World of Science.*

[47] Charles Hartshorne, *The Divine Relativity* (New Haven: Yale University Press, 1948).

[48] Pierre Teilhard de Chardin, *The Phenomenon of Man* (New York: Harper and Row, 1959). I have discussed Teilhard in "Five Ways of Viewing Teilhard," *Soundings,* **51** (1968): 115-145, and in "Teilhard's Process Metaphysics," *Journal of Religion,* **49** (1969) 136-159.

[49] Charles Birch and John B. Cobb, Jr. *The Liberation of Life* (Cambridge: Cambridge University Press, 1981).

[50] John B. Cobb, Jr. and David Griffin, *Process Theology: An Introduction* (Philadelphia: Westminster Press, 1976) 94. See also L. Charles Birch, *Nature and God* (London: SCM Press, 1965).

NATURAL SCIENCE AND BELIEF IN A CREATOR: HISTORICAL NOTES

ERNAN MCMULLIN, University of Notre Dame

Introduction

How does belief in a Creator relate to natural science?[1] It is a question almost guaranteed to cause discomfort for the believer. It would seem that if the natural order is in some way dependent upon a Creator, there ought to be some testimony to this within a science that aims at a comprehensive understanding of that order. Yet as that understanding has deepened, the signs of God's active presence in nature, once evident to all, have become equivocal. Many, indeed, argue that the sweep of evolutionary explanation, stretching backwards through a countless multitude of chance events to the first instants of time, leaves no room for Providence and eliminates the need for, perhaps even the possibility of, further explanation.

I am not going to face these questions directly, not right away at least. Instead, I am going to take an excursion through history; knowing where we have come from can often help us to know where we are going.[2] If one confines oneself to a review of the present situation, one may easily take too seriously the contingencies of the present mode of framing the question of God's relationship to the world. And one may easily miss ambiguities or presuppositions that were long ago laid bare in the historical record.

The "people of the Book," Jews, Christians, Moslems, have always seen their God's relationship with the world as an intimate one. I mean to sketch in broadest outline how their understanding of that relationship very gradually developed and to indicate the part played by the natural sciences in that development. There has been a tension almost from the beginning, we shall see, between two very different ways in which the believer might construe God's relationship to the regularities of nature that constitute the starting point for science.

Jerusalem

Let us go back, by way of introducing the topic, to the centuries when in one part of the Mediterranean world the Hebrew writings that would shape all later Western religion were in process of formation. In another part of that same world, the Greeks were groping towards notions of nature, of cause, of demonstration, that would, after two millennia of slow transformation, provide the matrix for what we have come to call the Scientific Revolution.

It would be generally agreed among Biblical scholars today, I think, that the primary focus of the writings that together comprise the Hebrew Bible, the Old Testament of the Christians, is on salvation history, on Yahweh's covenant with Israel, and not on cosmology, on Yahweh's role as cosmic creator. Indeed, it would seem that the Biblical references to creation were a later development in the Israelites' slowly dawning realization as to who the Yahweh who had led them out of Egypt really was.[3] We might easily be misled by the order in which the books of the Bible now appear into supposing that this was the actual order of their composition. If the two creation narratives with which the *Book of Genesis* opens were the first part of the Bible to have been composed, it would be plausible to suppose that they were intended to define the character of what would come after. Were this to have been the case, the Bible might seem to have been written as a sort of cosmic history, opening with an explanation of how it all began.

But the creation narratives in *Genesis* were, so far as we can tell, written much later than the accounts of the Exodus and the histories of David and Solomon.[4] Indeed, the majestic first chapter of *Genesis* was probably not composed until after the bitter experience of the Babylonian exile in the sixth century B.C., long centuries, then, after the historical chronicles. These much older writings celebrate Yahweh, the one who chose Israel as his special possession, dearer to him than were all other peoples.[5] They tell of a mutual promise between Yahweh and the people whom he favored, a promise often betrayed on the side of Israel, but constantly renewed by the one who had first extended his arm on their behalf. This was the Lord who had led a disorganized group of slaves out of Egypt, who had taken their side against their enemies and who had eventually confirmed them in the possession of the land he had promised them, a land from which, they were convinced, he had helped them dispossess the original inhabitants.

There is nothing cosmic in this story, quite the reverse, it would seem. Yet as it was told and retold, as generations of prophets and priests reflected on who their Yahweh must be, the story took on new dimensions. In perhaps the earliest direct statement of Yahweh's making of the universe, Jeremiah wrote: "Thus says the Lord of hosts, the God of Israel: It was I who made the earth, and man and woman and beast on the face of the earth, with my outstretched arm",[6] and went on to speak of a "new covenant", a much broader one that recognized Yahweh as the giver of "sun to light the day, moon and stars to light the night", thus linking him not only with the people of Israel but with the entire cosmos.[7]

It was in the Psalms that the dependence of the entire universe upon the mighty power of God first came to be celebrated in those ringing verses that have echoed down the ages. In the most eloquent of the creation psalms, Psalm 104, the writer addresses Yahweh:

> You stretch the heavens out like a tent;
> You build your palace on the waters above.
> Using the clouds as your chariot,

You advance on the wings of the wind.
You use the winds as messengers
And fiery flames as servants.
You fixed the earth on its foundations
unshakable for ever and ever.
You made the moon to tell the seasons,
the sun knows when to set:
You bring darkness on, night falls....
All creatures depend on you
to feed them throughout the year;
You provide the food they eat....[8]

The Yahweh of Mount Sinai is now the Lord of heaven and earth, responsible for making all things what they are. The psalmist announces the dependence of all things on Yahweh, their utter fragility. Even the earth, sun, and moon, eternal as they seem, owe their stability to his will: "The vault of heaven proclaims his handiwork" (Ps. 19). The world does not stand of itself; it needs his constant support.

The Psalmists were obviously not responding to a request for explanation. They did not write as they did in order to explain why the world is the way it is. When the Psalmist said, for example, that Yahweh wrapped the earth with waters that overtopped even the mountains and then caused the waters to retreat to a reservoir made for them beneath the earth, (Ps. 104), he was not proposing an explanation of the present relation of earth and sea. He was simply taking a belief about the waters beneath the earth which the Hebrews shared with other peoples of the Near East at the time, and using it with poetic force to help make his real point, which was the dependence of all things on Yahweh.

The shattering experience of the fall of Jerusalem in 587 B.C. and the loss of the land that Yahweh had given deepened this sense of dependence, of the need for redemption on the part of a forgiving Lord. The earlier easy confidence was gone. The writings of this time reflect this feeling of a collapse, a chaos over all the earth, and cry out to Yahweh as the one on whom all order depends, the one who first brought order from chaos. The opening chapter of *Genesis*, composed around this time, expressed confidence that the same Lord who had protected Israel from its beginnings as a people was the mighty creator, the fashioner of heavens and earth. It retold the story of creation presented in the older and more primitive account of the origins of man and woman that now stands as chapter two of *Genesis*, drawing perhaps also upon the creation stories of the Canaanites and of the other peoples with whom the people of Israel had had such intimate dealings.

The familiar opening lines of *Genesis* may not yet, however, be the creation from nothing of later Christian tradition. Though the best translation is still disputed, there would seem to be a preference for the reading that has God bring order to something pre-existent, to a waste of earth and waters: "When God set about to create the heavens and the earth, the earth was a formless void, there was darkness over the deep, and

God's spirit hovered over the waters."[9] And when the work of creation, of bringing order to this chaos, is done, the waste of waters still exists, surrounding the inhabited earth on all sides, held back only by the power and goodness of God. Were it not for this power exerted as gift, chaos would return.[10]

Much more should be said, but I must summarize. The central theme of the Old Testament is the covenant between Yahweh and Israel, the covenant that, for the Christian, is finally sealed in the life and death of Christ. The awareness that one can see growing among the Israelites that the earth is the Lord's, *their* Lord's, complements this earliest and more formative conviction. Their spokesmen, the prophets and leaders who brought this conviction into clearer and clearer focus, were not trying to *explain* anything. The creation narratives were not written as a cosmology but as an affirmation about the identity of the One who had redeemed them from the land of Egypt and who still sustained them. The *warrant* for these narratives, if one may use a notion that would have been alien to the writers themselves, was the continuing encounter of Israel with Yahweh. What they had learnt, what they had been helped to realize was that not only they but everything in the heavens and on earth depends utterly on God. They had come to appreciate, as their Near Eastern neighbors had not, the gulf that separates Creator and creature. Recall God's powerful reminder to Job, and through him to all creatures:

> Where were you when I laid the earth's foundations?
> Tell me, since you are so well-informed!
> Who decided the dimensions of it, do you know?
> Or who stretched the measuring line across it?
> What supports its pillars at their bases?
> Who laid its cornerstone
> when all the stars of the morning were singing with joy? . . . .
> Have you ever in your life given orders to the morning
> or sent the dawn to its post
> telling it to grasp the earth by its edges?[11]

The lesson could not be mistaken. God and God alone can give orders to the morning; he alone can mark the boundary of the seas and set the stars in their courses. He entirely transcends his world; he is in no way part of it, though everywhere present in it. There is not the slightest suggestion that he can be identified with any power that is immanent in nature, as the other creation-stories of the Near East had implied.[12] Nature itself, indeed, is his gift; it is not to be taken for granted but must be seen as contingent, as something that might not have been. Though Yahweh had sometimes been presented in very human terms in the earlier writings, in his dealings with the first man and woman, for example, the *Book of Job* leaves us in no doubt that he lies beyond all human naming. Yet it also conveys that there is still much we *can* say, and it is what Israel has darkly known from the beginning: that the God who holds all things in existence is, incredibly, a being to whom his creatures can confidently look for redemption.

Athens

I want to move now in imagination across the Mediterranean in order to bring out a striking contrast, a contrast (to use a time-honored phrase) between Athens and Jerusalem. The Biblical writers showed little or no interest in a causal explanation of natural process. But the Greeks were fascinated by it and constructed speculative but highly ingenious accounts of how water or fire or atoms in motion might explain the diversity of kinds and of changes they observed in the world around them. The "physicists" of Ionia took the world itself as a *given*. Even though they might speculate about cosmic origins, their world was a solid one and the only origins they considered were natural transformations of one kind of stuff into another. Some of them saw traces of mind working within the cosmic process, others did not. And those who did would, on occasion, link it with the "Divine". But this was a very different notion of the "Divine" to that of the Hebrews. It was needed in order to explain natural process, that was all. It was immanent within that process and thus accessible to the same sort of reasoning as any other aspect of nature.

Of course, the contrast here has been drawn too sharply. The rituals of the Orphic mystery-religions, the popular beliefs in the gods of Olympia, serve to remind us of other facets of that complex world. In a famous passage in the *Phaedo*, Socrates recalls his own disillusionment with the natural science that had been the enthusiasm of his youth, and sketches an alternative non-materialist account of causation that enables him to secure the reality of the "immortal realm" of God and soul. The *Timaeus* presents an account of cosmic origins in which both God and soul play a significant part: "When he was framing the universe, God put intelligence in soul, and soul in body, that he might be the creator of a work which was by nature fairest and best".[13] Plato did not believe that a *science* of the physical world was possible, strictly speaking. But provided one were satisfied with probability, the evidences all round us ought (he insists) to lead us to believe that God fashioned all things "by form and number".[14] Plato's was a voice that Jerusalem could be brought to understand.

But it was in Aristotle that Greek natural science attained its height, and it is to Aristotle that our attention must be devoted, since so much of our later story is already foreshadowed in his extraordinary intellectual achievement. He created whole fields, like physics, theoretical astronomy, logic, and above all, biology, his first love and lifelong passion. Usually unemotional in laying out arguments, he once introduced a work on physiology by speaking of the "immense pleasure" felt by "all those who can trace the links of causation," and went on:

> We must not recoil in childish aversion from the examination of humbler animals. Every realm of nature is marvellous . . . so we should venture on the study of every kind of animal without distaste, for each and all will reveal to us something natural and something beautiful. Nature's works exemplify, in the highest degree, the conduciveness of everything to an end, and the resultant end of Nature's generations is a form of the beautiful.[15]

The scientist of today would find no difficulty, I think, in recognizing and identifying with the spirit that animated those lines. Aristotle's sense of wonder, his admiration for nature in all its complexity, the excitement he so evidently felt in discovery — these assure us that natural science as we know it was already on the way. What would he and Jeremiah have had to say to one another, I wonder? Not very much, I suspect. A separation was beginning to open between two very different ways of addressing the world, a separation later to become a gulf.

Aristotle was not an irreligious man; indeed, if reverence for the natural world suffices to qualify a person as "religious", in that broader sense of the term often endorsed today, he could be called "religious". In the chapter from which I have already quoted, he notes that "of things constituted by nature, some are ungenerated, imperishable and eternal," and are thus "excellent beyond compare, and divine."[16] These are, of course, the celestial bodies, animated by intelligence and moving in their unchanging circular orbits. The evidence we have concerning them from sensation, the only source of evidence he allows, is scanty, and thus there is little (he reminds us) that we can know with certainty about them. But this knowledge, limited though it may be, gives us, he says:

> more pleasure than all our knowledge of the world in which we live, just as a half-glimpse of people we love is more delightful than a leisurely view of other things.[17]

Is there a hint here of religion in the more familiar sense, involving love and worship? I think not. Note that he situates these beings among the things "constituted by nature"; they are as much part of the world as the humbler animals whose study he also wishes to extol. What sets them above these others is only the character of their motions; these motions being circular, and thus returning on themselves are, in principle, eternal. There is a department of natural science devoted to the celestial bodies, the highest beings in Aristotle's world. And the eighth and last book of his massive work, *On Physics*, terminates in the famous proof of the existence of a First Mover, itself unmoved. The First Mover is required, he argues, in order to explain how motion, *any* motion, occurs. It is an indispensable part of the physical order, though itself pure actuality, without any liability to change, consequently immaterial.

If Aristotle speaks of "love" in this context, it is of a purely intellectual sort, of the kind he would also have had for the sea creatures he so painstakingly describes. Human happiness lies in the life of reason, in the pursuit and contemplation of truth:

> He who exercises his reason and cultivates it seems to be both in the best state of mind and most dear to the gods. For if the gods have any care for human affairs, as they are thought to have, it would be reasonable both that they should delight in that which was best and most akin to them, namely reason, and that they should reward those who love and honour this most.... and that all these attributes belong most of all to the philosopher is manifest. He, therefore, is the dearest to the gods.[18]

This notion of the gods as rewarding or caring for people is called into question elsewhere by Aristotle's characterization of God as pure thought eternally contemplating itself. Be that as it may, Aristotle's physical universe is entirely self-contained, capable of being fully understood in human terms. This is naturalism in as clear a form as it has ever taken. There is no reference to a power on whom man depends for his being, nor of one to whom he may turn in worship or in prayer. In his works on ethics, Aristotle showed remarkable insight into the varieties of moral weakness, but the notion of sin, of an action that is wrong because it offends a loving God, is entirely absent. Aristotle's world, in short, was in many ways remarkably like that of many scientists today.

There were differences, of course, and one of these I want to underline. Aristotle argued that there could be science, real science, only of the *necessary*. Knowledge at its best would have to be unchanging, definitive. (Once again, the preoccupation with changeability as defect.) For a true science of nature to be possible, the regularities of nature would themselves have to be necessary in character. In principle, one could argue that the essences of things could not be other than they are. Otherwise, it would seem, explanation would still be incomplete.

Demonstration took on a quite technical meaning in this system, influenced, as it very likely was, by the axiomatic geometry just then beginning to be perfected. To demonstrate was to move from premises, themselves seen on intuitive grounds to be unquestionably true, to conclusions that followed deductively. It is a demanding notion, obviously. To sustain such a science, the operation of nature itself has to be necessary, inexorable. Chance events can occur when lines of causality intersect. Acorns may be eaten by pigs and thus never attain their natural end of becoming oak trees. But if they are given the proper environment, they *necessarily* become oak trees. Nature not only operates with necessity, when not impeded, but it would seem that it could not in the first place be other than it is.

I hope you will forgive this excursion into what might seem like irrelevant detail. But you may perhaps have grasped already that a collision is now inevitable: Jeremiah and Aristotle cannot long go about their separate ways. Some day those ways will cross, and their descendants will be forced to join battle. But before we come to that dramatic moment, now ancient history since it occurred some seven centuries ago, I must return first to the world of Jeremiah, or rather to the world he prepared. I have not, after all, said anything yet about Christianity. What did it contribute? Remember that the thread that we are following is God's relationship to the world of nature.

Augustine

As we saw a moment ago, the doctrine of God as creator came only gradually into focus across the centuries as the Israelites struggled to understand the Protector to whom they had been bound by covenant from their beginnings as a people. The notion of God's action as a creation "from nothing", that is, an act of absolute bringing to be, and not just a making from pre-existent matter, is hinted at in a passage in the last historical writing of the Old Testament, *Second Maccabees*,[19] and again in Paul's *Letter to the Romans*.[20] But it took firm shape only in the first centuries of the Christian era, in part at least in response to the prevalent dualisms of the day that represented matter as evil, or at least, as resistant to God's action.

What ruled these out for the Christian was above all, perhaps, the central affirmation that God had redeemed his world by entering into it and himself taking on the reality of a man, Jesus of Nazareth. It was no longer possible to suppose that matter could somehow frustrate God's action. Since Jesus had taken on the full materiality of human existence, matter had to be entirely dependent on God's act of creation. The conviction deepened of the absolute transcendence of the God on whom the universe depends, and yet of his entrance into time in the person of Jesus, as well as his continuing action within the world, symbolized by the Spirit whom Jesus had promised would always be present. The doctrine of the Trinity thus meant a new and far more complex understanding of God's relationship to the world. It was not posited as a means of explaining otherwise inexplicable phenomena. Its warrant lay in the Scripture and ultimately in the long revelation of God that had taken place across the centuries in the life of ancient Israel as well as in that of the new Israel announced by Christ.[21]

Augustine was the one who finally brought the linked doctrines of Creation, Incarnation, and Trinity, into clear focus. He is, in a way, the crucial figure in my story. I will have to make an effort to be brief in his regard, since there is so much that could and should be said.[22] Augustine argued that Divine creation is a far more radical relationship than mere making from materials already there. It is a total bringing to be, an act whereby the very existence of the world and of each thing in the world is affirmed and sustained. God himself cannot be part of nature, as Aristotle's First Mover was. Nor can he without contradiction be said to create himself.

Since time is a condition of the creature, it too must be created in the act whereby the world itself is brought to be. The Creator himself is thus outside temporal process. He brings past, present, future (these are *our* terms, as creatures) to be in a *single* act. Creation is not just something that happened a long time ago when all of a sudden things began. It is also an action that at this moment sustains all things in being. We must understand, he says, that "God is working even now, so that if his action should be withdrawn from his creatures, they would perish."[23] This is the insight that the Psalmist had long ago expressed, but now it has been sharpened. God brings all to be in a single act within which temporal and causal connections can be discovered by us. The intelligibility we thus discover is a reflection of the Divine Mind. The stability of natures that makes a natural science possible is grounded in the relationship between these natures and the Ideas to which they are a witness.

Let me by way of preview recall for a moment Jacques Monod's *Chance and Necessity*,[24] which the author believes to have somehow undercut a theistic understanding of cosmic process by excluding the directive operation of mind within that process, as *part* of the process. Augustine's immediate response would be that chance and necessity are equally God's instruments. God achieves his purpose by bringing about the mutations and the random encounters in the same act whereby he brings about the regularities *we* interpret as "necessity". For God there is neither chance nor necessity; he knows the future not by knowing the present and inferring what will happen next, but in the same act by which he knows present and past (always remembering that these tensed terms reflect only the perspective of the created being). He brings about his ends not as a mind which directs cosmic process from within, so to speak, but as the *Creator* of the process, that is, the One responsible for there being a process in the first place.

Several other features of Augustine's thought are worth recalling. He argued that the *Genesis* account of creation in six days could not have been meant as literal history.[25] How could there be days in the literal sense before the sun was created? Yet it appeared only on what is called in the text the fourth "day". Further, the term "day" in its usual sense is relative to one's position on the earth; when it is day in one part of the earth, it is night in the other. Yet the six "days" of the *Genesis* account involve the entire earth. So, he concludes, the term clearly must be taken metaphorically, and he goes on to speculate what the significance of the choice of the "seven day" metaphor might have been.

He made use of a principle here that Galileo was to call on vainly in his own defence a thousand years later. Augustine asserts that, if there is a conflict between a literal reading of Scripture and a well-established truth about nature, this of itself is sufficient reason to seek a metaphorical interpretation of the Scripture passage. There cannot be a contradiction between Nature and Scripture since God is speaking to us in both. This principle of exegesis is remarkable in that it allowed natural science a role in determining the proper sense of Scripture. I say "remarkable" because Augustine is emphatic elsewhere that no one should worry if Christians are ignorant of the work of those he calls the "physicists" regarding the natures of things; it is enough for Christians, he says, "to believe that the cause of all created things, whether in the heavens or on the earth, whether visible or invisible, is nothing other than the goodness of the Creator, who is the one and true God."[26] Yet even though natural science ranks for him far below the knowledge of God, he allows it enough firmness to make it significant even for the interpretation of Scripture.

How, then, *did* the universe begin, in his view? God made all things together, and with them time itself began. In that first instant, the seeds of all that would come later were already present; there would be no need for later additions. Even the bodies of Adam and Eve, our first parents, were already present in potency in the materials from which the cosmos would gradually develop. One of his predecessors, Gregory of Nyssa, had already said the same thing: that a God who is truly creator, not just a shaper of pre-existent materials, would endow his creation from the beginning with all it needed to carry out his ends, as well as sustaining it at all times as those ends are being achieved. Gregory said in a wonderfully expressive passage:

> The sources, causes, potencies, of all things were collectively sent forth in an instant, and in this first impulse of the Divine Will, the essences of all things assembled together: heaven, aether, star, fire, air, sea, earth, animal, plant — all beheld by the eye of God.[27]

Gregory added that, because nature requires time and succession, the natures that were implanted in causal potency in that first instant would unroll only later in an order already implicit from the beginning. This was the famous doctrine of "seed-principles" that Augustine would later develop, and which was so often referred to by Christian defenders of the theory of evolution in those first decades after Darwin, when it still seemed a problematic theory from the Christian standpoint. Augustine's theory was not an evolutionary one, strictly speaking, since, for one thing, the natural kinds developed not from one another in a sequence over time, but each from its own proper seed-principle, when the material environment was propitious.[28] Augustine puts it this way:

> All things were created by God in the beginning in a kind of blending of the elements, but they could not develop and appear until the circumstances were favorable.[29]

In those "seeds", as Augustine graphically calls them, there were "invisibly present" (as he puts it) not only sun, moon, and stars, but even the immense diversity of living things that required only in addition to the seeds, the causal properties of water and earth.[30] What is striking about all this in the context of my theme here is Augustine's conviction that nature is complete in its own order. It does not need to be supplemented, adjusted, added to. He allows, of course, for the occurrence of miracle, noting that it is not, as is commonly said, contrary to *nature*, only contrary to our human expectations.[31] Nothing can happen which is strictly contrary to nature. The "nature" of each thing (and here his definition departs sharply from Aristotle's) is "precisely what the supreme Creator of the thing willed to be,"[32] which might, on occasion, include departures from the order normally observed by us.

What Augustine effectively did was to distinguish between two orders of cause or explanation; each is complete in itself, but each also complements the other in a distinctive way.[33] His way of reaching to God is not through gaps in the natural order, through the inability of natural science to explain certain phenomena. God is "cause", not as part of the natural order, not as intervening here and there to bring things about that otherwise would not happen, but as a primary creative cause of the entire natural order, as the agent responsible for its existence and its entire manner of being. It is one thing to call on God who alone builds and governs creatures from the summit, Augustine reminds us; it is quite another thing to explain why things happen in the way in which they do in the natural order. For this, it is sufficient to refer to the capacities that God has woven into the texture of the world from its first appearance.

The natural world is a *sign* of God — Augustine has much to say, in consequence, about the nature of signs — because it is his handiwork and therefore reflects his purposes. God can, as it were, be seen through it:

> I asked the heavens, the sun, the moon, and the stars: "We are not the God whom you seek," said they. To all the things that stand around the doors of my flesh I said, "Tell me of my God...." With a mighty voice they cried out, "He made us!" My question was the gaze I turned on them; the answer was their beauty.... Is not this beauty apparent to all men whose senses are sound and whole? Why, then, does it not speak the same to all men?[34]

This is, of course, the crucial question. Are not the invisible things of God "to be seen in the things that He has made", as Paul had insisted?[35] How, then, can it be that some simply do *not* see? Though the world may present the same appearance to the unreflective as to the reflective, Augustine remarks:

> it is silent to one, but speaks to the other. Nay, rather, it speaks to all, but only those understand who compare its voice taken in from the outside with the truth within them. Truth says to me: Your God is not heaven or earth or any bodily thing.[36]

God is not to be seen *in* the universe, then, but *through* it. And the seeing is not a matter of natural science, but requires an attending on the part of the individual to the truth that lies within himself or herself, within his or her own history.[37] This is, indeed, what the *Confessions* itself was intended to illustrate, as Augustine reflected on his own life in the light of the insight that all things come from God's hands.

How Could a Christian be an Aristotelian?

The first great confrontation between the Christian religion and the natural science that Augustine saw as complementary to it came in the thirteenth century in the new universities of Western Europe (it had happened earlier in Islam) when Aristotle's works on natural science became available in translation for the first time. Within a few decades, they had become the standard fare for Arts students in all the universities, notably in the two most renowned, Oxford and Paris. In scope and detail, these works had no rival. Plato's *Timaeus*, which had been the handbook for so long, was pushed aside. By the mid-1200's the natural science taught in the universities to all students, including theology students, was that of Aristotle.

But from the beginning an uneasiness manifested itself among the theologians. Aristotle's world was, after all, not a created world. It depended on nothing other than itself for its existence. Aristotle's science took the world as a *given*, and what was more, assumed its structure to be a *necessary* one. Indeed, the notion of scientific demonstration, as Aristotle had elaborated it, seemed to depend on this. The status claimed for the truths of Aristotelian physics presupposed that the world of nature could not be other in kind than it is. How, then, could a Christian be an Aristotelian? Natural science and Christian belief began to seem incompatible. The freedom of God in his act of creation, fundamental to the Christian understanding, appeared to be excluded by the structure of Aristotelian science. And there were difficulties about specific doctrines like the eternity of the world and the immortality of the soul.

The university teachers of natural philosophy made heroic efforts to reconcile their Aristotelian teaching with Christian doctrine. Two young friars, Roger Bacon in Oxford and Thomas Aquinas in Paris, were especially creative in that regard. But the theologians of Paris, the dominant school of theology in Europe, were for the most part unpersuaded, and they exerted pressure on the Church to ban the dangerous new views. In 1277, three years to the day from the untimely death of Thomas Aquinas, the Bishop of Paris condemned 219 propositions drawn indiscriminately from a variety of Aristotelian works, including those of Aquinas.

Historians of science have debated whether this condemnation may have marked a turning-point in the history of science. I am not sure that the condemnation itself was as influential as has sometimes been claimed. It did no more than reinforce objections that had already been fully formulated years before. The main issue was the necessitarianism underlying the entire Aristotelian notion of science, which seemed to compromise in a fatal way the Christian doctrine of God's freedom in creating.[38] Some of the Christian upholders of this freedom went so far as to challenge the entire Aristotelian framework of nature and essence, and to insist on the priority of the individual and on the conventionality of the way in which names are given to kinds of things, as though they had in common something called a "nature" or "essence" or "form". Denying this claim entailed that knowledge could be gained only of the singular; generalizations could at best only be probable. No demonstrations of the Aristotelian kind would then be available, since there would be no essences to be known in the way that Aristotle had supposed.

This new and controversial stress on the individual (what came among philosophers to be called "nominalism") was thus associated with a corresponding stress on the absolute freedom of God in the creating of each individual (what came to be called "voluntarism" by theologians). The reason why historians of science have of late spent so much time on an episode that hardly seems relevant to their interests is that some of them, at least, are convinced that the new stress on the primacy of the knowledge of the singular led to a novel conception of inductive science that was the immediate antecedent of (some have even said: necessary condition for) the new science of Bacon and Descartes.[39] The matter is still disputed; there have been, to my mind, some manifest exaggerations on the part of historians who have claimed that the origins of modern science lie in the Calvinism they see as the inheritor of both the nominalism and the voluntarism of the fourteenth century. But it does seem fair to say that in this instance Christian theology may have served as a corrective, a needed corrective, for Greek science. It is also worth celebrating the single occasion, perhaps, when an ecclesiastical declaration turned out to be perspicacious in regard to the presuppositions of scientific method!

Among those touched by the condemnation of 1277, Aquinas was not to stay in disfavor for long. Even before he was canonized in 1323, his mode of "Christianizing" Aristotle was widely accepted, and indeed Aristotle was well on his way to becoming "the Philosopher" for Christians. By 1346, Pope Clement VI could reprove those teachers and students at the University of Paris who showed a lack of respect for the "time-honored writings" of Aristotle, whose text (he said) should be followed "so far as it does not contradict Catholic faith."[40] What a dramatic change in only a few decades!

When arguing for the utility to the Christian of Aristotelian doctrines, Aquinas singled out the proof of an Unmoved Mover from the *Physics* as the "more manifest" way to prove the existence of a Being whom (he said) all Christians would understand as God. And he went on to draw further on the resources of Aristotle's natural philosophy to construct several other alternative proofs, setting five of these at the celebrated opening of his principal work, the *Summa Theologiae*.[41] There could, after all, be no more effective way to demonstrate the orthodoxy of Aristotelian natural philosophy from the Christian standpoint than to show that it could be made the basis for a multiple affirmation of the existence of God.[42]

But the first objection to this procedure was obvious. How is one to proceed from the Unmoved Mover of Aristotle to the Creator God of Augustine? The Unmoved Mover is at the end of a series which can be traced upwards from every single motion; each member of the series reduces from potency to act, i.e. moves, the member immediately beneath it. The outermost sphere must be self-moved; its motion can be explained only by supposing that the self in question is an intelligence which is moved through desire. The object of this desire is not itself required to be in motion (as a mechanical cause must); it is thus a mover, itself unmoved, the proper terminus for an argument of this sort. Though the causality at this last level is teleological, not mechanical, the argument still belongs to natural philosophy.[43] Without the Unmoved Mover, all physical motion would cease. Is it part of the order of secondary causality then? Can the distinction between the two orders, elsewhere so important for Aquinas, be maintained here? Could this Mover also be the Creator of the universe? It *could* be, of course. But as far as the argument itself is concerned, it need only be an ideal object sufficient to motivate desire. The step from such an object to a being responsible for the very *existence* of the intelligence that desires to emulate it is a very large one indeed.

Aristotle's argument depends on a fundamental principle of his physics: "Whatever is in motion is being moved by something other than itself," which, in turn, is based on a conceptual analysis of the notions of potency and actuality. The argument also assumes a hierarchy of movers; it implicitly excludes, for example, the possibility that A might move B while B moves A. The fifth "Way" likewise depends upon Aristotle's analysis of motion, specifically on the teleological directedness he took to be basic to the explanation of change. Aquinas does not look primarily to the living world for evidence of "design", as later writers would. Natural bodies, he says, act always or nearly always in the same way so as to bring about the best outcome, that is, they act for an end. Bodies removed from their natural place, for example, tend to return to it. And yet they lack any power of conscious planning on their own account. So they must be (constantly) directed by a being with intelligence. In this argument, there is no interesting hierarchy of causes between the Designer and the natural motions of the elements. The assertion of the Designer's existence seems, however, to belong once again to physics. The Designer is to be held responsible for continuously goal-oriented motion on the part of bodies incapable of purpose or conscious desire.

Conceived as flowing from Aristotle's natural philosophy, these proofs may prove vulnerable in several ways. There may be logical flaws in the proofs themselves;[44] if the natural philosophy from which they derive is abandoned, the proofs fall; the Mover to which they conclude bears little resemblance to the Creator of the Christian tradition. Gilson and many other modern commentators on the "Ways" argue that they have to be extracted from the matrix of Greek natural philosophy and formulated in metaphysical language, utilizing a broadened existential notion of efficient cause that leads to the affirmation of a First Cause and not just an Aristotelian Mover or Platonic Demiurge.[45] In this way, the weaknesses of the original formulation can (they believe) be overcome. The "Ways" then reduce, in a sense, to a single proof, one that begins from some observed general feature of the physical world, such as motion or efficient causal relationship, and infers to the necessity of a First Cause for the *existence* of such a feature. Gilson is at some pains to present the proofs, even the first Way, as being "independent of any scientific hypothesis as to the structure of the universe."[46] Whether in the end such a transformation is possible, while retaining the logical structure of the proofs, may be questioned. And whether the resultant argument ought be characterized as "natural theology" is also dubious. It would seem that an argument which relies on features like contingency and finitude, imputed to the physical universe as a whole on the basis of conceptual considerations, is more properly labelled "metaphysical". There can be no doubt in any case that Aquinas himself saw his proofs as being rooted in natural philosophy.

The importance of all this for us today is that with the revival of the work of Aquinas in the 19th century, the "Five Ways" took on the character almost of an inspired text for Catholic apologists.[47] Claims that the existence of God could be readily "demonstrated" nearly always referred back to the five favored proofs. And their cosmological character supported the assumption that the proper way to justify the rationality of the Christian belief in God is to begin from some specific feature of the natural world (like motion or biological adaptation) that "science alone can never explain."

The consequences have been in many ways lamentable. The proofs have been detached from their original complex context in Aquinas's theology and made to look as though they could serve, as they stand, as autonomous and entirely conclusive demonstrations. It has not been difficult for teachers of introductory college courses in philosophy to show their inadequacies when they are taken in this way; indeed, it would not be a large exaggeration to say that an indictment of the logic of the Five Ways has become a standard part of the formation of philosophy students at many American universities.

The Rise and Fall of Physico-Theology

The natural philosophy of Aristotle was widely believed to support a set of demonstrative arguments for the existence of the Christian God. What happened when this philosophy came under challenge from the "new science" of the seventeenth century? The practitioners of the new science were almost without exception strongly affirmative Christians, and they were determined to show (just as Aquinas had been) that their new science of nature did not have the dangerous implications for Christian faith that critics claimed it to have. The "mechanization of the world-picture," as one recent historian of science has termed it, seemed to some to suggest that the world could operate on its own, without need for any Divine intervention. But the "scientists" themselves (the label had not yet been invented,[48] but the professions of "scientist" and "philosopher" were already beginning to separate) were concerned to show that their science, far from leading to atheism, would in fact furnish new motives for belief.

The incentive to construct a new natural theology was felt more strongly in Protestant England than in Catholic Italy or France. Galileo made no attempt to derive a proof for God's existence from his new mechanics. He often speaks of God fashioning the universe in the "simplest" way, i.e. the way that conforms best to mechanical law, without need to call on miracle.[49] He took the *fact* of God's existence to be obvious, and in no need of scientific underpinnings.

Descartes' position was much more complex. Unlike Galileo, he was greatly troubled by the growing skepticism of the age, and found it necessary to base his new system on a proof of God's existence. But the proof was a metaphysical one; his physics was carefully designed to be self-sufficient. In the beginning, God set the cosmic matter in motion. The total quantity of motion thus imparted remains invariant, but its distribution constantly alters. All of the diversity of natural structures from planets to organisms developed later in accordance with the laws of mechanics alone. (Or, at least, they *could* have developed in that way; he prudently left open the possibility of a more traditional account that would invoke a special Divine intervention for every natural kind.) Descartes' physics provided no handholds for a natural theology. The continuing existence of the matter-in-motion could only be explained by supposing a Creator, but the argument in support of this was not a physical one. There were no features of the material world that demanded a *special* reference to God,[50] no gaps that the science of mechanics might not, in principle, at least, some day bridge.

Descartes did not deny that God had to conserve the universe in existence at every moment. Indeed, in the *Third Meditation* he argues that since the same power and action are needed for conservation as for the original creation, the distinction between conservation and creation is "solely a distinction of reason." This would later become the starting-point for Malebranche's argument that God is the true cause of all bodily motions; not only do we not have any clear idea of a force within the bodies themselves that would enable them to act upon one another, but his conservation of a body at different places in successive moments is, because of the efficacy of his will, equivalent to *causing* the motion. Natural science is not required in order to infer God's existence; the only way to account for our knowledge of material objects (he argues) is to suppose that we see them in God. The laws of mechanics are thus nothing more than the norms that God has freely chosen to govern his own action in the world. The occasionalism of Malebranche, and later of Berkeley, turns Cartesian naturalism on its head by challenging the notion of a world constituted by natures with their own powers.

More important for our topic was the response of those natural scientists who insisted that the testimony of science itself could be brought in evidence against a naturalism as sweeping in its scope as that of Descartes. Boyle, for example, writes:

> I confess I sometimes wonder that the Cartesians, who have generally, and some of them skilfully, maintained the existence of a Deity, should endeavour to make men throw away an argument, which the experience of all ages shows to have been the most successful (and in some cases the only prevalent one) to establish among philosophers the belief and veneration of God.[51]

He does not want to discount Descartes' more metaphysical approach to proving God's existence. But:

> I see not why we may not reasonably think that God, who as themselves confess, has been pleased to take care men should acknowledge him, may also have provided for the securing of a truth of so great consequence by stamping characters, or leaving impresses, that men may know his wisdom and goodness.

This argument to God's existence is to begin from certain unmistakable features of the living world:

> The excellent contrivance of that great system of the world, and especially the curious fabric of the bodies of animals and the uses of their sensories and other parts, have been made the great motives that in all ages and nations induced philosophers to acknowledge a Deity as the author of these admirable structures.[52]

Boyle is, of course, exaggerating the importance to his predecessors of the argument from design. His own version of the argument draws attention to the presence of means-to-end adaptation in the instinctive behaviors, as well as in the physiological structures, of numerous animal species. He documents his case with an abundance of references to natural history. The great naturalist, John Ray, in his *The Wisdom of God Manifested in the Works of Creation* (1691), was to add a great deal more supportive detail drawn from acute observation of insects, birds, fish and other creatures.

Boyle was sensitive to the Cartesian objection that his argument was not a properly "physical" one since it alludes to God's purposes in creation. He responds that though he may, by invoking a fashioning intelligence, be straying outside physics, strictly construed:

> To me it is not very material whether or no, in physics or in any other discipline, a thing be proved by the peculiar principles of that science or discipline, provided it be firmly proved by the common grounds of reason.[53]

And he was quite sure that his inference from biological adaptation to the necessity of a Designer *did* rest firmly on the "common grounds of reason." The conviction that the basic structures of the animal world required something more than a Cartesian Creator who would do no more than set matter in random motion deepened in the last decades of the century. William Derham in his *Physico-Theology*, aptly subtitled *A Demonstration of the Being and Attributes of God from His Works of Creation* (1713), is quite sure that the atheist has to be willfully blind to ignore the clear arguments for God's existence everywhere manifest in the living world.

Newton took a very different, but equally affirmative, approach to physico-theology. In the celebrated opening of his first letter to his disciple, Richard Bentley, who was eager to show how the new mechanics of the *Principia* could be called in support of belief in a Creator, Newton wrote in 1692:

> When I wrote my Treatise about our System, I had an eye upon such principles as might work with considering men for the belief of a Deity, and nothing can rejoice me more than to find it useful for that purpose.[54]

Actually, there was nothing in the first edition of the *Principia* which *would* have had this effect on "considering men". But in his letters to Bentley, in the General Scholium he appended to the second edition of the *Principia* in 1713, and in the *Opticks*, Newton outlined a set of arguments that he hoped *would* have the desired effect. They all took more or less the same form: pointing out some feature of the planetary system which "could not spring from any natural cause alone," but had to have been "impressed by an intelligent Agent."[55] Some of them pertained to cosmogony: "To make this system, with all its motions, required a cause which understood and compared together the quantities of matter in the several bodies of the sun and planets...."[56] Others had to do with the remarkable stability of the planetary system. Since the planets must affect one another's motions gravitationally, some sort of intervention on God's part must keep the system from collapsing. Newton's arguments all had a teleological overtone, since they related to the production and preservation of a suitable abode for human existence. But they relied ultimately on the claim that some feature of the world was in principle inexplicable in terms of the new mechanics and required therefore, a non-natural intervention on God's part.

This is the "gaps" type of argument in its purest form, and its vulnerability was amply demonstrated in the century that followed. Feature after feature of Newton's universe that he had thought to require for its explanation something more than a "natural cause" yielded to the probings of physicists like Kant and Laplace. Kant found it possible from the standpoint of the religious believer to accept the Cartesian postulate that our ordered world could have come to be gradually, simply through the operation of mechanical law (and, of course, God's constant conservation). God, he suggests, "put a secret art into the forces of nature so as to enable it to fashion itself out of chaos into a perfect world system."[57] The "art" here did not involve any departures from what our minds discover as the "natural" order. The adaptation of means to ends is not something that had to be added subsequently to the coming-to-be; teleology is a part of nature. Kant devoted a great deal of effort to showing that teleology is compatible with Newtonian mechanism, but, as he insists in the *Critique of Pure Reason*,[58] it cannot support a demonstrative argument for the Existence of a Creator. Physico-theology begins from a particular feature of the world of sense; it can at very best only imply a Craftsman, not a transcendent Creator. When it attempts to conclude to the existence of God, it covertly presupposes the (invalid) ontological argument of the metaphysicians. And even the inference to a Craftsman is suspect, since it either situates God within the chain of empirical causes or else it is forced to employ a dubious notion of cause to disengage God from the natural order and thus from possible experience.

Kant recognizes that it "would be utterly hopeless to attempt to rob this argument of the authority it has always enjoyed."[59] He does not object to its use, provided that it not be taken as providing demonstrative certainty. The human mind quite properly is raised by contemplation of the wonders of nature to the thought of the most perfect Being. But this, Kants insists, is not proof; valid proof cannot in principle begin from some feature of the world which empirical science seems unable to explain. Such a feature must be left to the science of the future.

And the future did not take long to arrive in regard to the remaining arguments of classical physico-theology. Darwin was able to take the evidence of adaptation that Boyle and Ray had thought to imply a Designer and reinterpret it in the light of a naturalistic theory of evolution. There was no longer any need to postulate any special intervention of Intelligence in the process, Darwin maintained, though his metaphor of *selection* was to remain a source of some confusion in that regard. The hostility that the new theory engendered among Christian thinkers, especially in England, has to be understood as being due not only to its drastic revision of the story of human origins but even more to its undermining of the physico-theology which had served as a secure bulwark of faith for so long.

The collapse of physico-theology in the latter part of the nineteenth century undoubtedly contributed to the growing crisis of religious faith at that time. In retrospect, it is easy to see where the trouble lay. The believer was too readily tempted, in the new scientific age, to seek for quasi-scientific validation of religious beliefs. God appears as the terminus of what purports to be a standard causal argument beginning from some feature of the natural world. Though this mode of argument attained its greatest popularity in the two centuries separating Boyle and Darwin, it has affinities, as we have seen, with the earlier Aristotelian and Thomistic traditions.

The God of physico-theology has often been called a "God of the gaps." But it is worth noting that the metaphor of a *gap* is somewhat ambiguous in this context. It was not a gap in the explanation of motion that led Aristotle, for example, to his conception of an Unmoved Mover. Similarly, Boyle might well have denied that his appeal to an intelligent all-powerful agency to explain certain features of the living world depended implicitly on a "gap" of some sort.

What is required for the conception of a "gap" is in the first instance the existence of a well-defined stock of principles and modes of inference constituting a natural "science". When the claim is made that "science" cannot explain a particular feature of the natural world, and that recourse must be had to a *different* level or type of explanation, a "gap" is being declared. But the founders of physico-theology did not usually see things in that way. They saw their explanations as being *part* of science, broadly construed. They relied on the claim that the "Craftsman" hypothesis was simply the best one in the circumstances. Inferring retroductively to the shaping action of a Craftsman seemed no different logically than arguing back from a secondary quality like color to the unobservable corpuscles believed to be responsible for it. Boyle for one realized that the form of the argument was in certain respects unorthodox. But he insisted that it lay on the side of "physics", not of metaphysics.

Many of Newton's arguments were, however, of a different sort. God had, for example, to *override* the "natural" tendencies of multi-planet systems to collapse in order to ensure the stability of our system. The "gap" here was an ontological one. God had, it seemed, to *intervene* in the natural order and bring about something that would otherwise not have occurred. (How much credence Newton himself gave these "considerations" is hard to say. His own faith in God quite certainly did not depend on them.) The distinction between "design" arguments that focus on features of the world that can best (only?) be explained by postulating a shaping Intelligence and "gaps" arguments of the Newtonian sort that point to apparent breaks in the natural causal order, is an important one.

The latter are much more vulnerable to "gap-filling" than the former. When Lecomte du Nouy claimed in our own day that the first living cell *could* not have come to be through the operation of the forces of nature only, he was gambling on the future of biological science, a gamble he seems likely to lose. But the less risky "Design" type of argument is also vulnerable: as science advances, the postulate of a shaping Intelligence in a specific context may prove unnecessary. The order that at first sight appeared to require an initial action of an intelligence that consciously harmonizes means and ends *may* prove explicable in another way.[60]

Contemporary Natural Science and Belief in a Creator

There are traces of physico-theology in many recent discussions of the role of God in the evolutionary process. Teilhard de Chardin divides the energies that propel the world forward into two radically different sorts, tangential and radial. Only tangential energies are accessible to the methodology of conventional natural science. To explain the evolutionary process itself, he argues, one must introduce a "radial" energy which is basically psychic in nature, and whose operation can be discerned only by employing a mode of understanding, a special "seeing" of pattern, which is very different in character from the modes of inference ordinarily recognized in biology.[61]

If the evolutionary process be carefully scrutinized in all its amplitude, he argues, the causal operation of mind-like energies will be discovered within it. These can be understood only by recognizing them as the manifestation of a creative mind acting within the process, steering it, as it were, towards goals that are set in advance. In a footnote in *The Phenomenon of Man*, Teilhard defines his disagreement with neo-Darwinism in a particularly clear way:

> I shall be accused of showing too Lamarckian a bent in the explanations that follow, of giving an exaggerated influence to the *Within* in the organic arrangement of bodies. But be pleased to remember that in the 'morphogenetic' action of instinct as here understood, an essential part is left to the Darwinian play of external forces and to chance. It is only really through strokes of chance that life proceeds, but strokes of chance which are recognized and grasped, that is to say, psychically selected. Properly understood, the 'anti-chance' of the Neo-Lamarckian is not the mere negation of Darwinian chance. On the contrary it appears as its utilization.[62]

Teilhard maintains that neo-Darwinian explanations in terms of natural selection, mutation, and the rest, are insufficient of themselves to account for some of the most basic features of the evolutionary process, notably its orthogenetic, or directional, character. One needs in addition a "psychical selection", an intelligence operating somehow within the process, capable of recognizing the opportunities offered by the mutation-induced alternatives and of choosing among these. The crucial issue separating the "neo-Lamarckian" and the neo-Darwinian is, therefore, the prospective completeness of the neo-Darwinian mode of explanation. The neo-Lamarckian has got to show that his opponent's approach is *in principle* inadequate if his own approach is to carry conviction. This may help to explain the violence of the opposition to Teilhard on the part of leading neo-Darwinian biologists.[63]

A further line of argument that Teilhard often relies on might be called a Principle of Homogeneity. He asserts that the Within must be present in *all* matter since it is present in some; the properties associated with consciousness could not have developed from a matter entirely lacking in them. They must, therefore, have been present in some rudimentary form right from the beginning. The implications of this principle are in one way sharply anti-evolutionary, since it excludes the possibility that fundamentally new properties can emerge from matter that entirely lacks them.[64] This implicit exclusion of emergence at the most basic levels encourages him to view his own system as the only viable alternative to a reductionist materialism.[65] But defenders of emergentist forms of evolutionary philosophy would, of course, insist that their view constitutes another alternative.

Did Teilhard intend *The Phenomenon of Man* to serve the function of physico-theology? That is, did he think of it as a means of *proving* the existence of God, beginning from the facts of evolutionary change? It is difficult to decide. It is much easier to make a case for such characteristic Teilhardian notions as the Within by appealing to a broader context, a sort of "meta-science" that would embrace metaphysics and theology, as well as natural science in the normal, more limited sense. Unless the epistemic credentials of metaphysics and theology are barred in principle, this would seem to be a legitimate manner of proceeding, though one that is (it must be admitted) out of favor, even in these post-positivist days. Whether, of course, this approach would be sufficient to validate Teilhard's system on the grounds of coherence is another matter.

But Teilhard himself, it would appear, did not want to fall back on this alternative, perhaps because it would have meant that his system could not then be expected to carry conviction with those who did not share its metaphysical and theological presuppositions. When he calls this system a "science", as he often does, he seems to be asserting that it possesses a broadly empirical warrant for its theoretical claims, above all for the claim that the radial energies of the universe are the manifestation of a "hyper-personal Omega Point" towards which the universe is both ascending and converging. The identification of this Omega Point with the Creator God of the Christian tradition Teilhard sets in an "Epilogue", but even here, "it is not the convinced believer but the naturalist who is asking for a hearing."[66] He seems to be saying that a science of cosmic evolution, if properly carried through, will end in an affirmation of the existence of a Being who is at once immanent and transcendent, spiritual and personal. Teilhard would not use the language of demonstration, but he surely wanted his science to be seen as a means of raising people's vision to God.

By now, we know what the dangers are of locating God as the terminus of explanation in a natural science. First, the sciences of cosmic evolution, as they develop, may find no place for Teilhard's "Within", for a psychic energy powering the universe in steady progress to a pre-set goal. And second, the God he reaches in this way may be no more than a world-soul, a cosmic mind, or the like. It is not at all evident that there *is* a way to reach the transcendent God of Jeremiah and Augustine through a "science" based on the energies of cosmic evolution.

Where does this leave us? Do the natural sciences bear on belief in a Creator in *any* way today? There has been one intriguing recent development which is worth mentioning; an adequate treatment of it is beyond the scope of my presentation here.[67] Until the early 1970's, cosmologists assumed that the kind of universe we have could originate from an initial state that did not need to be specified in any detailed way. Descartes had talked about an initial "chaos" of particles in motion out of which order gradually emerged, according to mechanical law. The details of this story proved far more difficult to fill in than Descartes or anyone in the seventeenth century could have anticipated, but by the 1950's, it seemed as though a plausible story could be told, in outline at least.

In the mid-1960's, the "Big Bang" model received strong confirmation from the discovery of a pervasive incoming microwave radiation of just the sort it had predicted long before. But the application of elementary particle theory, and of quantum theory generally, to the first moments of the "Big Bang" gave a most unexpected result. It turned out that a "life-bearing" universe, one that would allow life to develop, and thus for which the existence of planets, of elements heavier than hydrogen and helium, and of time for evolution to occur were among the necessary conditions, was *extremely* unlikely. Estimates varied of just how unlikely, but the Cambridge theorists who first developed these ideas (Hawking, Rees, Carr) thought it could be much less than one in a million.

What do probability estimates *mean* in this context, when we have only one universe to work from? The application of current physical theories to a "Big Bang" universe (specifically to the first few seconds of its existence) shows that a great many widely different lines of development are possible. Almost *none* of them lead to a universe in which complex life could develop, so far as we can tell. To get a life-bearing universe requires one to set very precise constraints on its initial state; to use a metaphor that has since become famous, it has to be "fine-tuned". The theories themselves do not limit the possible universes sufficiently, so the limitation must come from a *very* precise setting of the initial conditions.

A number of writers suggested that this limitation could be "explained" by adverting to the presence of *human* life in the universe, and thus the much-discussed "anthropic principle" was born.[68] Since we are here, the universe *must* be limited in this way: if it were not, we would not, after all, *be* here! But does this explain? Not, to my mind, as it stands. Some further supplement is needed. Quite a number have been suggested: for example, the possibility that this is only one of a large number of existing "parallel" universes, either serial or simultaneous, in which case it *would* become explanatory to say that we are in the only one we *could* be in.

But the most obvious way to convert the anthropic principle into a properly explanatory (but no longer strictly "scientific") one is to suppose that the "fine-tuning" is the work of a Creator who in some sense "intends" life to develop in the way it did. The Creator would choose one among all the physically "possible" universes (recall: "possible" in used in reference to current physical theory); that it should be the life-bearing one is no surprise to the Christian!

What makes this form of the design argument attractive to many (apart from the credentials it claims in cosmology) is that it does not require any *intervention*, strictly speaking, on God's part. There is none of the alteration of causal lines that we saw in classical physico-theology. It is just a matter of God's *choosing* a certain sort of universe in the first place, the universe in which human life will "naturally" develop.

The argument does, of course, rely on a "gap" still, namely, the inability of contemporary physical theory to explain the original tight specification of the initial cosmic state. And this has already, in the short time since the argument first appeared, proved to be its main weakness. More recent developments, particularly the so-called "inflationary" models of the initial cosmic expansion, have shown that the original puzzlingly "unlikely" specification *may* be at least partially explicable in broader theoretical terms.[69] The issue is much debated, and it is far too early to say where the debate may lead. And, of course, other sorts of possibility that would undermine the theological version of the anthropic principle must also be kept in mind — the possibility of life-forms quite different from anything we could presently imagine, for example.

This recent revival of the "design" type of natural theology is unexpected and raises a lot of intricate philosophical issues. The argument points to a transcendent Creator, not to a world-soul or even an all-powerful Craftsman. The agent who brings the universe to be is not itself limited to that universe. The argument, however, is of the classical "design" type: it relies on the discovery of an apparent means-end relationship that cannot (or at least *apparently* cannot) be explained in a non-teleological way. Even though the argument avoids many of the hazards of the older natural theology, its conclusion is at best a "consonance" one: a Being who "fine-tunes" the universe (if such there be) is *consonant* with the Creator God of the Christian tradition.[70] Consonance here is more than logical consistency, but much less than proof.

The Ways Divide

Are there any morals to be drawn from this long and complex story? One clear moral would seem to be that physico-theology is not to be trusted. Not only do the gaps not remain gaps, but even when they do, the "Filler of the Gaps" is hard to identify with the creator God of the Christian tradition. Must we give up all attempts to set up an explicit relationship between the content of scientific theory and belief in God? Christian thinkers are divided in their response to this question. They are agreed that God's action must permeate the history of both the physical and the human orders; a deism that would make the universe a structured whole independent of its Creator is unacceptable. But they disagree on whether natural science has the means to single out, within its own proper domain, some special evidence of an action distinctively "Divine" in character.

On the one hand, there are those who still would enlist the aid of science to search for traces of mind or life within cosmic process, and then to identify, or at least relate, that mind with the Divine. The "science" here is most often evolutionary science, and so this alternative has its roots in evolutionary philosophy, in the work of writers like Bergson, Morgan, and especially Whitehead. More recently, quantum theory has been invoked to support the presence of subjectivity even in basic quantum processes; though the mentalistic interpretation of measurement at the quantum level seems to many (I would be tempted to say most) physicists a mistake, some defenders of religion have seen in the consequent panpsychism not only a repudiation of materialism but an affirmation of a distinctively religious world-view. Distinctively religious, but also, it must be added, rather more Eastern than Western. Most of those who attempt to make quantum theory yield religious implications express their preference for a Hindu or a Taoist conception of religion which seems to them to fit this strongly idealist conception of nature rather better than the Christian perspective would.[71]

I am persuaded that this attempt to bring about a rapprochement between the quantum theory of measurement and Eastern religious cosmologies does justice to neither, but to argue this would draw us too far afield. My concern here is with Christian conceptions of the relationship between God and nature; though quantum theory may pose for the theologian some intriguing new questions in regard to causal action, it does not seem at all likely that the subjectivist interpretations of that theory would sustain a natural theology.[72] It must suffice, then to confine our attention to cosmologies inspired by the evolutionary metaphor, whose resources for Christian thought have been fairly thoroughly explored.

Theologians in the "process" tradition argue that the classic Augustinian and Thomistic notions of creation, which set God entirely outside temporal process, were based on Greek models that are inconsistent with the Biblical account of God's action in the world.[73] Their own proposal sets God firmly within cosmic process; it is through this process, indeed, that God is said to achieve self-realization. God is not identical with the world — Charles Hartshorne carefully separates "panentheism" from pantheism — but he includes the world in himself.[74] God's essence is prior to any actual world; it is what makes him what he is. But this is an abstraction; in the concrete, God's actuality derives from the world whose becoming is his becoming. He is thus neither omniscient, nor self-sufficient, not immune to suffering. But then (so the argument runs), this is just how the Bible would lead us to characterize him.

There are two separate issues here. One is how natural science bears on belief in a Creator; the other concerns the relationship of God to time and finitude. Those who argue that God is not immune from change, that he must belong to the realm of time and becoming as we do, do not necessarily see cosmic evolution as a manifestation of Divine evolution. Theologians who maintain that the creator God can still be a suffering God do not necessarily subscribe to process theology.[75] The two issues are not unrelated, of course. Those who, on theological or philosophical grounds, claim God to be a finite being, dependent in some respects on His creation, are more likely to discover traces of God's "special" action in certain aspects of cosmic process than are those who see God as timeless. My immediate concern here, however, is not with the larger theological issue, but with a certain way of relating science to belief in God.

Process theology tends to lean on the claim that its own explanation of cosmic process is superior to the conventional one given by the astrophysicist and the neo-Darwinian biologist. Though process philosophers differ in the detail of their accounts, they would agree in maintaining that notions like striving are required for the understanding of material process generally, and that evolutionary change testifies directly to the shaping action of mind. Whether to take this mind to be part of "nature" (and perhaps risk the charge of pantheism) or to set it in some way "above" nature (thus limiting the scope of a purely "natural" science) is the dilemma that process thinkers have wrestled with from the beginning. But whichever option they choose, it is clear that their approach presupposes a quite specific physico-theology, one that depends for its persuasiveness on the proposition that the categories of conventional natural science are inadequate for the explanation of evolutionary process. We have seen enough by now, perhaps, to lead us to be wary of any such proposal.

The alternative way is the one we traced earlier, the one that harks back to Augustine. The emphasis there was upon the transcendence of God, and the self-sufficiency in "natural" terms of the universe he created. There would, then, be no need to invoke a special not-quite-natural energy-animating cosmic process. The "chance" and "necessity" of the evolutionary story could be reinterpreted; they are such only in the eyes of the theorist. For the Creator, there is neither chance nor necessity: only a single Act in which *all* comes to be. Within that Act, particular actualities come to be out of a virtually limitless field of possibilities:

> The given natures of things are not forced, but out of them are drawn a particular world of life, consciousness and spirit, particular human histories and particular human destinies. Given the natures of elementary particles, atoms and molecules, the process of creation of life and of personal being was a long one. Nevertheless, God has drawn out of this evolving cosmos a world of persons. Given the nature of human beings thus rooted in the physical world, it was again a long process (not *so* long, however) before their complex history, including the history of religions, could become the vehicle of special revelation to the point of incarnation.[76]

The strengths and weaknesses of this approach are, as one might expect, almost the complement of those of the other. The transcendence of God is strongly emphasized; God is not made to depend on the universe for his own being. Yet he is present in the universe through his conserving power at every moment and in every place. He is not active in a different way in evolutionary process or in the working of the human mind; he works equally in all parts of his creation.[77] The traditional doctrine of the Trinity conveys something of this complex relationship between God and his world; God is present to the universe not only as creator and conserver but also in a quite special way through the history of a particular planet, a particular people, a particular person. Because nature in this view is conceived as complete in its own order, the integrity of natural science is in no way challenged. But, of course, the negative side of this is that, since there are no real "gaps" to fill, we may be left without an argument for God's existence of the kind that would convince a science-minded generation. God does not *seem* to make a difference, not at least of the sort that science can deal with on its own terms.[78]

There is, of course, one large difference that he *does* make on this view: it is he who brings it about that there should be a universe for the scientist to study in the first place. This is not a question that would ordinarily arise for the scientist; modes of explanation that rely on the regularity of natural process could not be deployed here. Yet the question is one that ought not be disallowed on *a priori* grounds. The appeal is not to a "gap" in scientific explanation but to a different order of explanation that leaves scientific explanation intact, that explores the conditions of possibility for there being *any* kind of scientific explanation.

The issues here are intricate and much debated. At the root of the disagreements are very different ways of relating the major ways of knowing: natural science, philosophy, theology, history, aesthetics, politics.... The question we have been tracing quite evidently requires a collaborative answer. Contemporary specialized scholarship offers little in the way of guidance as to how such collaboration is to be regulated.

NOTES

[1] An earlier version of this paper appears as "Natural science and belief in a Creator" in *Religion, Science, and Search for Wisdom*, ed. David M. Byers (Washington: National Conference of Catholic Bishops, 1987) 14-41. This revised and enlarged version owes much to the critique of my colleagues at the Center for the Philosophy of Religion at the University of Notre Dame, who afforded me the opportunity to discuss the paper in serial fashion at their weekly colloquium. I am also grateful to Dr. Robert J. Russell and the Center for Theology and the National Sciences at Berkeley. A stint at the Center as the J. K. Russell Fellow in February 1988 helped me complete work on the paper.

[2] For an example of how the historical approach can illuminate many of the issues dealt with here, see *God and Nature*, ed. David Lindberg and Ron Numbers (Berkeley: University of California Press, 1986).

[3] "In Israel's faith, redemption was primary, creation secondary, not only in order of theological importance, but also in order of appearance in the Israelite tradition." Bernhard W. Anderson, "The Earth is the Lord's," *Is God a Creationist?*, ed. Roland M. Frye (New York: Scribner's, 1983) 176-196; see especially 180.

[4] See Dianne Bergant and Carroll Stuhlmuller, "Creation according to the Old Testament," in *Evolution and Creation*, ed. Ernan McMullin (Notre Dame: University of Notre Dame Press, 1985) 153-175.

[5] *Exodus*, 19:5.

[6] *Jeremiah*, 27:5.

[7] *Jeremiah*, 31:31-34.

[8] Translation from the Jerusalem Bible.

[9] See the discussion in David Kelsey, "The doctrine of creation from nothing," in *Evolution and Creation*, 176-196; see 186.

[10] See *Genesis*, 7:11; 8:2.

[11] *Job*, 38:4-7; 12-13.

[12] Anderson, "The earth is the Lord's," p. 184.

[13] *Timaeus*, transl. B. Jowett, 30B.

[14] *Timaeus*, 53B.

[15] *On the Parts of Animals*, Oxford translation, Book I, chapter 5, 645a, 10-75.

[16] *On the Parts of Animals*, 644b 23-25.

[17] *On the Parts of Animals*, 644b, 34 - 644a, 1.

[18] *Nichomachean Ethics*, X, 8; 1178a, 23-31.

[19] *Second Maccabees*, 7:28.

[20] *Romans*, 4:17.

[21] See Kelsey, "The doctrine of creation from nothing," 184-192.

[22] For a fuller discussion, see McMullin, "Evolution and Creation," in *Evolution and Creation, op. cit.*, 9-16, and R. A. Markus, "Augustine," *Cambridge History of Later Greek and Early Medieval Philosophy*, ed. A. H. Armstrong (Cambridge: Cambridge University Press, 1967) 395-405.

[23] *De Genesi ad litteram*, V, 20; *The Literal Meaning of Genesis*, translated by J. H. Taylor (New York: Newman, 1982) 171.

[24] Jacques Monod, *Chance and Necessity* (New York: Knopf, 1971).

[25] Augustine used the phrase *"ad litteram"* much more broadly, but the point he was making about the *Genesis* text can be accurately rendered by the modern sense of the term, "literal".

[26] *Enchiridion*, adapted from the translation by Albert Outler, *Library of Christian Classics* (Philadelphia: Westminster Press, 1955) 342.

[27] *Apologetic Treatise on the Hexemeron, Patrologia Graeca*, ed. J.P. Migne, *44*, col. 72; quoted by E.C. Messenger, *Evolution and Theology* (New York: Macmillan, 1932) 24; translation modified.

[28] This is discussed more fully in McMullin, "Evolution and creation," 11-16.

[29] *De Trinitate*, III, 9.

[30] *Literal Meaning of Genesis*, V, 23, 175.

[31] See David Lindberg, "Science and the early church," *God and Nature*, 37-8.

[32] *City of God*, XXI, 8; translated by Gerald Walsh and Daniel Honan, *Fathers of the Church* (Washington: Catholic University Press) **8**, 362.

[33] It would be an exaggeration to say that he made this entirely clear. There are hesitations in his exposition, and the view will only be worked out in its fullness by his successors. But the essentials of it are there in Augustine, and this is what I want to stress here. See McMullin, "Evolution and creation," 15.

[34] *The Confessions of St. Augustine*,translated by John K. Ryan (New York: Doubleday, 1960) X, 6, 234.

[35] *Romans*, I, 20.

[36] *Confessions, X*, 6, 235.

[37] See Frederick Crosson, "Cicero and Augustine," to appear.

[38] See Edward Grant, "The Condemnation of 1277, God's absolute power, and physical thought in the late Middle Ages," *Viator* **10** (1979) 211-244.

[39] See, for example, R. Hooykaas, *Religion and the Rise of Modern Science* (Grand Rapids: Eerdmans, 1972).

[40] Cited by Etienne Gilson, *History of Christian Philosophy in the Middle Ages* (New: Random House, 1955) 471.

[41] Part I, q. 2, a. 3.

[42] Only the first of the five "Ways" is drawn directly from Aristotle; the third is inspired by Avicenna while the fourth is neo-Platonic in concept. Gilson remarks that Aquinas is drawing here on philosophies that are, at bottom, inconsistent with one another. Averroes, for example, maintains that the existence of God can be proved only through physics; hence for him, the proof from motion is the only valid one. Avicenna, on the other hand, insists that the proof must be metaphysical. Gilson argues that Aquinas took these different "Ways" from the works of his predecessors, cast all of them in the broadly Aristotelian language of natural philosophy, but transformed them in the light of his own distinctively Christian understanding of the notion of efficient cause. See Etienne Gilson, *Elements of Christian Philosophy* (New York: Doubleday, 1959) 80-87.

[43] In the much fuller exposition of this argument that Aquinas gives in the *Summa Contra Gentiles*, I, 13, he cites Aristotle's *Metaphysics* at the crucial point in sec. 28, but this does not remove the assertion of an Unmoved Mover from natural philosophy. Because the *Metaphysics* equated the study of being with the study of substance, and because the Unmoved Mover is the sole example of unmoved substance, its study belongs both to physics (where its existence is required for the completeness of the explanation of motion) and metaphysics.

[44] Anthony Kenny has summarized the critical case against the Five Ways, considered as logical demonstrations of God's existence. See *The Five Ways* (Notre Dame: University of Dame Press, 1980).

[45] Gilson, *Elements of Christian Philosophy*, 85.

[46] Gilson, *Elements of Christian Philosophy*, 67.

[47] The noted Thomist author, Fernand Van Steenbergen, tells of the crisis that a paper of his, mildly critical of the logic of the Five Ways, caused at the

International Thomist Congress in Rome in 1950. His paper merely argued that Aquinas's proofs had to be reformulated to make them cogent for the modern reader. But even this implicit criticism was enough to elicit outrage. The encyclical, *Humani Generis*, had just appeared and was repeatedly invoked in the Congress discussions against those who would tamper with the venerable text of the Five Ways. See *Dieu Caché* (Louvain: Presses Universitaires, 1961) 169.

[48] It was introduced by that fertile inventor of terms, William Whewell, in 1840 in his *Philosophy of the Inductive Sciences* (London, 1840).

[49] Winifred Wisan, "Galileo and God's Creation," *Isis* 77 (1986) 473-486.

[50] There was, of course, the human soul or mind, but this was not, strictly speaking, part of the material world. Descartes nowhere attempted to utilize the existence of the soul as a starting-point for an argument that would conclude to the necessity of a Creator of souls.

[51] *A Disquisition about the Final Causes of Natural Things*, from *Works* (London, 1744) vol. 4, 522.

[52] *Op. cit.*, 521.

[53] *Op. cit.*, 520.

[54] *Isaac Newton's Papers and Letters on Natural Philosophy*, ed. by I. Bernard Cohen (Cambridge, Mass.: Harvard University Press, 1958) 280.

[55] *Op. cit.*, 284.

[56] *Op. cit.*, 286.

[57] *Universal Natural History and Theory of the Heavens* (1755), transl. by W. Hastie (Ann Arbor: University of Michigan Press, 1969) 27. See William Shea, "Filled with Wonder: Kant's Cosmological Essay," in *Kant's Philosophy of Physical Science*, ed. Robert Butts (Dordrecht: Reidel, 1986) 95-124.

[58] The argument is found in the section of the Transcendental Dialectic called "The Ideal of Pure Reason."

[59] *Critique of Pure Reason*, transl. by J. M. D. Meiklejohn (London: Bell, 1908) 383.

[60] There was an unresolved ambiguity in the Design arguments of the seventeenth century. Did they require a causal *intervention* in the natural sequence (a "miracle") or only an *initial* shaping at the first creation? Boyle and Ray assumed that God had formed the animal world at the beginning; although they are not clear on this, they seem to suppose that the later evidences of adaptation in each generation would not require a *continuing* "non-natural" intervention on God's part.

[61] See McMullin, "Teilhard as a philosopher," *Chicago Theological Seminary Register* **60** (December 1964) 15-28.

[62] *The Phenomenon of Man* (New York: Harper, 1965) 149.

[63] In a recent sympathetic account of Teilhard's work written from a biologist's standpoint, Edward Dodson notes that there are difficulties with Teilhard's notions of the Within, of critical thresholds in biology, of the orthogenetic-development of complexity-consciousness, and of the Omega Point. His own recommendation, more favorable than that of most of his colleagues, is that since these concepts can neither be verified nor disproved on empirical grounds, it is best, for the moment, to suspend judgement on them. See *The Phenomenon of Man Revisited* (New York: Columbia University Press, 1984) 239-243.

[64] The principle may rest on a misunderstanding of the notion of potency. To say that A has the potency to give rise to something with property B does *not* mean that A must already possess property B in some diminished way. See McMullin, "Four senses of potency," in *The Concept of Matter in Greek and Medieval Philosophy* (Notre Dame: University of Notre Dame Press, 1965).

[65] This appears to be one of the major arguments in support of Teilhard's cosmology for N. Max Wildiers, *The Theologian and His Universe* (New York: Seabury, 1982) chap. 8.

[66] *The Phenomenon of Man*, 292.

[67] There is by now a very large literature on the so-called "anthropic principle". The most thorough account is that of John Barrow and Frank Tipler, *The Anthropic Cosmological Principle* (New York: Oxford University Press, 1986). Their treatment of some topics, particularly of what they call the "Final Anthropic Principle," is however exceedingly speculative. See also John Leslie's numerous articles on this subject, for example, "Anthropic Principle, World Ensemble, Design," *American Philosophical Quarterly* **19** (1982) 144-151 and in this volume, 297-311; and E. McMullin, "How should cosmology relate to theology?", in *The Sciences and Theology in the Twentieth Century*, ed. Arthur Peacocke (Notre Dame: University of Notre Dame Press, 1981) 17-51; see especially 40-46.

[68] The term was first used by Brandon Carter, "Large number coincidences and the anthropic principle in cosmology," in *Confrontation of Cosmological Theory and Astronomical Data*, ed. M.S. Longair (Dordrecht: Reidel, 1974) 291-298.

[69] See Alan H. Guth, "Inflationary Universe: A possible solution to the horizon and flatness problems", *Physical Review D* **23** (1981) 347-356.

[70] In an essay that excited considerable controversy, Freeman Dyson reaches a carefully-stated conclusion: "I conclude from the existence of these accidents of physics and astronomy that the universe is an unexpectedly hospitable place for living creatures to make their home in. Being a scientist, trained in the habits of thought and language of the twentieth century rather than the eighteenth, I do not claim that the architecture of the universe proves the existence of God. I claim only that the architecture of the universe is consistent with the hypothesis that mind plays an essential role in its functioning," *Disturbing the Universe* (New York: Harper, 1979) 251; see *The Argument from Design*, 245-253.

[71] The book that set the fashion here was Fritjof Capra's *The Tao of Physics: An Exploration of the Parallels between Modern Physics and Eastern Mysticism* (Berkeley: Shambhala, 1975). For a critical review of this genre of literature, see Sal P. Restivo, "Parallels and paradoxes in modern physics and Eastern mysticism," *Social Studies of Science* **8** (1978) 143-181.

[72] Gordon Kaufman makes this point in his *God the Problem* (Cambridge, Mass.: Harvard University Press, 1972).

[73] See, for example, William Temple, *Nature, Man and God* (London: Macmillan, 1934); John Cobb, *A Christian Natural Theology* (Philadelphia: Westminster, 1965). For a useful review of the varieties of the "process" approach to philosophy and theology, see Eric Rust, *Evolutionary Philosophies and Contemporary Theology* (Philadelphia: Westminster, 1969).

[74] Charles Hartshorne's voice has been the most influential of those who have attempted to adapt the system of Whitehead to make it the foundation of a Christian world-view. See, for example, *The Divine Relativity* (New Haven: Yale University Press, 1964). For other representatives of this school, see *Process Philosophy and Christian Thought*, ed. Delwin Brown, Ralph James and Gene Reeves (Indianapolis: Bobbs-Merrill, 1971).

[75] For an illustration, see a recent essay by Langdon Gilkey where he argues that the "fundamental puzzles of Christian theology — and of the texts on which that theology is based — are more appropriately understood if we view the divine mystery as a polarity of being and non-being rather than as a mystery of absolute being." The explanation of evolutionary process plays no part in his argument, "Creation, Being, and Non-Being," to appear in *Creation in Interfaith Perspective*,

ed. David Burrell and Bernard McGinn (Notre Dame: University of Notre Dame Press, to appear).

[76] Brian L. Hebblethwaite, "Providence and Divine Action", *Religious Studies* **14** (1978) 223-236; see 231.

[77] Howard Van Till calls this the "creationomic perspective", and defends it ably in *The Fourth Day* (Grand Rapids, Mich.: Eerdmans, 1986) chap. 12.

[78] Austin Farrer tries to meet this objection in his *A Science of God?* (London: Bles, 1966).

THE NEWTONIAN SETTLEMENT AND THE ORIGINS OF ATHEISM

Michael J. Buckley, S.J., National Council of Catholic Bishops, Washington, D.C.

Introduction

The nature of the sciences and their relationship one with another have engaged the educated intellects of Western civilization since the centuries of the ancient Greeks, but neither the definitions of these disciplines nor the texture of their interrelationships or mutual adjustments have achieved a fixity that would allow either for a stable curriculum within institutions of higher learning and inquiry, or for a settlement which one generation would, without radical modification, accept from those who had gone before. To study the history of intellectual culture is to encounter with some fair degree of regularity such issues as what constitutes the physical sciences and mathematics; how these relate to the biological sciences and (even more) to the newly designated *Geisteswissenschaften*; how all of these bear resemblances or distinguishing differences to what passes for the arts of a given period or for "the more humane letters"; and how validly one can differentiate or dialectically identify theoretical, practical, and productive inquiries, and successfully delimit from these the arts of interpretation, persuasion, and creativity.

These common issues of culture assume an increasing complexity when one asks about the relationship of such sciences or arts or disciplines with theology. Helmut Peukert has remarked that "in our century theories of science have radically challenged the very possibility of theology," while arguing that a "certain convergence can be established between contemporary reflection on the fundamental principles of theology on one side and the results of research into the theory of science on the other." Peukert lays the basis for this fundamental theology in Habermas' theory of communicative action.[1] But over the centuries one can trace a similar reformulation of theology with its commensurate recasting of appropriate issues and methods as one or another of the academic disciplines assumes an hegemony in systematic reflection. The idealization of geometry, for example, brought Alanus de Insulis to frame the first apologetic work of the Middle Ages within a set of definitions, postulates, and axioms from which he would deduce theorem after theological theorem to demonstrate the truth of the Catholic faith against the Albigensians, the Waldensians, the Jews, and the Muslims. The legal and rhetorical methods of Roman controversy, moving dialectically through oppositions to their resolutions in the tracts of Hincmar of Rheims and Gratian, advanced into theology through the *Sic et Non* of Abelard and then into Peter Lombard's

Book of the Sentences, quite literally the theological textbook for all of Western Europe until the Reformation. Grammar gave its methods of exegesis and interpretation to constitute a monastic theology which was prior to this scholastic revolution and which countered its advance, while the rhetorical arts of invention issued in the theologies of Augustine and Luther. It is not an idle piece of information to discover that Thomas Aquinas never made a formal study of rhetoric, while Augustine gave over his early career to the teaching of this art.[2]

What constitutes the sciences, or any disciplined human inquiry, and how they bear upon theology are critical questions, and the variety of answers given in the history of culture have specified much of the structure and value of the theology of a particular period. The understanding of a serious intellectual discipline frames the settlement within which theology will be understood and attempted.

This paper proposes to explore one such settlement, that which dominated the latter half of the seventeenth century and much of the eighteenth and so accompanied the rise of atheism in the Western world. Its explorations ask whether there is any connection between two remarkably important and mutually contradictory developments within modernity: the *Newtonian Settlement* and the rise of atheism. To do so entails some framing of the *Newtonian Settlement,* by reflecting somewhat upon Newton's work, upon the world that preceded him, and upon the world of followers that turned Newton into Newtonian. The paper proposes to investigate only something of this complex question because neither the limits of a single paper nor, for that matter, of a single conference here at Castel Gandolfo, could do such a subject justice.[3]

Posing the Issues: Mechanics, Mathematics, and Theology

At the juncture where the *Two New Sciences* introduces the generation of circles either from the center of a horizontal plane or from the top of a sphere, Simplicio attempts to raise this last and greatest dialogue of Galileo about mechanics to a higher subject. This geometrical wonder, he suggests, "leads one to think that there must be some great mystery hidden in these true and wonderful results, related to the creation of the universe (which is said to be spherical in shape) and related also to the seat of the first cause (*prima causa*)." But Salviati and Sagredo will have none of this peripatetic enthusiasm, whatever its influence from the *Timaeus.* Mechanics is not the place to bring up theological considerations. "Profound considerations of this kind belong to higher sciences than ours [*a più alte dottrine che le nostre*]," says Salviati, the interlocutor who expresses the teachings of the Master. "Now, if you please, let us proceed."[4] Proceed they did, generating by their conversations Galileo's last work and the one which he considered "superior to everything else of mine hitherto published."[5] Galileo was writing about mechanics and local motion, and there was no place there for the theological aspirations of Simplicio.

Mechanics was a limited science. Sagredo had made it initially depend directly upon mathematics, laying its foundations securely in geometry.

Salviati would, in some contrast, praise those who had constituted it a science distinct from mathematics, even those "cutting loose from geometry" and so able to deal with "the imperfections and the variations of the material."[6] But Salviati would insist that resistance was so fixed and constant a property that it could be treated with the same rigor as mathematical figures, even if it prevented material objects from being simply reduced to mathematical figures. There might be disagreement about the nature of mechanics, abstract and concrete subject-matters, but both Sagredo and Salviati would agree that to introduce considerations of a *prima causa* was to introduce a foreign element.

Now that is precisely what Isaac Newton denied. It seems curious to line up Newton with Simplicio, but on this last issue there is no escape. From the *Two New Sciences* one can abstract three issues: (1) the relationship between mechanics and mathematics; (2) the extension of mechanics as a *dottrina*; and (3) the appropriateness of theological considerations within that ample extension. Newton's resolution of the first two issues will tell both in the philosophy of science and in the philosophy of religion, but we are concerned with them here only because they form the necessary context to understand his resolution of the third issue, i.e. the relationship between mechanics and religion. It is upon this last problem that this inquiry principally devolves.

On the first issue, Salviati's grudging praise is misplaced and Sagredo is exactly wrong. Geometry is neither distinct from mechanics nor is it foundational. In the preface to the first edition of the *Principia*, Newton insisted — in direct contradiction to the Cartesians — that mechanics does not obtain its principles from geometry, but that geometry receives its principles from mechanics, "for the description of right lines and circles upon which geometry is founded belongs to mechanics. Geometry does not teach us to draw these lines, but requires them to be drawn." The relationship is precisely the opposite to that formulated by Sagredo, for the student must first learn to describe geometrical objects mechanically before geometry can begin. "Hence geometry is founded in mechanical practice, and is nothing else but that part of universal mechanics which accurately proposes and demonstrates the art of measuring."[7] How directly such a position posed an alternative to the going Cartesianism can be gauged by contrasting it with a single sentence in which Descartes summarized his own thought for Père Marin Mersenne: "My entire physics is nothing else than geometry."[8] Newton's claim remains an extraordinary one. The foundations of mathematics do not lie in intuition [as, much later, Poincaré argued] nor in logic [following Russell and Whitehead], but "in mechanical practice."

On the second issue, Newton gives a comprehension to mechanics that allowed him to include not only mathematics within its compass, but all the phenomena of nature and its philosophy. Mechanics comprises "the whole burden of philosophy." Newton would extend this sense of Universal Mechanics beyond Galileo and beyond Descartes. *Mechanica* would become *universalis* by comprehending both mathematics and also theology, needing no first philosophy either to furnish a justification for his mechanics or to establish the existence of God in order to guarantee the extra-mental world.

Startling as such a claim may have been at the time, so universal a view of mechanics continues a tradition reaching back to Heron and Pappus of Alexandria. The first line of Newton's preface to the first edition calls attention to the eighth book of Pappus' *Synagoge*. It is in Pappus that this massive extension of the mechanical imperium can be located. Mechanical theory surpasses Aristotle's physics, contended Pappus, because mechanics deals not just with the natural movement of bodies, but also with those locomotions effected by violent efforts contrary to natural, "physical" tendencies. It further overmatches the Aristotelian science because it treats the subject-matter of physics mathematically, a procedure which the Second Book of the *Physics* indicates is incommensurate with the kind of hypothetical necessity intrinsic to natural things. One can mathematize the study of natural things, of course; one can do mechanics. As a matter of fact, Aristotle is credited with a book by that name now lost. But if one confuses this kind of inquiry with physics properly so called, one will lose the uniqueness of nature. Precisely wrongheaded, countered Pappus. This change of the subject-matter to include all movement and the introduction of mathematics directly into the heart of the method unleashed the genius of mechanics. The School of Heron made mechanics into a universal discipline, and it divided this mechanics precisely as Newton would later do into the rational (λογικὸν) and the manual (χειρουγικὸν). The sweep was enormous: physics and astronomy, geometry and arithmetic — all of those sciences which would later form the medieval quadrivium and constitute the disciplines that dealt with things as opposed to the disciplines that dealt with words. The manual included a mastery in painting and carpentering, metal work and architecture, and "anything that would involve manual skill," or all of the "arts" which would be called, in the most narrow sense, the mechanical arts. Placed together, they bestowed a universality on mechanics which would give it "perhaps first place (σχεδὸν πρώτη) among the natural inquiries which deal with the matter of the elements in the world."[9] Mechanics did bestride the Alexandrian world of Heron and Pappus like a Colossus. Newton explicitly aligns himself with this intellectual tradition, changing the somewhat modest σχεδὸν πρώτη to the prouder claim of *maximi* and elaborating a mechanics which would not only give mathematics its foundational elements but theology its subject-matter.

From Mechanics to the First Cause

And this takes us to the third issue: to the scandal of the Cartesians. Newton did not formulate or indeed have any patience with a first philosophy that would establish the existence of God and eliminate the skepticism of Montaigne. Newton's scientific temperament possessed its own liabilities, but skepticism would not find a place among them. He did not need God to guarantee the existence of the world. Quite the contrary. He needed the world to demonstrate the existence of God. And as for this demonstration, it was to be brought under the competencies of a mechanics now coming into its universality.

Newton concludes the extensive discussion of the divine existence conducted in the General Scholium with this assertion: "And thus much concerning God: to discourse of whom from the appearance of things certainly does belong to Natural Philosophy."[10] He considered this demonstration to be included within the "main business of natural philosophy," and the defense of his theological interests usually occurs in those rare places in his published works where he sets the method of his mechanics in sharp contrast with that of the Cartesians:

> The main Business of natural Philosophy is to argue from Phaenomena without feigning Hypotheses, and to deduce Causes from Effects, till we come to the very first Cause, which certainly is not mechanical; and not only to unfold the Mechanism of the World, but chiefly to resolve these and such like Questions.

Note the questions that Newton maintains belong to mechanics "chiefly to resolve:"

> What is there in places almost empty of Matter, and whence is it that the Sun and Planets gravitate towards one another, without dense Matter between them? Whence is it that Nature doth nothing in vain; and whence arises all that Order and Beauty which we see in the world? To what end are Comets, and whence is it that Planets move all one and the same way in Orbs concentrick, while Comets move all manner of ways in Orbs very excentrick; and what hinders the fix's Stars from falling upon one another?

Newton continues in a manner that will allow Mechanics to incorporate the discoveries of William Harvey and Marcello Malpighi in providing evidence for the divine existence:

> How came the Bodies of Animals to be contrived with so much Art, and for what ends were their several Parts? Was the Eye contrived without Skill in Opticks, and Ear without Knowledge of Sounds? How do the Motions of the Body follow from the Will, and whence is the Instinct in Animals? Is not the Sensory of Animals that place to which the sensitive Substance is present, and into which the sensible Species of Things are carried through the Nerves and Brain, that there they may be perceived by their immediate presence to that Substance?

Such questions inevitably force mechanics into consideration of a first cause that is non-mechanical:

> And these things being rightly dispatch'd, does it not appear from Phaenomena that there is a Being incorporeal, living, intelligent, omnipresent, who in infinite Space, as it were in his Sensory, sees the things themselves intimately, and throughly [sic] perceives them, and comprehends them wholly by their immediate presence to himself.

Mechanics is not theology, but it serves to ground theology, especially by establishing the existence of its fundamental subject-matter. Mechanics is not religion, is distinct from revelation, but it renders both credible:

> And though every true Step made in this Philosophy brings us not immediately to the knowledge of the first Cause, yet it brings us nearer to it, and on that account is to be highly valued.[11]

To Richard Bentley, the great classicist and the first of the Boyle lecturers, Newton had written some five years after the first edition of the *Principia*:

> When I wrote my treatise about our Systeme, I had an eye upon such Principles as might work with considering men for the beliefe of a Deity; and nothing can rejoice me more than to find it used for that purpose.[12]

One does significant violence to the history and the achievement of ideas, if one takes "mechanics" as a word possessing a single meaning and designating one obvious subject-matter. Isaac Newton endows mechanics with a comprehension of meaning and an extension of subject that neither Galileo or Descartes before him nor many in contemporary science would admit. For Newton, it was mechanics which both provided the foundations of geometry and also established the existence of God.

This theological interest was intrinsic to the universal mechanics for at least two reasons: the nature of its subject-matter and the progress of its method. If absolute motion entailed absolute space and time for its existence, it entailed the infinite and the eternal, the immutable and the impassible. Inevitably the question arose whether such realities were not divine, whether Newtonian mechanics were not covertly dealing with God when dealing with these absolutes. Secondly, if the mechanical method of analysis, relentlessly following the procedures indicated by the parallelogram of forces, had to resolve motions back to their aboriginal forces, and if any system could be treated in all of its complexity as such a motion demanding such a resolution, and if the Cartesian insistence upon a final resolution through mechanical principles was unwarranted and, indeed, led to the illegitimate feigning of hypotheses, then it was equally inescapable that the continuation of analysis in the calculation of motions, masses, geometrical patterns, and balances which composed the system of the world would lead "to a first cause which certainly is not mechanical."[13] I have argued this case at the Cracow Conference earlier this year against a previous article by Professor Edward Strong, and I shall not repeat my argument here. What I should like to do, however, is to build upon it.[14] I want to suggest that Newton saw mechanics not only as corroborating theology, but as serving for its foundation. Here we pass in contemporary discussions to an issue posed recently by such books as *God and the New Physics* by Professor Paul Davies and by theories such as that of the much contested anthropic principle of the past five years, associated with such distinguished names as John Barrow, Frank Tipler, and John Leslie. It is the issue of the use of science to ground religious affirmation. That issue was a live one also for Isaac Newton.

Mechanics as a Foundation for Theology

In an extraordinary paragraph, the last in the *Opticks*, Newton claims that moral philosophy will be enlarged by the methods of analysis and composition which structure mechanics or natural philosophy. How so?

> For so far as we can know by natural Philosophy what is the first Cause, what Power he has over us, and what Benefits we receive from him, so far our Duty towards him, as well as that towards one another, will appear to us by the Light of Nature.

And this, in turn, corrects the polytheism of the pagans and teaches authentic worship:

> And no doubt, if the Worship of false Gods had not blinded the Heathen, their moral Philosophy would have gone farther than to the four Cardinal virtues; and instead of teaching the Transmigration of Souls, and to worship the Sun and Moon and dead Heroes, they would have taught us to worship our true Author and Benefactor, as their Ancestors did under the Government of *Noah* and his Sons before they corrupted themselves.[15]

Religion itself, Newton had written in his *Short Scheme of the True Religion*, is "partly fundamental and immutable, partly circumstantial and mutable." This distinction obviously parallels in many ways that of absolute and relative motion, space, and time. Fundamental religion "consists of two parts, our duty towards God and our duty towards man, or piety and righteousness, which I will here call Godliness and Humanity."[16] Now these are precisely the two areas which the *Opticks* called "moral philosophy," and which the mechanical examination of the phenomena of nature was to purify or even to establish. It is not a great leap to assert that Newton saw mechanics as providing for theology what it provided for mathematics: its foundation. This does not mean that Newton collapsed any distinction between religion and mechanics. He expressly stated that they are to be kept distinct when religion is identified with revelation: "We are not to introduce divine revelations into Philosophy, nor philosophical opinions into religion."[17] But natural philosophy or mechanics could give the foundation for the credence which one extended to the objects of revelation. Mechanics could dispose of the objections of the atheist:

> Opposite to the first [Godliness] is Atheism in profession, and idolatry in practice. Atheism is so senseless and odious to mankind that it never had many professors. Can it be by accident that all birds, beasts, and men have their right side and left side alike-shaped (except in their bowels), and just two eyes and no more, [one] on either side the face, and just two ears, [one] on either side the head, and a nose with two holes and no more between the eyes, and one mouth under the nose, and either two fore-legs, or two wings, or two arms on the shoulders, and two legs on the hips, one on either side and no more? Whence arises this uniformity in all their outward shapes, but from the counsel and contrivance of an Author? Whence is it that all the

eyes of all sorts of living creatures are transparent to the very bottom and the only transparent members in the body, having on the outside a hard transparent skin and within transparent layers with a crystalline lens in the middle and a pupil before the lens — all of them so truly shaped and fitted for vision that no Artist can mend them? Did blind chance know that there was light and what was its refraction, and fit eyes of all creatures after the most curious manner to make use of it?[18]

The argument from design lies at the foundation of Newton's affirmation of the existence of God.

These and suchlike considerations, always have, and ever will prevail with mankind to believe that there is a Being 1) who made all things and 2) has all things in his power, and 3) who is therefore to be feared.[19]

These are precisely the three persuasions which the *Opticks* had claimed that natural philosophy would instill. And the "main business" of Newtonian mechanics was to move carefully either by analysis or by synthesis to provide for religion its fundamental basis.

What one sees in the universal mechanics of Newton is not the mechanics of Descartes or even of Leibniz. Rational mechanics can be generally understood as "the science of motions resulting from any forces whatsoever, and of the forces required to produce any motions, accurately proposed and demonstrated." And this very easily identifies with philosophy itself: "For the whole burden [*difficultas*] of philosophy seems to consist in this — from the phenomena of motion to investigate the forces of nature and then from these forces to demonstrate the other phenomena."[20]

Motion in Newton seems much less puzzling than force. For force became very quickly in the definition either *potentia resistendi*, differing only in perspective from the inactivity of mass, or impressed force, the *actio* upon the body which induces mutation of any kind. In Books I and II of the *Principia* impressed force, the influence behind any change, become particularized as motive force to deal with the motion of bodies in and out of a resisting medium. To explain the inner coherence of the system of the world, it becomes gravity in Book III. And finally, to explain the existence of "this most beautiful system of the sun, planets, and comets," Newton no longer spoke of motive force or gravity; he spoke of dominion [*dominium or dominatio*] and he makes this dominion the principal theological attribute of God.

This Being governs all things, not as the soul of the world, but as Lord over all; and on account of his dominion, he is wont to be called Lord God, παντοκράτωρ, or Universal Ruler; for God is a relative word, and has a respect to servants; and Deity is the dominion of God, not over his own body, as those imagine who fancy God to be the soul of the world, but over servants. . . . It is the dominion [*dominatio*] of a spiritual being which constitutes a God: a true, supreme, or imaginary dominion makes a true, supreme or imaginary God."[21]

Dominion makes God to be God. Dominion is to all things what impressed forces are to change or what gravity is to the systematic coherence of the planets.

Newton's mechanics reduces all things, bodies or motions or systems back to force. But force is ultimately — that is, in its farthest reaches as "first cause" — not a mechanical reality. It is the dominion of God. Newton has crafted a mechanical method which scorns the Cartesians who "banish the consideration of such a Cause out of natural Philosophy, feigning Hypotheses for explaining all things mechanically, and referring other Causes to Metaphysicks."[22] For these causes do not belong in metaphysics. As a matter of fact, nothing that Newton prized belonged in metaphysics! These causes belong here, in a natural philosophy that was a universal mechanics, supplying the foundations of geometry and of religion, an interrelationship which comprises the *Newtonian Settlement*.

It was not strange that science should take up the cudgels against a putative atheism. The great Robert Boyle had yielded to no one in the seventeenth century in his insistence that the new science was providentially oriented to disclose the divine footprints in nature. It was Boyle's legacy that provided for an annual series of public lectures "to prove the truth of the Christian religion against infidels, without descending to any controversies among Christians."[23] Richard Bentley had inaugurated this series with his *Confrontation of Atheism,* and Newton had aided him in extensive correspondence with the mechanical data that could serve as foundational to the natural theology he was to argue. Some twelve years later, Samuel Clarke, the greatest of all of the immediate followers of Newton, continued the series with *A Demonstration of the Being and Attributes of God.* In building his lectures upon a Newtonian model, Clarke proposed that he would employ "One only Method or continued Thread of Arguing; which I have endeavoured should be as near to Mathematical, as the Nature of such a Discourse would allow."[24] Clarke labored to bring theology into the world of Newton, whose mechanics he had introduced into English text-books, whose *Opticks* he would translate into Latin and whose defender he would prove in the celebrated correspondence with Leibniz. As did Newton, Samuel Clarke argued in propositions and theorems. The *Demonstration* developed through twelve propositions to establish the existence of the necessary Being and his essential, personal attributes. The mechanics of Newton together with its assimilation of new discoveries in anatomy furnished foundational evidence. To prove against Spinoza, for example, that such a necessary being could not be material, Clarke invoked Newton's doctrine of the *vis inertiae,* resistance, as essential to matter. The lack of this resistance in many "places" indicates that all space is not material, that there must be vacua, indeed that much of space is empty of matter. "It follows plainly, that *Matter* is *not* a *Necessary Being*. For if a *Vacuum* actually be, then 'tis evidently more than possible for Matter *not to Be*." Newton's mechanics had done more than indicate that matter may not be necessary; it had shown that it was actually not necessary at all. There are any number of places — indeed, over the greatest extent of space — in which it *de facto* does not exist.

> "If a Being can without a Contradiction be absent from one Place, it may without a Contradiction be absent likewise from another Place, and from all Places: And whatever Necessity it may have of Existing, must arise from some External Cause, and not absolutely from itself."[25]

So successful were these lectures, that Dr. Clarke was invited to deliver the subsequent series.

Two factors emerge as of critical importance here. First, Clarke, unlike Bentley the classicist, was a theologian. In his acceptance of Newtonian mechanics, he signaled the reliance which subsequent theology was to place on Newton. Roger Cotes, in his celebrated "Preface to the Second Edition," promised that "Newton's distinguished work will be the safest protection against the attacks of atheists, and nowhere more surely than from this quiver can one draw forth missiles against the band of godless men."[26] Here was a theologian who realized that promise.

Clarke built his case with twelve propositions or theorems. He employs the analysis-synthesis methodology of Newton. The analysis of phenomena back to aboriginal forces entailed the demonstration of the Being of God; synthesis conjoined the divine existence with the impersonal attributes; analysis again accounts for the design within the phenomena by a divine intelligence and freedom, and then synthesis follows with a demonstration of the personal and moral attributes. What refutes either Lucretius and Epicurus or Spinoza and Hobbes is contemporary science, "the *Late* Discoveries in Anatomy and Physick, the Circulation of the Blood, the exact Structure of the Heart and Brain, the Uses of Numberless Glands and Valves for the Secretion and Motion of the Juices in the Body..." And in celestial mechanics, one could cite "the *Exquisite Regularity* of all the Planets' Motions, without Epicycles, Stations, Retrogradations, or any other Deviation or Confusion whatsoever." Clarke brings to court the evidence of the "*inexpressible Nicety* of the Adjustment of the Primary Velocity and Original Direction of the *Annual* Motion of the Planets, with their distance from the Central Body and their force of Gravitation towards it."[27] One after another, the elements and findings of Newton's celestial mechanics are listed in their proportions, balance, and central gravitational unity. Newton has given the theologians, as no one before him, asserts Clarke, the grounds to demonstrate the existence and nature of God.

Thus, in the consideration of these factors that constitute for Clarke "the First Foundations of Religion,"[28] Newtonian mechanics has furnished both the evidence and the methodology with which the existence and attributes of a personal God could be demonstrated: the constitution of the world and the structure of the human body. There were others which could have been used, wrote Clarke, but they were not so integrated into the natural theology which gave foundation and solidity to his own theology. Clarke strengthens and draws out the lines of the *Newtonian*

Settlement, establishing the ground or the foundations of religion. This does not reduce theology to physics nor does it elevate natural philosophy to religion. But it does maintain that there is a radically important relationship between religion and science, here mechanics; that the fundamental cognitive claims about the existence and nature of God are established by science or natural philosophy or (here) a universal mechanics; that religion itself does not offer evidence of comparable cogency — indeed, it may offer little evidence at all for such a basic religious commitment. The *Newtonian Settlement* as such does not argue so extensive a position, but it does maintain that the arguments that come out of the new mechanics furnish religion with its first foundations.

The Historical Context of the Newtonian Settlement

It is not strange that this settlement should have come out of the intellectual ferment of Europe at this time. At the beginning of the century, two major works were written against the atheists. The one was by the Jesuit theologian of Louvain, Leonard Lessius, *De providentia numinis et animi immortalitate* (1613), introduced into English dress under the engaging title of *Rawleigh His Ghost* (1631). The other was by a Franciscan polymath, Marin Mersenne of Paris: *L'Impiété des Déistes, Athées, et Libertins de ce Temps (1624)*. In some respects, the two books could not have been more different, but they were symptomatic of the age. Lessius' atheists were either from the classical lists of antiquity (Diagoras of Melos, Theodore of Cyrene, Bion of Borysthenes, etc.) or those contemporaries who "only secretly among their familiars do vomit out their Atheisme."[29] Mersenne, who claimed that there were 50,000 atheists in Paris alone, gives honor of preeminence to Pierre Charron, Geronimo Cardano, and Giordano Bruno.[30] Lessius could name no atheists, and Mersenne swept into his ample net thinkers who would have rejected his characterization with horror. The fact of the matter is that there were no significant thinkers in Europe who thought of themselves as atheists. This type of person would only emerge in France in the middle of the eighteenth century. Like witches, atheists were discerned everywhere, refuted, run to the earth, and put to death. The only problem is, it is not certain that they existed.

If Lessius and Mersenne agreed upon the danger, they agreed even more upon its remedy. Religion, even the Christian religion, could not establish or defend the existence of God. This was the job of philosophy. Lessius mentions revelation, but dismisses it as belonging to the realm of faith, and turns to the *rationes philosophicae*. Mersenne builds out of the methodology and mechanics whose origins lie with Epicurus to end with the final arguments taken from Augustine and Anselm. The irony of the situation is palpable. The dawn of modernity was raising for religious Europe the great questions of the existence of God, and typically Catholic theologians stood ready to engage these questions as philosophers. The classic arguments and loci of Cicero's *De natura deorum* are all represented: the skeptical dialectic, the Epicurean mechanics, and the Stoic judicative logic. These provided the devices, the "topics", by which the wealth of new facts discovered in the Renaissance could be taken in as further evidence. In 1613, the *fistula dioptrica* was revolutionizing astronomy, and in *De providentia numinis* — published only two years after Galileo's *Siderius nuncius* had announced the discovery of the uneven face of the moon, Jupiter's four Medicean planets, the waxing and waning of Venus — Louvain's Lessius was using these findings as further evidence of the design that justified belief: "Saepe haec omnia ipse instrumento conspexi cum summa admiratione divinae sapientiae et potentiae."[31] Mersenne, whom Professor James Collins called "the clearing house for scientific and philosohical information in the decades just prior to the appearance of the first learned journals," still figures in mathematics for a formula that initially attempted to derive all of the prime numbers, for his investigation of the cycloids, and for his suggestions to Christian Huygens that the pendulum be used as a timing device, "thus inspiring the pendulum clock."[32] Both men used the methods of variant philosophies to incorporate the emerging sciences as their evidence within the world to indicate the existence of God.

The arguments of both are constructed as if sixteen hundred years of Christianity had never occurred. The facts are new, but the topics in which they are used or with which they are argued could be found in Cicero or Plato or, perhaps, even Aristotle. Christianity as such contributes nothing. Both treat the atheistic question as if religion had nothing to say to this issue short of categorical faith, that before this radical challenge to its cognitive claims religion stood empty-handed. The question could only be settled philosophically. By the opening of the seventeenth century, there was widespread conviction that the atheists were everywhere, a fifth column in the religious culture of Europe, and that the defense of the religion had passed to philosophy.

The philosophers had accepted the charge, even leaped to it. In his dedicatory letter to the dean and the theological faculty of the University of Paris, Descartes informed these "sapientissimis clarissimis viris," that such questions were the *opus* of philosophy: "I have always been of the opinion that two questions — those dealing with God and with the soul — were among the principal ones which should be demonstrated by philosophy rather than by theology."[33] How so? Theology is limited by the faith out of which it comes, and this is useless to persuade those who do not believe. Again there is no question that religion might possess its own intrinsic evidence and methodologies, no question of a pneumatology of religious experience or of the historical weight of Jesus. Faith needs the external support of philosophy — less modestly, of the *Meditationes* which Descartes was dedicating to these learned theologians. It was a conviction that he would live with all his life: "I make bold as to say that never has faith been so strongly supported by human reasons, as it can be if my principles are followed."[34] In so many ways Descartes and Newton would divide as superpowers the seventeenth century world between them, but on this they would implicitly agree. The Christian religion is intrinsically powerless to establish or justify the existence of God to those who deny it;

the claims it makes are subsequent to the commitments of faith. To demonstrate the existence of God can neither entail the evidence of religion nor constitute the work of theology.

Whose work was it then? Here Descartes lay down his own particular path which Newton would never walk. For Descartes, it was the work of first philosophy, a metaphysics which needed the existence of God to justify its own assertions that there was a world at all. The contrast with Newton here is total. In Newton, the world served as warrant for the existence of God; in Descartes, God served as warrant for the certain existence of the world. In Newton, a universal mechanics which would deal with all of the phenomena of nature could then incorporate the divine existence; in Descartes, metaphysics would have to establish the divine existence from the proportions given in cognition, and from that deduce the truth of God which would guarantee the existence of the world. Metaphysics served as the foundation for physics; the world itself provided no evidence for God. The study of the world could then reduce all of nature to mechanical principles, to matter upon which motion had been conferred and to causes that move all things through pressure. This became a cardinal principle in Descartes. Since first philosophy had so adequately treated the existence of God, one could have a thoroughly mechanical physics deduced from mechanical principles and excluding any considerations that were theological. The world and its study could be this secular because the enemy of religion along with the skeptics had been destroyed in the wars of metaphysics. By reducing all mechanical problems to mechanical principles, Descartes offered physics something to compensate for universality; he offered it its own internal independence from theological concerns.

Cartesian physics enjoyed but a little century in which to monarchize, yielding even on the Continent before great Newton. By the time of the Enlightenment, *la physique expérimentale* had become synonymous with Newton, and Descartes was dismissed in the new enthusiasms as a metaphysician. Theology followed suit in its justification of the existence of God. The young Denis Diderot could observe that "it is only in the works of Newton, of Musschenbroek, of Hartsoeker, and of Nieuwentijt that satisfactory proofs have been found of the existence of a reign of sovereign intelligence."[35] Bernard Nieuwentijt's massive *L'existence de Dieu, démonstrée par les merveilles de la nature* (1714) incorporated biology into this all-encompassing physics, and presented learned Europe with an encyclopedic description of the argument from design. Petrus van Musschenbroek's inaugural address, *Oratio de certa methodo philosophiae experimentalis* (1732) had celebrated the repeated victories of Newtonianism in the West, and Redi and Leeuwenhoek had expanded the field of design to include "le mécanisme de l'insecte le plus vil." The influence of Newtonian mechanics was everywhere, pouring into a world prepared for it by such works as John Ray's *Wisdom of God Manifest in the Works of Creation* and reaching its *mot juste* in William Derham's *Physico-Theology* (1713). Mechanics became increasingly more universal and each new discovery seemed to support the wisdom of the Newtonian settlement with theology.

Beginnings of the Divorce: Physics and Theology

What is crucial for the theologian to note is not only what is present as evidence for the divine existence, but what is absent. Religion offers nothing as warrant for its most central assertion. Religious experience of whatever dimension or character counts for nothing, neither the interior claims of an absolute, nor the disclosures of "limit experiences," nor the movements and attractions towards the transcendent. Or, if one looks not for the witness of subjectivity but for the historical or external witness within human tradition, one will look in vain for the history of holiness as a perpetual manifestation of mystery, the testimony of mystics, the depth of human religious practice over thousands of years, and — even more remarkably for a Christian culture — anything of the reality and meaning of Jesus of Nazareth. Religion either in its internal, intuitive, affective dimensions or in its historical, institutional, external, traditional dimensions has nothing to offer to the question. It is presumed, though this statement is never made, that religion stands empty before such an issue. That is why it looks to physics to sustain its truth.

Ernst Mach credits Lagrange with the next pertinent development of physics, the elimination of theological concerns:

> After an attempt in a youthful work to found mechanics on Euler's principle of least action, Lagrange, in a subsequent treatment of the subject, declared his intention of utterly disregarding theological and metaphysical speculations, as in their nature precarious and foreign to science. He erected a new mechanical system on entirely different foundations, and no one conversant with the subject will dispute its excellences. All subsequent scientists of eminence accepted Lagrange's view, and the present attitude of physics to theology was thus substantially determined. The idea that theology and physics are two distinct branches of knowledge thus took, from its first germination in Copernicus till its final promulgation by Lagrange, almost two centuries to attain clearness in the minds of investigators.[36]

The effect of this new autonomy of physics from theological concerns is often symbolized in the famous interchange between Napoleon and Laplace, recorded by William Herschel from a visit by the First Consul on August 8, 1802. The conversation turned to celestial mechanics and Napoleon asked: "'And who is the author of all this?' M. de Laplace wished to show that a chain of natural causes would account for the construction and preservation of the wonderful system. This the First Consul rather opposed." Subsequent legend has shortened the story by having Laplace reply to Napoleon's Newtonian theology: "Je n'avais pas besoin de cette hypothèse-là."[37] Neither, of course, did Descartes for the design of the universe! Given matter and motion, the universe would of necessity have eventually arranged itself in its present configurations. Laplace is not denying the existence of God. He is only insisting [with Descartes] that mechanical problems must have mechanical principles as their solution, not theological ones. Laplace had assisted Newtonian celestial mechanics as no other figure in French astronomy, but his efforts entailed both the restoration of mechanical principles as ultimate in science and the consequent elimination of theology. In Laplace, Cartesian methodology had its partial revenge.

Whether one awards the palm to Lagrange or to Laplace, both of them bring to completion a dialectical revolution that had begun much earlier and gathered strength during the Enlightenment: physics needs nothing beyond physical principles to explain itself. No one better exhibits the change than Denis Diderot, from the Newtonian *apologia* of the *Pensées philosophiques* to the elimination of all such physico-theologies in the *Lettre sur les aveugles* and finally to the open atheism of *La Rêve de d'Alembert*. Perhaps better than any single figure, this great genius of the Enlightenment develops the internal alienation of this "first foundation of religion" into its contradiction.

In Diderot's *Lettre sur les aveugles,* the Reverend Mr. Holmes attempts the argument of Newton, Leibniz, and Clarke on the blind Cambridge mathematician, Nicholas Saunderson. How would or could Saunderson explain the design even "dans le mécanisme admirable de vos organes?" Saunderson counters that his own blindness must also be explained as well as the broader history of deformed monsters and of the diseased, lingering away into half-death. The intelligent Author of design is not adequate to explain all of this, and so Saunderson introduces another principle, collapsing the disjunction between matter and motion, and insisting against the mass of Newton and the extension of Descartes "le matière se mouvoir et le chaos se débrouiller."[38] Matter, blind but dynamic, evolving form after form in the gradual establishment of those organic beings which are self-sustaining, was no longer the passive extension of Descartes. Neither was it the mass of Newton that "was unable to initiate any action itself, passively dominated by external forces but endowed with a power to resist them."[39] Newton had allowed that mass had the *potentia resistendi,* and in its very resistance had the power to change the impressed forces brought against it. Newton had also acknowledged the necessity of active principles to be found in Nature: fermentation, magnetism, and the cause of gravitation, and Leibniz gave matter the dynamism of *vis viva* [momentum] and the *conatus* for continuance which could be awakened on contact. But both needed a cause other than matter to initiate action.[40] No longer. Taking his understanding of matter and the evolution of organic forms from Lucretius, Saunderson had something that explained all natural phenomena better and was itself commensurate with it: "La matière faisait éclore l'univers."[41]

Diderot has taken something from both Newton and Descartes and turned these weapons against those who forged them. From Newton, he takes the universality of mechanics and its competence to handle definitively the existence of God. He accepts the *Newtonian Settlement*. From Descartes, he takes not the nature of his method, but the nature of his principles; they must be mechanical, reflexively commensurate with the subject-matter to be explored. Experimental physics remains universal [Newton], but contains within itself mechanical principles [Descartes]. Diderot's understanding and use of dynamic matter was coordinate with

Toland's *Fifth Letter to Serena* and even with the theories of the tortured Giordano Bruno, but here they are introduced into the tradition of Newton, into the *Newtonian Settlement*. In this development, physics has obtained its autonomy — its emancipation, Marx will later write — and theology has lost its foundations.

The Beginnings of Atheism

It was out of this loss that the first major atheistic tractates of the Enlightenment issued: Diderot's own *Dream of d'Alembert* and a series of works by the Baron Paul d'Holbach that culminated in his massive *System of Nature*. For the first time in Western Civilization, "atheist" became more than an invective, an epithet hurled at an alien thinker. With these few of the Enlightenment, it became a signature. Hegel insisted upon the novelty of what was occurring in this cirlce around d'Holbach and Diderot:

> We should not make the charge of atheism lightly, for it is a very common occurrence that any individual whose ideas about God differ from those of other people is charged with a lack of religion, or even atheism. But here it really is the case that this philosophy has developed into atheism, and has defined matter, nature, etc., as that which is to be taken as the ultimate, the active, and the efficient.[42]

Hegel was correct in recognizing that something new had occurred. What he could not know was that this persuasion of a very few would wax through his own century to become increasingly the mark of an intellectual elite, augmenting its members in our own century through its identification with mass movements and totalitarian ideologies.[43] This growth constitutes a religious development unprecedented in the history of the world. What had reached its first articulation in Paris as the Enlightenment among so few obtained this massive increase in a little more than two centuries.

It would be wide of the mark to accuse the Enlightenment figures of indifference to religion. It would be far more accurate, maintains Ernst Cassirer, to say that they were obsessed with it — and to add that they would discuss it in terms of the *Newtonian Settlement*, in terms of physics supplying or denying the foundations of religion.[44] Atheism came out of a turn in the road in the development and autonomy of physics. So many of the theologians had appealed to physics. Now let them be content with its development even if it meant their contradiction. Descartes had needed only motion and matter to have the universe emerge out of chaos to its present state. Now merge matter with motion: "Motion," wrote d'Holbach, "is a manner [of existence] which matter derives from its own existence."[45] Newton's definitions had elaborated mass and impressed force, one resisting change and the other causing it. But mass differed from inertial force only in our way of conceiving it. Instead simply of a *vis inertiae*, make the *vis insita* of mass also a *vis motiva* which combines absolute and accelerative forces, and you need nothing more than time and chance out of which the system of the world can form. The analysis of phenomena can terminate now in something which Newton would never have recognized: a dynamic matter reflexively responsible for itself and for all of the phenomena of nature.

With this principle, d'Holbach can bring the dialectical history we have been detailing to its conclusions. Each paragraph of Newton's General Scholium is cited separately in the *System of Nature* and followed by an argument that its exigencies can be satisfied more adequately by dynamic matter. The twelve theorems with which Clarke's *A Demonstration of the Existence and Attributes of God* had proven the existence and nature of God were subsumed by d'Holbach as a progressive, concatenated demonstration of the comprehensive validity of dynamic matter as a universal explanation. As for Descartes and his first philosophy, this field had long since been swept clean by the experimental physics out of England. D'Holbach is content with a few condescending paragraphs, concluding: "We might then with great reason accuse Descartes of atheism, seeing that he destroys in a very effectual manner, the feeble proofs which he gives of the existence of God."[46] Finally one is left with nature, and d'Holbach's final conclusion about Newton: "This God is nothing more than nature acting by necessary laws necessarily personified, or destiny, to which the name of God is given."[47]

One might contend that Toland's *Fifth Letter to Serena* had prepared the way for both Diderot and d'Holbach to move in this atheistic direction. Certainly their work had many predecessors. Toland argued against the merely static extension of Descartes and the inert mass of Newton, insisting, as Giordano Bruno before him, that motion was essential to matter. But neither Toland nor Bruno dominated Europe. One might study physics as it came from Descartes or from the universal mechanics of Newton, but the choice was between Descartes and Newton, not between Bruno and Toland. On the records of d'Holbach, Descartes was "the restorer of philosophy" and Newton the "father of physics." D'Holbach's atheism, and the circle which this *System of Nature* represented, took its power essentially from its dialectical character. The strength and enduring dynamism lie not so much in its arguments — they really are not that impressive — but in its dialectical gathering of strength from its contradictions. This work had its influence in the history of ideas not because it completes Toland, but because it completes Newton. It is so devastating a dialectical reversal of Clarke, because it uses Clarke's own pattern of argument, and it obtains a presence in Europe that Anthony Collins could never have obtained. The strength of what Diderot and d'Holbach initiated lies in this dialectic: "it purifies and then confirms as its own what should have been its contradiction." Atheism emerges, generated dialectically, out of the very efforts used to counter it, out of a physics made "its first foundations."

The Consequences for Religion

It would be wonderful to record that Catholic reflection learned its lesson from this profound reversal, but the history of ideas does not afford that comfort. The National Assembly of the French Church in 1770 engaged that universal genius, l'abbé Nicolas-Sylvain Bergier, to deal intellectually with the deluge of atheistic literature pouring into the capital. In 1771, to fulfill his mandate from the Assembly, he wrote as his refutation of d'Holbach's *System of Nature* his own *Examen du matérialisme,* "one of the best pieces of critical writing of the century." [48] Felicité de Lamennais in the subsequent century would evaluate l'abbé Bergier as "le plus grand apologistes des siècles passés, et peut-être de tous les siècles." [49] And how did Bergier intend to deal with the atheism of the Baron?

> As soon as it is evidently proven that movement is not essential to matter, that the latter is purely passive by its nature and without any activity, we are forced to believe that there is in the universe a substance of a different nature, an active being to which movement must be attributed as it is to the first cause, a Mover that is not itself matter." [50]

Bergier represents the relationship between science and religion that continued the *Newtonian Settlement* and that still found obvious favor in the Church and among its major apologists. The Benedictine scholar, Louis-Mayeul Chaudon, carried out the same high hopes that the conflict could be waged and won through a reading of nature. He concludes with two axioms that can bring our own précis of the *Newtonian Settlement* to its conclusion:

> The study of Physics is quite properly the cure of two extremes: Atheism and Superstition.... It proves that there is an intelligent first cause, and it makes us know the particular mechanical causes of this and that mechanical effect. Physics augments admiration and diminishes astonishment." [51]

Over the centuries the theologians had become philosophers in order to counter a putative atheism. What is more, philosophy had become Newtonian physics, and its apologetic value lay in the ability to do physics better than one's opponents. If physics declared that it wanted and needed only commensurate principles to explain the mechanical phenomena it was investigating, then the existence of God was left without the foundations upon which the theologians had been counting and to which they had made appeal.

The book, *At The Origins of Modern Atheism,* contextualized and explored in considerably greater dimensions the argument and the history touched upon in this paper. Thomas V. Morris maintained as the thesis of this book that "religion turned to philosophy for its defense and philosophy betrayed it." [52] It would more accurate to express that thesis as religion betraying itself. In turning to philosophy for its foundations, for its "first foundations" as Samuel Clarke's revealing phrase had it, religion was implicitly confessing its own intrinsic lack of warrant, confessing that it did not possess the proper resources to deal with the existence of the God upon whom it reflected, that it must look elsewhere for the substantiation of its most profound claim. It was only a question of time until that confession would become public. The last sentence of that book expresses again its basic position: "The origin of atheism in the intellectual culture of the West lies thus with the self-alienation of religion itself."[53] Obviously there is nothing intrinsically inimical between the philosophic enterprise and Christian faith. Centuries of Christian wisdom have proven just the opposite to be the case. Etienne Gilson remarked on more than one occasion that, in contrast with Islam, philosophy flourished in Western Europe among the theologians of greatest orthodoxy. From the earliest stages of philosophy in the West, philosophers have asserted the existence of God as essential to their study of the nature of things. To assert that philosophy can neither substitute nor provide the foundations for religion is not to assert that it is the enemy of religion or has no properly theological role. Philosophy does not betray religion. Religion can only betray itself. The problem with the *Newtonian Settlement* is not that philosophy was present, but that religion was absent.

Perhaps what was basically wrong with the *Newtonian Settlement* only emerged when it generated its own denial. In turning to some other discipline to give basic substance to its claims that God exists, religion — or that reflection upon religion for its evidence that we have been calling theology — is admitting an inner cognitive emptiness. If religion does not possess the principles and experiences within itself to disclose the existence of God, if there is nothing of cogency in the phenomenology of religious experience, the witness of the personal histories of holiness and religious commitment, the sense of claim by the absolute already present in the demands of truth or goodness or beauty, the intuitive sense of the givenness of God, an awareness of an infinite horizon opening up before inquiry and longing, an awakening jolted into a more perceptive consciousness by limit-experiences, the long history of religious institutions and practice, or the life and meaning of Jesus of Nazareth, then it is ultimately counterproductive to look outside of the religious to another discipline or science or art to establish that there is a "friend behind the phenomena." Inference cannot substitute for experience, and the most compelling witness to a personal God must itself be personal. To attempt something else either as foundation or as substitute, as did the *Newtonian Settlement*, is to move into a progress of internal contradiction of which the ultimate resolution is atheism.

NOTES

[1] Helmut Peukert, *Science, Action, and Fundamental Theology: Towards a Theology of Communicative Action,* translated by James Bohman (Cambridge, Mass.: The MIT Press, 1986) xxiii.

[2] See Michael J. Buckley, S.J., "Toward the Construction of Theology: a Response to Richard P. McKeon," *The Journal of Religion,* Supplement, **15**, 58 (1978) 52-63.

[3] This paper is reworked and integrated into a much more extensive study of its thesis by the author in a work that was published in 1988 by Yale University Press, *At the Origins of Modern Atheism.* Because of its later date of publication this paper was able to incorporate the suggestions of the members of the conference at Castel Gandolfo and to respond to some early questions about the book.

[4] Galileo Galilei, *Dialogues Concerning Two New Sciences,* Third Day, Theorem VI, Proposition VI, Corollary III, translated by Henry Crew and Alfonso de Salvio with an introduction by Antonio Favaro (New York: Dover Publications, 1914) 194. As Stillman Drake remarks, the name "Simplicio" comes from the sixth century Greek commentator on Aristotle. See Galileo Galilei, *Dialogue Concerning the Two Chief World Systems,* translated with revised notes by Stillman Drake (Berkeley: University of California Press, 1974) 467-468*n*7.

[5] Galileo Galilei to Elia Diodati 9 June 1635, in *Commercio Epistolare di Galileo Galilei,* edited and illustrated by Eugenio Albèri (Florence: Società Editrice Fiorentina, 1859) II:57. Cf. Favaro, *op. cit.,* xi.

[6] Galileo, *Two New Sciences,* First Day, p. 2: "Sagredo: 'Now since mechanics has its foundation in geometry, where mere size cuts no figure, I do not see that the properties of circles, triangles, cylinders, cones and other solids will change with their size.'... Salviati: 'The common opinion here is absolutely wrong.... There are some intelligent people who maintain this same opinion, but on more reasonable grounds and cut loose from geometry...." Salviati will insist that his contribution, which makes this a new science, is the introduction of demonstration by the geometrical methods proving his conclusions "in a rigid manner from fundamental principles." See *ibid.,* 6.

[7] Isaac Newton, *Philosophiae Naturalis Principia Mathematica,* "Actoris Praefatio ad Lectorem." The third and final edition (1726) of this work has been assembled and edited by Alexandre Koyré and I. Bernard Cohen with the assistance of Anne Whitman (Cambridge, Mass.: Harvard University Press, 1972) [henceforth cited as K-C], 1:15. All citations from the *Principia* have either been translated by the author or checked against the Latin text in his use of the standard English translation of Andrew Motte (1729), revised by Florian Cajori: *Sir Isaac Newton's Mathematical Principles of Natural Philosophy and His System of the World* (Berkeley: University of California Press, 1962) [henceforth cited as Cajori], xvii.

[8] James Collins, *Descartes' Philosophy of Nature* (Oxford: Blackwell, 1971) 61. Collins maintains that this statement is accurate enough for Descartes' general physics, but must be modified in the more differentiated and specialized portion of his scientific study.

[9] Pappus of Alexandria, *Synagoge* [*Collection*], 8, preface 1-3, in *Selections Illustrating the History of Greek Mathematics,* English translation by Ivor Thomas, Loeb Classical Library (Cambridge, Mass.: Harvard University Press, 1941) 2:614-620. Pappus makes quite clear that the understanding he is reporting is not one that is general among Greek theoreticians of mechanics by attributing it to the

school of Heron of Alexandria. For Heron, it is the movement of the point which generates the line, the movement of the line which generates the surface, the movement of the surface which generates the solid. The science which studied or described movements would lie at the basis of geometry. See Heron of Alexandria, *Mensauration: Definitions,* 14: 1-24 in *op. cit.* 2:468-469. Newton gives a similar description of the generation of the "quantitates mathematicas, non ut ex partibus quam minimis constantes, sed ut motu continuo descriptas, hic considero. Lineae describuntur, ac describendo generantur, non per appositonem partium, sed per motum continuum punctorum; superficies per motum linearum; solida per motum superficium; anguli per rotationem laterum; tempora per fluxum continuum; et sic in caeteris. Hae [sic] Geneses in rerum natura locum vere habent, et in motu corporum quotidie cernuntur." "Introductio ad Quadraturam Curvarum," in *Isaaci Newtoni Opera quae exstant omnia,* with commentary and illustrations by Samuel Horsley (London: Joannes Nichols, 1779-1785) [henceforth cited *Newtoni Opera Omnia*], 1:333.

[10] *Principia* 3, General Scholium, K-C 2:764; Cajori, p. 546.

[11] Isaac Newton, *Opticks, or a Treatise of the Reflections, Refractions, Inflections and Colours of Light,* 3, query 28, based on the 4th ed. (London, 1730/New York: Dover, 1952) [henceforth *Opticks*], 369.

[12] Newton to Bentley, December 10, 1762, in *The Correspondence of Isaac Newton,* edited by H. W. Turnbal (Cambridge: Cambridge University Press, 1959) 3:233.

[13] *Opticks* 3, query 28, 369.

[14] See Michael J. Buckley, S. J., "God in the Project of Newtonian Mechanics," in *Newton and the New Direction in Science,* eds. G. V. Coyne, M. Heller and J. Zycinski (Vatican City State: Vatican Observatory, 1988) 85.

[15] *Opticks* 3, query 31, 405-406.

[16] Isaac Newton, "A Short Schema of the True Religion," in *Sir Isaac Newton. Theological Manuscripts,* selected and edited with an introduction by H. McLachlan (Liverpool: University Press, 1950) [henceforth NTM] 48.

[17] Isaac Newton, "Seven Statements on Religion," *NTM,* 58.

[18] Newton, "A Short Schema of the True Religion," *NTM,* 48.

[19] *Ibid.,* 49 [enumeration added].

[20] *Principia,* preface to the first edition, K-C 1:16; Cajori, xvii-xviii.

[21] *Principia* 3, General Scholium, K-C 2:760-761; Cajori, 544-545. Newton puts his position precisely: "Deitas est dominatio Dei."

[22] Newton, *Opticks,* query 28, 369. It is critical to note that for Newton the only way in which one could posit a mechanical principle for the ultimate explanation of natural phenomena was by "feigning Hypotheses."

[23] James P. Ferguson, *An Eighteenth Century Heretic: Dr. Samuel Clarke* (Kineton: Roundwood, 1976) 23-25.

[24] Samuel Clarke, *A Demonstration of the Being and Attributes of God* (London: J. Knapton, 1705) [henceforth *DBAG*], "Preface."

[25] *DBAG,* 26-27 and 44-45 [italics his].

[26] Cajori, xxxiii.

[27] *DBAG,* 111-112ff [italics his].

[28] *DBAG,* 126.

[29] Leonard Lessius, *De providentia numinis et animi immortalitate, libri duo adversus Atheos et Politicos* in the *Opuscula Leon. Lessii, S.J.* (Paris: P. Lethielleus, 1880). ET: *Rawleigh His Ghost: Or a Feigned Apparition of Syr Walter Rawleigh, to a friend of his, for the translating into English, the Booke of Leonard Lessius (that most learned man) entitled De providentia Numinis, et Animi immortalitate; written against Atheists, Polititians of these days,* 1558-1640, edited by D. M. Rogers (London: Scholars Press, 1977) 1.1, 5.

[30] Marin Mersenne, *L'Impiété des Déistes, Athées, et Libertins de ce temps* (Paris: Pierre Bilaine, 1624), preface, unnumbered, page 12. The number of atheists in Paris is delivered by Mersenne a year previous in his *Quaestiones in genesim* (Paris: Sebastiani Cramoisy, 1623), cols. 669-674. This was at a time when the entire city numbered 400,000.

[31] Lessius, *De providentia numinis*, 1.2.16-19, 235-238.

[32] James Collins, *God in Modern Philosophy* (Chicago: Regnery, 1959) 51.

[33] Rene Descartes, *Meditations on First Philosophy*, translated by Laurence J. Lafleur (Indianapolis: Bobbs-Merrill, 1978) 7. For the Latin text, see the standard edition of Charles Adam and Paul Tannery (Paris: Leopold Cerf, 1897-1909) VII, 1-3.

[34] Descartes to Vatier, AT I, 564. See Jacques Maritain, *The Dream of Descartes*, translated by Mabelle J. Andison (New York: Philosophical Library, 1944) 205.

[35] Denis Diderot, *Pensées philosophiques* #11 as in *Oeuvres philosophiques*, edited by Paul Venniere (Paris: Garnier Frères, 1961) 17-18.

[36] Ernst Mach, *The Science of Mechanics*, translated by Thomas J. McCormack, 6th ed. (LaSalle, Ill.: Open Court, 1960) 552.

[37] Roger Hahn, "Laplace and the Vanishing Role of God in the Physical Universe," *The Analytic Spirit: Essays in the History of Science in Honor of Henry Guerlac*, ed. Harry Woolf (Ithaca: Cornell University Press, 1981) 85-86.

[38] Denis Diderot, *Lettre sur les aveugles à l'usage de ceux qui voient*, in *Oeuvres philosophiques*, 121.

[39] Richard S. Westfall, *Force in Newton's Physics: The Science of Dynamics in the Seventeenth Century* (New York: American Elsevier, 1971) 450.

[40] Ernan McMullin, *Newton on Matter and Activity* (Notre Dame: University of Notre Dame Press, 1978) 29-56.

[41] Denis Diderot, *Lettre sur les aveugles*, 123.

[42] Georg Wilhelm Friedrich Hegel, *Lectures on the History of Philosophy*, translated by H. S. Haldane and Frances H. Simson (Atlantic Highlands, N. J.: Humanities Press, 1983) III, 387.

[43] David B. Barrett, *World Christian Encyclopedia. A Comparative Study of Churches and Religions in the Modern World 1900-2000* (Nairobi: Oxford University Press, 1982) 6.

[44] Ernst Cassirer, *The Philosophy of the Enlightenment*, translated by Fritz C. A. Koelln and James P. Pettergrove (Princeton University Press, 1951) 135-136. Cf. Peter Gay, *The Enlightenment: An Interpretation*, vol. 1, *The Rise of Modern Paganism* (New York: Norton, 1966) 18.

[45] Baron Paul d'Holbach, *Système de nature ou des loix du monde Physiques et du Monde moral*, new edition (London: 1771) 30.

[46] *Ibid.*, 150.

[47] *Ibid.*, 230.

[48] R. P. Palmer, *Catholics and Unbelievers in Eighteenth Century France* (Princeton: Princeton University Press, 1939) 215.

[49] Alfred J. Bingham, "The Abbé Bergier: An Eighteenth-Century Catholic Apologist," *Modern Language Review*, **54**, no. 3 (July 1959) 349.

[50] Nicolas-Sylvain Bergier, *Examen du matèrialisme, ou Réfutation du Système de la nature* (Paris, Chez Humbolt, 1771) I. 154, 174-176, cited in Alan Charles Kors, *D'Holbach's Coterie: An Enlightenment in Paris* (Princeton: Princeton University Press, 1976) 65n.

[51] Louis-Mayneul Chaudon, *Anti-Dictionnaire philosophique* (Paris: Saillant and Nyon, 1775) II, 125-129, as cited in Kors, *op. cit.*, 56n.

[52] Thomas V. Morris, "A Door that Proved Most Narrow," *Commonweal* January 29, 1988) 58.

[53] Buckley, *Modern Atheism*, 363.

IS A NATURAL THEOLOGY STILL POSSIBLE TODAY?

W. Norris Clarke, S.J., Fordham University

Introduction

The enterprise of Natural Theology (or the Philosophy of God) is a particularly difficult one to carry out in our day. Philosophically it has come under heavy attack from empiricists and Neo-Kantians, from analytic philosophers tinged with both of the above, from historical and linguistic relativists appealing to hermeneutics, and more recently from Deconstructionists. We shall take up these philosophical roadblocks presently. But first, given the context of this book, we turn to the relations between natural theology and contemporary science, in particular theoretical physics and cosmology.

Relation to Science

Natural theology is, from one point of view, on better terms with contemporary science than it has been for a long time. The notion that mind has a place in nature, that nature points to mind as its completion, is much more acceptable, even plausible, to many scientists today, especially theoretical physicists and cosmologists. One example is that advanced by Fred Hoyle in his recent book, *The Intelligent Universe*. Many scientists are favorably impressed by the now famous Anthropic Principle, which seems to point to an extremely precise fine tuning of the four basic forces of the material universe, with its enormous statistical improbability, as a sign that the universe was planned from the beginning in view of the appearance of conscious observers like ourselves in it. Indicative is the comment of the physicist Freeman Dyson:

> I conclude from the existence of these accidents of physics and astronomy that the universe is an unexpectedly hospitable place for living creatures to make their home in. Being a scientist, trained in the habits of thought and language of the twentieth century rather than the eighteenth, I do not claim that the architecture of the universe proves the existence of God. I claim only that the architecture of the universe is consistent with the hypothesis that mind plays an essential role in its functioning.[1]

Two points are noteworthy here. The first is the openness to, or "compatibility" of the scientific picture with, the theistic hypothesis, rather than the former closedness that used to predominate. But the second is the warning that from inside the scientific outlook this hypothesis is only *compatible* with the results of contemporary science, not authorized or established by them. As Ernan McMullin's fine paper in this volume shows,[2] theistic philosophers in the past have persistently tried to argue to the existence of God from some gap in the existing scientific picture of the universe, from some need discovered within the web of scientific explanation for a further grounding that the scientific explanation itself could not supply. Thus Newton believed that God's intervention was necessary to maintain the constant motion of the heavenly bodies. Paley and others argued from the marvelous adaptation of the various species of living organisms to their environment — given the common pre-Darwinian acceptance of the fixity of species — to the hypothesis of a Cosmic Planning Mind that had thus ordered them; and so on. But in each case science eventually closed the gap in its web of explanation, and in so doing undermined the argument for the existence of God based on this gap. The "God of the gaps" has been progressively put out of a job.

The same kind of process seems to be at work again today. Despite the initial plausibility and strong suggestiveness of arguments for the need of a world-ordering Mind from unfilled gaps in the current scientific picture, especially those based on the statistical improbability of our present world-order,[3] this foundation does not seem to me a secure one for building a cogent natural theology. The figures are indeed impressive: for example a Princeton scientist, Don Page,[4] recently calculated that the odds against our present universe are something like 1 in 10^{133}. But opinions continue to vary as to the basis for making calculations, given the unique situation or "singularity" of the earliest stages of the cosmic system. Others have put forward ingenious hypotheses, such as that no choice is needed for the peculiar initial conditions of our universe, since an infinite number of all possible universes actually exist, so that ours is bound to turn up somewhere without the need for any calculus of probabilities or for a selective agent. Others try to argue that ours is in fact the *only possible* universe that can be actualized, assuming that in quantum physics many of the conjugate properties of subatomic particles can only be actualized by conscious observers like ourselves. Others weaken the basis of the impressive argument of the fine tuning of the four basic forces and other precisely balanced constants of the universe, by reducing the four forces, first to three, then to two, then hopefully, in the light of some as yet incomplete hypotheses like superstring theory, etc., to one simple all-embracing force from which all else can be deduced. Some suggest tracing the beginning to a mere chance fluctuation of a primordial quantum field, emerging out of a pure formless high-energy vacuum state or pre-space-time "foam," which they ambiguously identify as "nothing."[5] In view of the intense ferment of speculation going on at this time in high-level theoretical physics, it does not seem to me possible yet to find any secure foundation within the exigencies of scientific explanation for the postulation of a Transcendent Mind as the only adequate cause of the origin and structure of our cosmos.

Others from within the biological community, or philosophizing on its data, suggest there is even stronger evidence for the need of a Cosmic Planning Mind to explain the origins of life and the large jumps to new species in the course of evolution, in view of the huge statistical improbability of the passage from a non-living molecule to a living cell (Fred Hoyle and his associate have calculated it as 1 in $10^{40,000}$), and the failure thus far of all attempts to reproduce successfully the conditions of such a passage, plus the widely conceded breakdown of Darwinian chance selection as an adequate explanation for the passage from one species to another of a different order.[6] But again, such gaps in current explanations might possibly be filled in by some future hypothesis.

So, somewhat reluctantly, and without denying the powerful suggestiveness of inferences from the apparent enormous improbability of our present universe, both in its origins and in the evolution of life in it, I think it wiser to agree with Erman McMullin, at the end of his paper in this volume, that natural theology today should avoid any attempt to build its foundation on apparently unfillable gaps in the scientific picture of the universe. The "God of the gaps" has so often been put out of a job that I think he should be, if not permanently, at least for the time being, retired. Only a radically metaphysical argument, from the very existence of a determinate world, or the existence of any dynamic order at all, has a fair chance of succeeding.

Philosophical Obstacles

There have been many attempts in modern and contemporary philosophy to block any project of constructing valid philosophical arguments for the existence of a Transcendental Reality. There are, of course, both older and newer more sophisticated forms of empiricism and Kantianism, whether in scientific, linguistic, or phenomenological versions, that are still tenaciously pervasive in contemporary thought. All these are fundamentally anti-metaphysical, in the sense that it is impossible to move, by philosophical reason, beyond the world of human experience, inner or outer, to affirm legitimately the existence of some reality transcending this experience. Then there are the newer movements of historical, cultural, linguistic, or hermeneutical relativism, together with the latest "demolition squads," known as Postmodernists and Deconstructionists.

For the relativists, all our expressed knowledge is history, culture, and language-bound, meaningful only within a given historical and linguistic framework of inquiry and expression, but never allowing any unconditional truth statements transcending such frameworks, which would be true for all times, places, and cultures. This seems, at least, to cripple any attempt to construct an objective natural theology with any cogency outside its own narrow tradition — if even there. For the Postmodernist there is no "meta-narrative legitimation of first-order narratives." You have your story; I have mine (or my group has mine and your group has yours). But there is no norm beyond the individual stories by which to judge their truth or value. We must allow neither political nor conceptual tyranny; both are functions of power, not truth. "Let all flowers bloom."[7]

The Deconstructionist calls on us to resist — and sabotage — the arrogant *logocentrism* of the West, with its pretensions to capture reality adequately in an all-inclusive, totalizing conceptual system, transparently reflecting the non-linguistic real, à la Hegel. They propose a "heterology" (championing the Other, the different, the exception, the marginalized) in opposition to a "henology" (the reduction of the many, the different, to some all-inclusive, all-explanatory One) as has been customary in Western metaphysics. The more radical versions, which Jacques Derrida, the most visible "father" of the movement, often makes gestures of repudiating, maintain that no expressed signifiers ever connect up unambiguously with the truth, or non-linguistic reality, that there is no unambiguous dividing line between metaphor and objective concept, literature and philosophy, that all signifiers trail off into an endless labyrinth of reference to other signifiers and these to others, into traces of traces.... In place of so-called truth claims, they unveil the hidden pretensions of the philosophers to impose their metaphorical schemas on others by the "will to power" (the influence of Nietzsche is clear here, and often explicity avowed). In addition, all texts can be cracked open to reveal a hidden subtext which works against the surface text to undermine it. The radical Deconstructionist is a "double-agent and a nomad," who infiltrates one system to blow it up from within, then, with no "home" (or position) of his own, moves on to blow up another. It is obvious that an effective natural theology — or any kind of theology, it turns out — is, in such a context, a *logocentric* illusion.[8]

Clearing the Obstacles

Let me indicate briefly how I would go about removing or circumventing the above philosophical roadblocks to the positive construction of a natural theology. First, the contemporary relativists and Deconstructionists. I think it would be a serious mistake — an intellectual loss of nerve — to allow ourselves to be intimidiated by these movements, with their often strident proclamations of the end of Western *logocentric* reason. My argument is this: Whenever these positions move to a really radical stance blocking all access to objective truth, they promptly self-destruct and become inoperative as a critique. For *either* they are claiming to be informing us of some *truth* about *all* linguistically expressed human thought, and then their assertive performance contradicts the content of their assertions, namely, that all such assertions are culture and language-bound so that they cannot connect up unambiguously with the truth. *Or*, if they are not really claiming to tell us some significant truth about all of us, then their own position immediately becomes relativized. In this case it turns into just another late 20th-century culture-bound opinion, perhaps even localized to a thin veneer of thinkers in a few large cities. If this is so there is no reason for the rest of us to bother our heads about it; we are free to go on in our own contexts happily asserting our objective truth claims. In a word, the natural, spontaneous, and in the last analysis inextinguishable drive of the human mind to discover and give recognizable expression to the truth, to what *is the case* in reality, cannot tolerate for long any attempt at systematic self-sabotage of its own natural drive and innate cognitive structure of experience-insight-judgment. As

Derrida himself has well said somewhere, understanding this as a pragmatic necessity of actual human living: "Il faut la vérité" (There must be truth).

If the above movements are taken in moderation, however, they can lead us to an important, more realistic understanding of what in fact *is the case* about our human reason. There is no going back to a pre-hermeneutic understanding of human thought and language. What has really been demolished is the old and indeed arrogant Cartesian and Enlightenment ideal of human reason as pure, impersonal, autonomous, self-sufficient Reason, independent of any tradition, culture, historical perspective or authority. Such human reason would, in principle, be able to gather into itself unaided and with perfect transparency all that is real, knowable or worth knowing, with special priority given to the scientific method as the ideal method of reaching any truth available to us.

Accordingly, a self-aware contemporary thinker in *any* field should be willing to admit that our human reason must always see the world from some limited (and hence incomplete) historical perspective or vantage point and that what is seen from other vantage points is complementary, not contradictory. We cannot come intellectually naked to understand the texts of humanity and its world, but we must go through some apprenticeship in a living hermeneutical tradition. The reliable knowledge we can indeed attain about the real is not the Cartesian ideal of absolute certitude such that the opposite can be shown to be a *logical* contradiction, but rather the "reasonable affirmation," as Bernard Lonergan puts it, achieved not by impersonal, automatic, clearly specifiable rules for correct thinking, but by *personalized responsible* thinking. This includes striving for intelligent *insight* into the meaning latent in the data and for personally responsible *judgment* based on evidence recognized as sufficient for its purposes. All of our perception, concepts, and understanding are, as Polanyi has shown so well, a synthesis of focal and peripheral (or background) knowledge, such that it is neither *possible* nor *necessary* to make formal and explicit all that is in this background knowledge. Ours is a mode of existential, *lived* knowledge acquired by sharing in a practical "form of life" never fully susceptible of explicit conceptual formulation, and this is not, as Deconstructionists so often overlook, crippling to our capacity to understand, but positively *enabling*.

On the other side of the picture, however, it should always be remembered that no matter how limited or incomplete a perspective may be, it is still an opening *onto* something beyond the viewer. A perspective which opens onto nothing, or only inward into the viewer, is not a perspective at all, but a hall of mirrors. Similarly, no matter how much one must start within a hermeneutical tradition to learn the tradition and the skills of inquiry and interpretation, a hermeneutics that effectively does its job is one that *enables* us to understand a situation or text that needs interpretation, and, by a sensitively intelligent fusion of horizons, to come to understand significantly a different or older tradition. A hermeneutical viewpoint is a *vantage point* from which we discover and understand something — although not *all* that can be seen; it is not like a labyrinth or prison in which one comes to know only the prisoners.

As for the Deconstructionists, Polanyi's theory of local and peripheral knowledge, appropriately amplified, already takes care of most of their significant warnings, without the skeptical consequences. As for the claim of the Other, the different, the absent, which deserves equal status with the One being, the present, it is my impression that *Deconstructionists* often exhibit a systematic blindspot to what St. Thomas was so well aware of, namely, the distinction between the *mode or path of discovery* of a concept and the *content signified* by it. Thus in a realistic metaphysics like that of St. Thomas the metaphysical notion of *being*, like most metaphysical concepts, is intrinsically analogous, pregnant with the one and the many, sameness and difference, remaining systematically vague so all that is in it can never be made fully explicit. The notion itself is brought to explicit possession by contrast with partial absences, instances of non-being, etc. But all these differences, absences, partial non-beings, etc. are always enveloped within the overall horizon of *being*, differences *within* being, not outside of it in a radical and unqualified sense. There is no warrant to conclude, from the mode of discovery of our basic concepts, that in the real order unqualified non-being and absence can claim independent equality or priority with respect to being, presence, unity, etc. All truly analogous concepts contain the many within their womb from the outset.

As for empiricism, it cannot make secure its claim to block any ascent of rational human intelligence to transcendent reality either within us or above us. It is in essence an arbitrarily restrictive theory of knowledge, attempting to constrict the natural drive of the mind to know, holding it merely to the realm of experience. It allows description and correlation of the data of experience but no explanations reaching beyond experience to fill the gaps of intelligibility within experience. But the radical weakness in this procedure is that the knower cannot be caught adequately in the empiricist's net of all that is knowable. The intelligent knower who is looking at the data, striving to understand it, interpreting and judging it, discerning value or disvalue, oughts and ought nots within it, is not out there among the sensory data. The knower cannot "look" at him or herself as a self-conscious, self-possessing, self-governing I, if empiricism is true. But we do just this all the time. The knower transcends all empirical data, is not reducible without remainder to all or any part of empirically given data. This surplus between the knower and empirical data opens the way to a non-empirical (i.e., a meta-physical) ascent of the mind through the exigencies of intelligibility to whatever transcendent reality is needed to fill the gaps in intelligibility found within our empirical data.

But there is one principle of explanation that must be explicitly rescued from the straightjacket of empiricism, if our ascent is to be viable. That is the principle of efficient causality. The empiricist would have us believe that the foundational and only legitimate meaning of causality is simply the regularly observed succession in time of empirically observable antecedent and consequent events, such that from the first one can predict the second according to some law. Any further intrinsic link between cause and effect, such as the active production or bringing into being of the second by the first, the fact that the cause is responsible for the effect, which therefore has an ontological link of dependence on its cause — all this is declared to be later, unnecessary, and unjustifiable baggage added on by the metaphysicians.

The opposite is in fact the case. Our modern natural sciences have indeed good reason, for methodological purposes, to restrict the meaning of efficient casually in practice to "predictability according to law," whether deterministic or, more often today, statistical. But the original meaning of the term, deriving from the Greek law courts[9] and its flourishing use today in ordinary life situations for explaining why things happen, is the notion of "that which is responsible (originally 'guilty', in court room use), for" the given occurrence of some event, the presence of some entity, etc., which of itself needs explanation, is judged not to make sense by itself alone.

The efficient cause, thus understood in ordinary life and more self-consciously and abstractly in realist metaphysics, is simply a function of the inquiring mind at work, with a flexible analogous application just as wide as the reach of the latter. It is really nothing more, but nothing less, than the reaffirmation of the basic commitment of the working human mind to the unrestricted intelligibility of the real, tailored to fit a particular situation. When allowed to operate without arbitrary restrictions, the search for the efficient cause is simply the search for whatever is needed to fill a discerned gap of intelligibility in the data of our experience. Wherever this search leads, to whatever is shown to be indispensable to fill this gap, whether a cause in the empirically given world, or beyond, this can be affirmed legitimately under pain of allowing the initial data we are trying to explain to remain with a declared unfillable hole or "wound" in its intelligibility. In its wide-open scope, as wide as being itself, this principle contains no restrictions such as empiricism would force upon it. Realist metaphysicians should reclaim without intimidation the principle of efficient causality as the natural birthright of the innate drive of the mind to know and the indispensable instrument for carryng out the mind's natural commitment to the intelligibility of being — a commitment of "natural faith," as Einstein and other great scientists have put it, that is really the inner dynamism and soul of all serious intellectual inquiry, scientific or otherwise.[10]

As to the last of our roadblocks, Kantianism, two brief remarks will have to suffice. Kant is indeed a great thinker, especially in ethical matters. But we have been too long intimidated by his long shadow in epistemology, in particular his anti-realist and anti-metaphysical stance, which claims to bar the way to a rational affirmation of anything beyond empirical appearances. In the first place, his refutation of the so-called Cosmological Argument for God is flawed by a serious misreading of the traditional argument as presented by realist metaphysicians like St. Thomas. In the last crucial step of the argument Kant distorts it, so that it becomes an attempted deduction of the existence of a Necessary Being (I would prefer to call it a Self-sufficient Being) from the idea of the *Ens realissimum* (or infinitely perfect being). St. Thomas would indignantly repudiate such a procedure, all too easily refuted by Kant. The traditional procedure is precisely the opposite. Once the reality of a Self-sufficient Being has been established from causal arguments, it is then argued that such a being could not be at once self-sufficient and finite, for the latter by nature requires a cause of its being as finite. Therefore the Self-sufficient Being must be infinite, and so, by an easy step, only one. There is no deduction from the idea of perfection or from any idea in itself, although some such procedure may have been invoked by some of the rationalist metaphysicians of the Wolffian type just before Kant.

As to Kant's attempt to bar access to any valid affirmation about a real world beyond the knower, it suffers from a fatal flaw, a massive blindspot that has also plagued most of modern Western epistemology since Descartes, as pointed out insightfully by John Dewey as well as by Thomists: namely, overlooking the key role of *action* as the self-revelation of being in our human knowing — an absolutely central theme in the epistemology of Aristotle, St. Thomas, and Dewey himself. For on the one hand Kant must admit action coming from the real world of things-in-themselves into the human knower, since he insists that he is not an idealist, that we do not create by thought the objects of our knowledge. On the other hand, he will not admit that this action is revelatory of anything objective in real things, anything true of them — not even their real existence, since being itself represents only the position by the mind of its own synthesis of the unordered appearances in the sense-manifold with its innate *a priori* forms of sense and intellect. But action which is totally non-revelatory of the nature of the agent-source from which it comes is itself unintelligible and cannot be truly *action*.[11]

Kant cannot have it both ways. Either there is no real action of the real world upon us and he is forced into idealism, which he rejects vehemently, or he accepts the fact of real action of the real world upon us and then this action is necessarily revelatory, a manifestation of its real agent-source. As a thing acts, so must it be — *agere sequitur esse*, as the ancient Scholastic adage goes.

The root of the trouble lies, I suspect, in Kant's implicit rationalist ideal of knowledge of the real as it is in itself, *independent of any action* upon others by a detached uninvolved pure knower. Of course such a knowledge is impossible save for a purely creative knower, which we are not. But the whole key to an action-based realist epistemology is that our knowledge, involving incoming action from the known, received according to the mode of receptivity of the knower, is indeed through and through *relational*; but this relation itself is thoroughly real, necessarily revealing something significant about both ends of the relation. It reveals the known as actor, as in itself *this kind of actor*, which is precisely in the last analysis just what an essence should really mean in a realist epistemology — not only in that of St. Thomas, but also in any successful realist epistemology. Thus St. Thomas is not in the least reluctant to admit — in a text which astonishes many contemporary epistemologists when they are shown it — that "the substantial forms of things in themselves are unknown to us, but shine forth through their accidents" (i.e., operations, etc.), ". . . as through doors placed around it (*quasi ostia circumposita*)." So that the mind points back to the hidden nature through its manifestations in a kind of "discursive movement" — which I interpret as the intentionality of judgment, not as a direct intuition.

Such relational knowledge through action is necessarily perspectival and incomplete, proportional to the limitations and conditions of the receiver. Still it is a genuine perspective on the known as agent — and in the last analysis isn't that what we most want to know about other real beings: what characteristic actions we can expect of them? Given this umbilical cord to the real world through action on us, we can pursue any gaps in the intelligibility of the world thus revealed to us to affirm as real whatever is needed to fill these gaps, empirical or trans-empirical, as the case may be. This is precisely the path of efficient causality.

To sum up now what has gone before, all attempts to lay mines that will definitively block the modest access of the human mind to the real, and to whatever is needed to fill the gaps in its intelligibility, succeed finally only in blowing up the mine-layers with their own mines, leaving the rest of us free to navigate with critical care between the rocks and through the rapids. The metaphysical hypotheses worked out along the way, including arguments for the existence of God such as I will now present, are not Cartesian absolute certitudes, but explanatory hypotheses which recommend themselves as worthy of reasonable affirmation because they fill the gaps in the intelligibility of the real world we experience. They do so in a more illuminating and adequate way than other competing hypotheses, which either leave out something significant from experience or leave gaps in its intelligibility in principle unfillable from their perspective. Now to our positive task.

Constructing a Natural Theology

From all that contemporary philosophical discussion has taught us, it should be clear that it is not realistically possible to construct a purely objective philosophical argument for the existence of God floating free from all personal roots, one that is capable of convincing by its pure impersonal cogency any intelligent hearer whatever, irrespective of all predispositions and presuppositions, moral and intellectual, and of all cultural and conceptual frameworks. As Polanyi, Gadamer, and others have shown, there is no presuppositionless thought, in any field. We do presuppose, therefore, in anyone who is willing to give sympathetic consideration to the arguments we are going to propose, a certain familiarity with what I would call a metaphysical type of thinking. By this I mean a way of thinking that is open to asking radical questions about the very existence and intelligibility of the world we live in, of following the discovered exigencies of intelligibility wherever they may lead, and not cutting short, *a priori* or arbitrarily, the innate drive of the mind to understand the real as fully as possible.

Our metaphysical procedure will be, first, to identify significant gaps in the intelligibility of our universe as a whole, if we can find them, then to propose in a kind of branching technique the main options for filling these gaps, and to try to eliminate all of these options, or explanatory hypotheses, save one. By "gap in intelligibility" I do not mean merely something I do not yet understand, some mystery. I mean that one must show positively that, given the nature of the data, there is something in them that *excludes* any adequate explanation of them if taken by themselves alone. In a word, they just cannot, because of some built-in deficiency of being, be self-explanatory, but demand the help of some further real being, which fulfills the role of efficient cause, i.e., that which is responsible for their actual being or their coming into existence. In a metaphysical type of inquiry, because of its vast generality, it is possible to reduce the revelant options for explanation to a very few at one step, then move onto the next and do likewise — something that is rarely possible in the natural sciences because of the complex details to be explained and the wide ranges of hypotheses left open. The elimination of all options but one can rarely be done by purely logical means, but usually requires metaphysical insight into intrinsic idea-connections, which cannot command but which can afford, if carefully checked, sufficient grounds for reasonable affirmation of what it reveals.

I agree with Charles Hartshorne in his later works that it is more accurate these days to speak of "arguments" rather than "proofs" for the existence of God, since "proof" as understood today has become so rigorous in its requirements that it is impossible, properly speaking, to prove the existence of any real being (outside the knowing "I"), let alone the existence of a transcendent reality like God. Secondly, I agree with him that such arguments (or "reasonable ascents of the mind to God") exhibit a certain cumulative effect, one argument opening one side of the intelligibility needed; another, another side, or perhaps one argument for one type of mind, another for another. I believe, too, that a well-rounded and effective natural theology should proceed in a two-pronged approach: one approach is what I would call the "Inner Path," through the exigencies of the inner life of the human person having intellect, will, and moral sensitivity; and the other approach is the "Outer Path," through the exigencies of the entire cosmos (including humanity). The Inner Path would proceed from reflection on the innate drive of the human spirit toward the unlimited horizon of being as truth and goodness. In this way one is faced with the option that either an Infinite Fullness of Being as truth and goodness actually exists as the only adequate goal of this innate drive, a drive that is constitutive of the human — hence not capable of being substituted for by any finite goal or set of them — or our human spirit is radically absurd, oriented towards what does not and cannot exist. And since there is no good reason for opting for the absurd, the unintelligible, and every good reason for opting for the existence of the Infinite as closing the gap in intelligibility in us, it is uniquely reasonable to opt for the latter, though the opposite can never be shown to be a logical contradiction. I fear we shall have no further space to develop the Inner Path argument in the present paper.[13] The reason I choose the Outer or Cosmic Path for fuller development here is that the Inner Path can indeed reach an Infinite Good as *my* final good, *my* God, but not the Ultimate Source of *all* being, which is necessary for a fully adequate notion of God.

Argument from any caused being to a single infinite source of all being

Let me start with my own adaptation of a classical argument of St. Thomas, combining three essential steps, which he stretches out over some nine questions in his *Summa Theologica*, Part I (beginning at question 2, article 3, and finishing at question 11 — a point too often overlooked by those who look only to the Five Ways in question 2 for his complete proof). This argument is the longest one I shall give, which will at least serve the purpose of initiating the reader into a metaphysical type of inquiry, even if it is not convincing.

The basic question we are raising about the beings of our experience is not, "What are they? What are they like? How do they operate?" but the radical question about their very existence: "Why do they exist at all in this way that they do exist? What is the ultimate intelligibility, or sufficient reason, why they in fact exist at all?" It is important not to shortcircuit this question from the start, as Bertrand Russell and many empiricists are wont to do today. "Explanation" for them, as one told me, means to relate the parts to each other within the system as a whole. But you can't raise any questions about the system itself as a whole. As Bertrand Russell put it in a nutshell in his famous BBC debate with the Jesuit philosopher, Frederick Copleston, some years ago, when the latter pushed him on this point: "The world just is, that's all." Now it is true that our explanations start there. *Scientific* explanation, indeed, must start there; a science can use its methods only on some subject matter already given to it in existence. But it is an intellectual cop-out, an arbitrary restriction of the natural drive of the mind to know, to refuse to ask the question *philosophically*. The question of existence itself is one of the basic and most natural questions for the inquiring mind to ask when it is allowed to work at its full scope.

The three steps in the argument are as follows: (1) Given any caused being, there must exist at least one self-sufficient being; (2) No being can be self-sufficient unless it is also infinite in perfection; (3) There can be only one such being infinite in all perfections — which therefore must be the Ultimate Source of all other beings.

Step One: There must exist at least one self-sufficient being

1. There must always have been something existing (not necessarily the same being all the time). This is obvious: for if at any time absolutely nothing existed, then nothing now existing could ever have emerged from it. Existence itself is the ultimate matrix of all possible explanations. It is deficient existence that needs explanation.

2. Among existents there must be at least one self-sufficient for its own existence (or self-explanatory of its own existence, with no gaps in its intelligibility). Take any real being, e.g., yourself. Now either this being is self-sufficient, or it is not. If it is, this part of the search for intelligibility is finished. If it is not self-sufficient, it will require an efficient cause to bring it into existence (either in whole or in part). Now

we can raise the same question about this cause: is it self-sufficient for its own existence, or not? If not, we must go on to another cause; and so on....

Now it is impossible to close the gap in intelligibility in the caused being we started with if we go on to infinity in an infinite series of causes all of whose members are non-self-sufficient for their own existence. It is not a question of there being any impossibility of an infinite series as such. Mathematicians and physicists today seem quite at home with them. It is a question only of the self-sufficiency of an infinite series of real causes, all of which are themselves caused. For none of them contains within itself the sufficient reason either for its own existence or for its own effects, since the conditions for its own existence have not yet been adequately fulfilled, but are indefinitely in the process of being fulfilled — a process that can never in principle be completely fulfilled, that is endlessly being put off. Thus it turns out that nowhere in this series is there any fully adequate sufficient reason for the existence of any one of the members, since its necessary conditions for existence are never actually unconditionally fulfilled, but are always being postponed, waiting for other conditions to be fulfilled, which in turn are waiting for their conditions to be fulfilled, and so on in an endless series of unfulfillable necessary conditions. The yawning gap in intelligibility can never be filled this way. It follows, therefore, that somewhere along the line, either at the head of the series or outside of it, supporting the whole, there must be at least one self-sufficient being, which is the initiator (not necessarily in time) of the causal flow of existence into all the others in the series.

Many arguments stop here — as for example do three of St. Thomas's Five Ways (four in fact, since the third gives no reason for its final step). But there is much more to be done. What sort of being will qualify as self-sufficient? Perhaps some primordial atoms? And how many such can there be? The great Aristotle himself thought finally that there must be 55 unmoved, uncaused Prime Movers. So we must put an appropriate question to the self-sufficient being we have discovered that will further highlight its significant attributes. Its "job qualifications" will reveal its nature.

Step Two: Any being self-sufficient for its own existence must be infinite in perfection, i.e. unlimited in its qualitative fullness of all perfections. Or: *No self-sufficient being can be finite*. Why? Let us suppose it were finite. This means it would be one determinate, limited mode of being (limited in qualitative intensity of perfection) among at least several other possible modes such that at least one higher mode were possible. Otherwise it would not be finite or limited. Now there must be some sufficient reason why the being in question exists in this particular limited and determinate mode of being and not in some other possible mode. Why *this* being, or this whole finite world-system, in fact, and not some other? A principle of selection is needed to select this mode of being from the range of possibilities and give actual existence (energy-filled existence) to it according to this limited mode (or "essence," as the metaphysician would say). But no finite being can select its own essence and confer existence on itself. For then it would have to pre-exist its own determinate actual existence (in some indeterminate state), pick out what it wills to be, and confer this upon itself. All of this is obviously absurd, unintelligible. It follows that no determinate finite being can be the self-sufficient reason for its existence as *this determinate being*. Therefore it requires an efficient cause for its actual existence as this being. But, since we cannot go on to infinity in finite caused causes, we must eventually come to some Infinite Cause of these finite beings.

Thus every finite being, not only each one in particular, but any system as a whole that is finite and determinate in its mode of existence, as ours clearly is, needs a self-sufficient infinite being to draw it out of the range of possibilities and make it to be in this particular way and no other. It does not matter, in fact, how many other modes actually exist, or even all possible ones. Each one needs to be given actual existence according to its determinate mode, and no one can do it for itself.

Step Three: There can be only one such being infinite in all perfections. This is a quick and easy one, admitted by just about all metaphysicians, I believe, once the existence of an absolutely infinite being is granted. For suppose there were two such. Then one would not be the other. But this is impossible unless at least one of the two lacks something the other one has. Otherwise they would coincide in total indistinguishable identity. But if one lacks some positive perfection it could not also be absolutely infinite in all perfections. Also, as Duns Scotus has pointed out, at least one would be unable to know the other — a great imperfection; for either one would have to create the other or be acted on by the other; and in either case one of the two would have to be dependent on the other, and hence not self-sufficient — which we already established.

We conclude, therefore, that if anything at all exists, then there must exist one and only one Infinite Source of all being. This we may call an apt philosophical definition of "God."

Argument from any finite being to a single infinite source of all being

It may have occurred to the reader that it is possible to condense the above long argument into a much simpler and more elegant one beginning with Step Two and proceeding directly from any finite being to the necessity of a single Infinite Source of all being. This is perfectly true.

Instead of starting from some particular finite being, one can also make this into a more powerful and impressive argument, simply by stepping back and taking as our starting point the entire system of our material cosmos as a whole. This is clearly a determinate limited system, whose basic contants are precisely limited — e.g., the four basic forces, the speed of light, Planck's constant, etc.. Therefore the entire system as a limited whole can provide no sufficient reason why it actually exists as this determinate system and no other. Physicists have been able to show, as other papers in this volume indicate, that it is possible to vary the values of the basic constants of our universe (e.g., the speed of light) and still get a consistent system. Why, then, *this* determinate one rather than some other? The system itself can provide no answer. The only conceivable way one could do so would be if one could show that this particular material universe was absolutely in every respect the only universe possible — that is to say, that there could be, not only no other type of material universe, but also no other type of universe at all, including all possible modes of immaterial being. It is clearly impossible to make any such all-inclusive claim, particularly from the point of view of any natural science, or science of any kind we know. The argument holds firm, I believe. There is no way for the system itself to fill the gap in its own intelligibility, to illuminate the sheer brute fact of its own limited existing *thisness*. Note, too, that this argument is a purely metaphysical one, quite independent of any changes or progress in the content of the sciences. For the natural sciences by their very nature must always be dealing with new patterns and systems of determinate, finite entities, of whatever sort. Science and the determinate finitude of its objects are by necessity always linked together.

Argument from any determinate essence (or system of essences) as contingent to a necessary ultimate cause of existence

I am well aware that many contemporary philosophers, including especially scientists thinking as philosophers, have trouble dealing with the concepts of finite and infinite perfection, in particular with the infinite aspect. When talking with various scientists, especially in Berkeley at the Colloquium on this topic at the Center for Theology and the Natural Sciences, it also became clear to me that it is difficult to find any secure starting point for an argument to God from particular aspects of order in the universe, in view of the genetic character of all recent speculation about its origins, in which scientists constantly strive to reduce the present order to some simpler earlier state. It then occurred to me that what was really needed, and might be most effective, was a very simple, radically existential argument, which would reveal the need of a cause of the existential energy of the universe, a cause of existence for the universe of a totally different order than all the patterns and modes of order revealed within it. I am very grateful for those discussions, for they precipitated one of those metaphysical breakthroughs in which an old, not well-known argument of St. Thomas (briefly sketched out by him in his *Summa contra Gentes*, Bk. I, ch. 15, no. 5), suddenly came alive for me in a new form and seemed to match up well with the current discussions between scientists, philosophers, and theologians on cosmic origins. At any rate I have developed a special liking for this argument, for its simplicity and depth, and for what it reveals to us about the very inner nature of God, and I am going to take the risk of presenting it.

There is one essential initial step for getting inside the inner "logic," or perhaps better, the inner movement of this argument. It is to bring clearly into focus the actual existence of this universe in all its fresh wonder and originality. This means getting beyond the mere epistemological recognition of the fact that this universe exists. We must look deeper into the very intrinsic actuality, the active presence within the existing beings themselves, which is the ground for our epistemologically true judgment about them. This is what St. Thomas calls by his own original term, "the act of existing" (the *actus essendi* or the *esse*, the to-be, of things). And this actual existence within real beings is not to be looked upon as some minimal, static factual state, but as an *active presence*, *presence-with-power*, as power-filled, energy-filled presence. For it is characteristic of every real being, as St. Thomas brings out so insistently and forcefully, that it is a center of power, of active energy, that it naturally pours over into self-manifesting, self-communicating activity toward other real beings, generating a web of mutual relations from their interaction. Real existence is active, energy-filled presence-with-power, (what Aquinas terms the *virtus essendi* or power of being).[14]

Now let us turn to any determinate essence or system of essences, i.e., a determinate pattern or mode of existing, and ask it if it contains within itself any explanation of why it actually exists with the determinate quota of energy and mode of operation that it does have.[15]

The crux of the argument is that no determinate essence (or system of essences), no model of reality, can specify or prescribe its own actual existence, its own real instantiation. It is especially obvious with a mathematical model, the essential tool of scientific inquiry. No mathematical model can specify its own instantiation as a real system with energy. It is only a web of intelligible relations, a blueprint for the flow of action, but in itself it is absolutely static; nothing in it provides for its actual existence, filled with power to act. As St. Thomas puts it, all such determinate essences are radically "contingent," i.e., neutral or "indifferent" (not the happiest term) to existence or non-existence — they can either be or not be. Hence it is that they need a cause of a totally different order from that of essence or model, a cause in the existential order that is a source of energy-filled existence and can communicate it to others, a cause that can bring them out of their contingent neutrality to exist — which really means out of their pure intelligible possibility — and "real-ize" them into the real order of energy-filled existence. Of themselves they have no necessary link with their existence; it is a sheer brute fact that they *are*. But they have no resources anywhere inside them to explain why in fact they do exist. They need an actualizing-energizing cause outside of themselves to link their essence-models to real existence.

If all such determinate essences or models need an actualizing cause, is there any sort of essence which can specify its own existence? Yes, there is one and only one such: one that contains actual existence as constitutive of its very essence, whose essence *is* energy-filled existence itself in all its unrestricted fullness, unreceived from any other, in a word, existence itself in its very source. It is clear there can be only one such essence. This is for St. Thomas the very essence and "proper name" of the unique God. "Who Is," or, in the language of Yahweh to Moses in the Book of Exodus, "I am who am." Each of us can only say, "I *have* existence; I am, but only so much, in such and such a way," but *not* "I am" — unqualifiedly, "I am existence in all its energy-filled fullness." Such a source, of course, of its nature could not be limited, imperfect, or incomplete existence, since it is the very source, the unique source of all possible modes of existential perfection. "Infinity" in the metaphysical qualitative sense simply means unrestricted, unparticipated fullness of possible perfection.

This argument needs to be meditated on so that its point and its force gradually emerge. It seems to me about the most radical and cogent one I can think of, certainly the most existential. Whether it is really just another way of putting the argument from finitude or a distinctly different approach, I am not sure yet myself. One advantage of it in the context of scientific discussion about the origin of our universe is that it undercuts all objections to the effect that this might be the only possible universe in an absolute sense. For one thing, all that science could possibly show — if it can, which is at present highly controversial — is that this might be the only possible *material* universe, not that it is the only possible universe absolutely, with respect to all possible modes of being, material and otherwise. But the point of the argument is that, whether this is the only possible material universe or not, it still of itself remains only *possible*; it contains nothing within its model that provides that it must actually exist.

The same holds true of the suggestion that an infinite number of all possible universes might actually exist, thus eliminating the need of a choice between them by a Cosmic Planner. This hypothesis already begins too late, supposing that all these possible universes actually exist. But the point is that just because they are all possible gives no grounds for asserting that any or all actually exist. The passage from essence to existence, from model to instantiation, is the crucial one, the gap in intelligibility that needs to be filled in from another source, in possession of energy-filled existence itself. No matter how simple the original state suggested by scientific hypothesis, pure quantum energy field or whatever, it still remains that one must first have some energy-filled existent to start with.[16]

Argument from order in the world

The ancient argument from order or design in the world, now commonly known as the teleological argument, St. Thomas calls the most widespread and the most efficacious path to God at all times and in all cultures. And I think he is right. Such an argument, however, needs special adaptation to be effective in the light of contemporary science, both evolutionary and cosmological.

I would like to present what I think is the most powerful version of the argument, fitting more closely the way we have come to see nature today. (St. Thomas himself laid down the basic principle, namely, that when many non-rational agents cooperate together to form a single world order, some unifying ordering mind must be the source of the unified order. But he developed it differently in detail.[17]) It runs thus. Take any dynamically ordered system of active elements, such as our own universe, in which the various active elements are ordered towards regular reciprocal interaction with each other. For example, when forming water, hydrogen atoms are ordered to combine with oxygen atoms in the proportion of 2 to 1. Similar orderings hold for all the molecular, atomic, and subatomic elements which form our unified cosmic-wide order. Now in such a system, where each active element's basic properties (their natures, in metaphysical terms) are defined by relation to the others in the system, no one element can explain its own nature or be the sufficient reason for its own active nature as existing and operating, unless it is *also* the sufficient reason for all the others reciprocally related to it. But this is impossible. For then this element would have to be both prior (in causal, not necessarily temporal, priority) in its activity to the others and responsible for them, and at the same time presuppose them, since its active properties are all ordered to interaction with them according to law. It would thus in its very nature as active presuppose the others as reciprocally constituted in relation to itself, and yet be independent of and responsible for these correlated properties in others by which its own active nature is defined. Clearly this will not work.

Such a cosmic-wide order, therefore, is one in which many are brought together under the unity of great overarching laws of mutual interaction. But this order can have its ultimate sufficient reason, its intelligible grounding, only in some pervasive unifying cause, capable of thus ordering many active agents into a single unity. Such a unifying, ordering agent, which must set up the unity of reciprocal ordering prior (causally) to the actual operation of these agents, in terms of not-yet-existent future effects, can only be a Mind.[18] In fact, this is almost a definition of intelligence: the power to creatively construct new order out of mere possibility. Such a Cosmic Ordering Mind must, obviously, transcend in its own being and activity the system that it orders — otherwise it could not itself operate until the system were already set up; it would be both prior to, and independent of, yet dependent on the same system.

Metaphysically speaking, to be complete one could still raise the question of whether such a World-Ordering Mind is finite or infinite, one or many — perhaps one for each possibly independent universe? Recourse would have to be had again to the earlier — and in the last analysis, it seems, indispensable — argument from finitude to one Infinite Source. But even without this step, which most people take for granted, we have still reached a Transcendent Reality upon which our whole cosmos and ourselves in it depend. And this gives a sufficient foundation for a basic religious attitude of gratitude, love, obedience, etc. to the Author of our nature and destiny.

The beauty of this argument is that it works for any basic dynamic order in our or in any universe whatever. For without some primal ordering of the basic active elements in a system of mutual interaction, prior to their actual operation, nothing could happen at all, not even by chance; the elements, if any, would just pass through each other in total mutual atomic isolation. There would be no such thing as a world at all. And only a mind, capable of creatively thinking up a determinate order out of possibility, could establish the basic mutual ordering of these active natures to each other prior to their actual operation in the existential order.

Does this argument, beginning from any existing dynamic order, lose its power when set in the context of current discussions in theoretical physics and cosmology, all of which are attemping to reduce the present complex order to an original simpler order, or possibly to an absolutely simple state? This does shift the ground of the argument significantly, and one must think carefully about the implications. It seems to me, though, that it still holds firm. For, no matter how simple the original energy state may be, it does evolve into determinate active elements which exhibit a built-in dynamic orientation to combine together through mutual interaction in regular determinate ways. Some prior dynamic orientation must have existed within the original energy state to thus evolve into a determinate dynamic order; otherwise it would emerge totally out of nothing, be in principle totally unintelligible, without any sufficient reason whatsoever — which is not an explanation at all, but an intellectual cop-out. Note, too, that even if the particular order that emerges from the original simple energy state emerges purely by chance — which is highly controversial — any order whatever would have to be internally ordered within itself, if anything determinate is to result, for it to be a discernible cosmic order at all.

It would seem, then, that the principle holds firm: any determinate order, whether stable in itself or originating from a prior physical state, must be grounded ultimately in an Ordering Mind transcending the system itself.

With regard to the evolutionary development of living organisms, let me add this note: no matter how much chance there may be in the external conditions of the environment which these organisms exploit to evolve, what remains as the essential prior condition for the whole process is the built-in inner dynamism, the unflagging dynamic drive of the organisms toward interacting in determinate ways with the other agents in the universe around them, toward actively exploiting the opportunities offered them by chance. This innate positive drive to survive, to act, to interact, cannot be supplied by any random exterior conditions. It must be built into the active potentialities (or dispositional properties) of the very natures of the organisms themselves, prestructured from the beginning to interact with one another in basic determinate ways. It is this innate drive that is not supplied by evolutionary theory, but must ultimately be predetermined by some creative ordering Mind, that alone can transpose intelligible possibilities of order from creative idea to actual existence with focussed power.

Postscript on Deriving the Divine Attributes

Let me indicate with extreme brevity just the general procedure for determining which attributes (or predicates) can legitimately be applied to the God we have discovered at the term of our arguments. Two basic principles are involved. One is a corollary to the nature of efficient causality — understood, of course, in its active productive ontological sense, not as a mere Humean succession of events. This is the similitude that must exist in some at least analogous way between an effect and its cause. For, since the effect as effect derives from its cause, receives its being from its cause, and the cause cannot give what it does not possess at least in some higher or equivalent way, there must be some bond of real similitude between them. This can then serve as the bridge by which we can link our knowledge of the effect with what we can affirm of the cause.

The second principle is that we cannot without further analysis transfer any attribute found in the effect directly and literally to its cause. In the case of God, for example, any attribute (or predicate) containing in its very meaning some limit or imperfection must be winnowed out as implying some contradiction if literally applied to God — this is what St. Thomas calls "the negative moment" in the process. The attribute in question must be reduced to some broader, more universal one that is purely positive, containing no limits or imperfections in its core meaning, including the lower limited attribute as only one of its limited modes or degrees, and calling for our unqualified approval, such that, were God to lack it, God would be less perfect than we are. Thus visual power in a creature would be transposed into knowledge in God, and so on. A small number of basic attributes survive this purification process and can be applied literally, though analogously, to God, with the index of infinity added. We can and must legitimately affirm these of God, but we cannot grasp directly and clearly just what these are like as possessed by God. Such viable attributes turn out to be: existence, unity, activity-power, goodness, intelligence, will, love, and a few others derivative from these. God's essence in itself (his proper mode of possessing these perfections) remains totally unknown to us, St. Thomas insists. Our own natural dynamism toward the Infinite helps us to illumine, though obscurely, this mystery-shrouded essence of God through a certain "connatural affinity" with him as his images, by a knowledge of the heart through longing and love as magnetized by the Infinite Good, a knowledge through longing and love, not through vision — at least not in this life.

Let me conclude by suggesting that the fascinating task-challenge-opportunity for natural theology today is to speculate imaginatively as to what the "personality" or "character" must be like of a Creator in whose image this astounding universe of ours is made, with its prodigal abundance of energy, its mind-boggling complexity, yet simplicity, its fecundity of creative spontaneity, its ever-surprising fluid mixture of law and chance, etc. Must not the "personality" of such a Creator be one charged, not only with unfathomable power and energy, but also with dazzling imaginative creativity. Such a creator must be a kind of daring Cosmic Gambler who loves to work with both law and chance, a synthesis of apparent opposities — of power and gentleness, a lover of both law and order and of challenge and spontaneity.[19]

NOTES

[1] Freeman Dyson, *Disturbing the Universe* (New York: Harper & Row, 1979) 251.

[2] Ernan McMullin "Natural Science and Belief in a Creator," in this volume.

[3] A valuable survey of details on this can be found in L.S. Betty and B. Cordell, "God and Modern Science: New Life for the Teleological Argument," *International Philosophical Quarterly*, **27** (1987) 409-35.

[4] See D.E. Thomsen, "The Quantum Universe: A Zero-point Fluctuation?" *Science News*, **128** (Aug. 3, 1985) 73.

[5] More accurately, it should be described as a nothingness of form, not of energy. See in this volume: John Leslie, "How to Draw Conclusions from a Fine-tuned Universe"; C.J. Isham, "Creation of the Universe as a Quantum Process"; Frank Tipler, "The Omega Point Theory: A Model of an Evolving God."

[6] See the article cited in note 3 above, and C. Thaxton, W. Bradley, R. Olsen, *The Mystery of Life's Origin: Reassessing Current Theories* (New York: Philosophical Library, 1984).

[7] For typical examples, see K. Baynes, J. Bohman, T. McCarthy, eds., *After Philosophy: End or Transformation?* (Cambridge: MIT Press, 1987), in particular the essay by J.F. Lyotard, a leading French Postmodernist, "The Postmodern Condition."

[8] On Deconstructionism, see the essay by Derrida in the book cited in note 7; also John Caputo, *Radical Hermeneutics: Repetition, Deconstruction, and the Hermeneutical Project* (Bloomington, IN: Indiana University Press, 1987).

[9] Cf. H. Boeder, "Origine et préhistoire de la question philosophique de l'AITION," *Revue des sciences philosophiques et théologiques*, **40** (1956) 421-43.

[10] For a fuller development of this whole question of efficient causality, see my book, *The Philosophical Approach to God: A Neo-Thomist Perspective* (Winston-Salem: Wake Forest University Publications, 1979) Chap. 2: The Metaphysical Ascent to God through Participation.

[11] For the central importance of action in St. Thomas's metaphysics and epistemology, and its relevance to Kant and modern epistemology, see my essay, "Action as the Self-Revelation of Being: A Central Theme in the Thought of St. Thomas," in Linus Thro, ed., *History of Philosophy in the Making: Essays in Honor of James Collins* (Washington: University Press of America, 1982) 63-81. For John Dewey's reference to action as the key to epistemology and its neglect in modern philosophy, see his 1897 Lecture, "The Significance of the Problem of Knowledge," *Early Works of John Dewey* (Carbondale: S. Illinois University Press, 1972) 295, 297.

[12] The first text is from *Summa Theologiae*, I, q. 77, art. 1, ad obj. 7; the second quoted is from *Comment, in III Sent.*, dist. 35, q. 2, art. 2, sol. 1.

[13] I have developed this argument in my book cited in note 10, Chap. 1: *The Turn to the Inner Path in Contemporary Neo-Thomism.*

[14] See *Sum. c. Gentes*, Bk. I, ch. 28; and my essay cited in note 11.

[15] Some scientists at the conference claimed that they are uncomfortable with the metaphysical term "essence," as though it were some kind of "black box" hiding within it strange entities like immutable Platonic forms, or who knows what. When I asked what term they would be more at home with, I was told, "model"; that this is what scientists are dealing with all the time, especially mathematical

models. The term "model" is fine with me. The force of the argument becomes even clearer, more pointed, I believe.

[16] As quoted in the New York Times (May 4, 1988, p. 22) Stephen Hawking in his new book, *A Brief History of Time* (New York: Bantam Press, 1988) speculates that: "There was in fact no singularity at the beginning of the universe... no edge of space-time at which one would have to appeal to God or some new law to set the boundary conditions for space-time.... The universe would be completely self-contained and not affected by anything outside itself. It would neither be created nor destroyed. It would just BE."

If by "the universe" he means to include *all that is*, then it is of course true that the universe is self-contained. And, if he means that the material universe need not have begun in time or space, this is controversial, but could well be absorbed by philosophers as such. But, if he is implying that our material universe, with the determinate ordered existence that it has, without any grounding Intelligence and Source of existence not itself limited, is completely self-sufficient for its own determinately ordered (this and not that) energy-filled existence, then he is cutting short his inquiry prematurely, leaving us with the existence of this universe as a brute fact or surd, with a gaping and unfillable hole in its intelligibility, with no answer to the ultimate "why it is and is so" — "a why" to which our human intelligence, with its unrestricted drive to know and understand, has every right to seek an answer. Our arguments outlined above had nothing to do with beginning in time or space, but only with finitude and determinate existing order as such. Contrary to common belief, a beginning in time is not the only reason — in fact not a decisive reason at all — why our material universe needs a Transcendent Reality as cause.

[17] Some samples are *Sum. c. Gentes*, Bk. I, ch. 13, no. 35; ch. 42, no. 7; *De Potentia*, q. 3, art. 6.

[18] Charles Hartshorne develops a similar argument, using the felicitous phrase, "a cosmic-wide order needs a cosmic-wide orderer." See *A Natural Theology for Our Time* (La Salle, IL: Open Court, 1967) 58-67. See Donald Viney, *Charles Hartshorne and the Existence of God* (Albany: SUNY Press, 1985) Chap. 6: "The Design Argument." For some further highly stimulating ideas about the Cosmos as Revelation, see Brian Swimme, *The Universe Is a Green Dragon*, (Santa Fe: Bear and Co., 1985).

[19] See Robert J. Russell, "Quantum Physics in Philosophical and Theological Perspective," in this volume.

CHRISTIAN BELIEF AND THE FASCINATION OF SCIENCE

OLAF PEDERSEN, University of Aarhus

Science Itself

In the following sketch I shall try to argue that (1) there is something which may be properly called the "fundamental scientific experience"; (2) that this experience can be located in certain areas of scientific activity and at all its levels; (3) that this experience can be interpreted in a way which reflects certain elements of Christian belief; and (4) that it is more a call to conversion than a prop for some kind of "natural theology". However, a corollary of these theses must be an explanation of the fact that there always have been and still are good scientists who are committed Christians without being naive, hypocritical or schizophrenic. Since I am not a theologian or philospher of science, I am unable to explore these matters in a systematic way; all I can offer is a series of personal reflections based on such disparate elements as my own impressions of scientific research done long ago in relativity and cosmology, some years' experience as a science teacher, and a long standing interest in science as an historical phenomenon. I shall also leave all particular sciences and all special scientific theories on one side. It may well be that some areas of scientific investigation — quantum mechanics, cosmology, genetics — present problems that are of particular interest to the ongoing dialogue on science and religion. However, I shall leave such matters to more competent minds and, for reasons that will appear later, restrict my reflections to some very general aspects of science. Both science and Christian faith have been with us for a very long time and it is at least a reasonable hypothesis that the proper locus of their interaction resides in features which have been there right from the beginning. This has a number of consequences.

One must remember that "science" in modern English usage often means science and technology viewed as a single entity. They are not clearly distinguished, or at least they are regarded as an inseparable partnership. But without denying the existence of a scientific-industrial complex in modern society or closing one's eyes to the serious problems it creates, it is still important to realise that throughout the ages technology was largely independent of science, and that even today there are huge areas of scientific investigation devoid of technical applications without for that reason being deprived of their true scientific character. Therefore, my general point of view implies that, although there are genuine matters for discussion in fields like "science and ethics" or "science and politics," the core of the problem of "science and Christianity" has something to do with science "itself" as a permanent activity which transcends more time-dependent connections with other fields of human concern.

"Science itself" is an extremely complex area of human activity. It applies a great variety of methods of observation and interpretation and more than one "language" for expressing its insights in the form of a public discourse about nature, which is the principal aim and final outcome of all scientific work. By scientific practice I shall now understand everything which scientists themselves actually do when they perform the work of erecting both the whole structure of science and its component parts. This structure is distinct from the discourse itself, which consists of scientific statements connected by their characteristic rules of grammar. We shall be concerned more with certain features of scientific performance than with its concrete and public results: What are scientists doing when they do science?

This question is of course partly answered in the philosophy of science, which investigates the methods by which scientific statements are established. But such problems are largely outside the pale of this paper, in which we are more concerned with the way that science is done than with results. All we need is a brief, and admittedly very incomplete and perhaps naive, survey of a few of the more conspicuous types of scientific statements. This survey contributes nothing to the philosophy of science, but may help us to discover such areas of scientific practice which are located at the boundary of Christian experience.

Scientific Statements

One way of characterising scientific statements is to place them within one or another of the linguistic traditions of that discourse about nature which emerged in Greek antiquity. We know much about a previous type of discourse in which the appearance of individual phenomena was ascribed in a personalistic language to gods and spirits of nature who provoked them in an arbitrary way. This gave way to an attempt to connect the phenomena by the assumption of some kind of immanent necessity which would be described in an impersonal and rational language. This change of the discourse generated a linguistic crisis, since the established language of mythological discourse did not accommodate the idea of immanent necessity. Out of this crisis arose three new traditions distinguished by their different linguistic apparels. The Aristotelian tradition used a metaphysical language shaped in terms of cause and effect, matter, substance, form, and the like, all of them originating as metaphors borrowed from ordinary language. In this tradition science was regarded as the quest for causal explanations of the phenomena. Against this arose two other traditions in which the phenomena were connected by mathematical relationships between non-metaphoric terms and where causal explanations were rejected, as even Aristotle had noticed with regard to final causes. One of these was the Platonic tradition, in which the mathematical relations appeared as *a priori* conditions, often with purely numerological speculations as the result. The other was the Archimedean tradition, in which these relations were established *a posteriori* through a direct appeal to experience. These three traditions have co-existed in science ever since, competing for the allegiance of scientists and philosophers, and often leading to unnecessary confusion.

Another way of approaching this matter is to sort out the scientific statements according to their more or less intimate connection with the world of phenomena. This leads to three principal orders of statements which may be very roughly described as follows.

In the first we meet with very simple references to something which is found by direct observation and, "when found, make a note of," to quote that sharp and systematic observer Captain Cuttle. Such statements are bound to a certain time and a certain place. They may be framed in ordinary language or in numerical terms. "This morning it was raining at Castel Gandolfo" is an example of a statement of this kind, and so is "Today at 8^h15^m the temperature of the air in the market at Castel Gandolfo was 18° Celsius," although this latter statement contains numbers read off two different instruments.

There is a second order of statements which grammatically are equally simple, but logically of a very different nature. They are not concerned with single observations of individual phenomena, but affirm correlations, valid always and everywhere, between two or more whole classes of phenomena. Thus the saying, "After thunder follows rain," expresses in ordinary language a relationship between the phenomenon of thunder and the phenomenon of rain in general. More complex is the statement that, "The boiling point of alcohol is 78° Celsius at a pressure of one atmosphere," which affirms a relationship between three classes of phenomena — the state of alcohol as a liquid or vapour, its temperature, and the pressure to which it is subjected. Statements of this type may be said to form the primary relations of science. They are just simple correlations without causal connotations; so when Mrs. Nickleby wondered that so many cobblers died from drowning and guessed that it might have something to do with leather, she not only expressed a correlation but went further into the realm of theory.

A scientific theory may be regarded as a network of statements constructed according to some kind of logical principles which sort the individual statements into certain classes, although this cannot be done in an unambiguous way. This cannot be discussed here in any detail; but it is useful to remember that special problems are raised by theoretical statements which refer to theoretical entities such as force, energy, gravitation, ether, and the like, which have no direct counterparts in the world of phenomena. Another special class is represented for instance in physics by the general laws of conservation of matter, energy, momentum and several other entities. Such statements enjoy a life of their own, being incapable of proof within the theory in which they function as hermeneutical principles or tools for interpretation and discovery. Reflecting on the apparent lack of energy conservation in beta-radioactive decay, Wolfgang Pauli did not draw the conclusion that the energy principle was violated in nuclear physics, but that a previously unnoticed elementary particle (the neutrino) was involved in the process in such a way that the energy balance was preserved.

The Fascination of Science

This extremely sketchy survey of some of the principal types of scientific statements raises a number of questions: How are such statements established? How can they be interconnected? In what sense can they be called true or false? What is the ontological status of theoretical entities? — and several other genuine and obvious problems which the philosophy of science must try to elucidate. However, here we must ask a question which is of a very different nature and much more delicate, since it seems to imply subjective considerations of a kind which science usually tries to avoid. What is the reason why scientists are devoted to and perhaps even fascinated by science?

Scientists may of course be influenced by all kinds of motives. They may be urged on by vanity and the desire to become known or even famous. They may suffer from ambition and the desire for a more prosperous career. They may give way to greed and try to exploit economically the results obtained. Or, passing to the credit side of the ledger, they may be motivated by benevolence towards a colleague or by altruism to society. This tells us no more than that scientists are human beings and exposed to all the moral forces to which our race is subject. All the motivations hinted at here are externally connected with the results of scientific activity, but they are not related to scientific practice in the sense in which this term was used in the introduction to this paper. Moreover, a great many scientists are more fascinated precisely by the scientific practice than by public results. They would surely continue to do science even if all external motivations ceased to operate. There seems in fact to exist a much more primitive fascination and a more deep-rooted satisfaction in doing science than all external incitements to pursue it can provide. In which area of scientific practice does this particular kind of fascination reside?

The elementary statements of the first of the orders mentioned above lead a rather precarious and shadowy life. They are the immediate outcome of some kind of observation, and many of them originate with a brief glance at a pointer on a scale or similar snappy procedures which are not apt to leave any lasting impression on the mind apart from very exceptional cases in which, for instance, the result of *an experimentum crucis* is eagerly awaited. They are stored in laboratory journals or other records from which a few among them are resuscitated to a more glorified after-life in a publication, while the rest are drowned in the mire of data processing and doomed to oblivion or even physical extinction, as many historians of science deplore. Moreover, as soon as such a statement is made, its accompanying circumstances disappear into the unretrievable past with the result that they cannot be verified or falsified. Everything considered, such statements are not really exciting even to scientists, just as they do not attract the attention of the general public.

At the other end of the spectrum is the complex variety of statements which are joined together into a scientific theory. They are usually surrounded by much more interest in all circles. The proliferation of popular literature shows how fascinated the public is by expanding universes, black holes, genetic codes, or theories of descent. Similarly, philosophers of science seem to be devoting most of their work to questions raised by theory. Also, any historian of science knows that most of the great battles within the scientific community itself have been fought over theoretical questions. On the other hand the historian must also be impressed by the changing nature of theories. Even if a new theory can be said in some sense to "comprise" its predecessor, this generalisation is usually achieved at the cost of a radical change of the meaning of basic concepts like mass, energy, space, time, etc. Such changes certainly form one of the most fascinating parts of the life of science. However, in the following we shall abstain from any attempt to analyse what scientists do when they erect new theories or adjust old ones . We are concerned with a more fundamental type of fascination, which can be described without regard to the complex philosophical questions that arise out of the art of erecting theoretical constructions.

This leaves us with statements of the second order, the primary relations of science. Here we meet with a picture which is radically different. Primary statements are more or less doomed to oblivion and theories are subjected to change, while primary relations show a remarkable resilience to the passage of time. Once critically established they are preserved as a treasure from which very little is again discarded. Of course, I am not here speaking of the idea that all natural constants might be slowly changing over the cosmological time scale, nor am I referring to the special class of statements which are true at one time and false at another, as when one states that, "The French coast passes two miles north of St. Malo," which is true at regular intervals of $12^{h} 25^{m}$ and false in between. What I have in mind is primarily that kind of statement which is listed in a work like the well known *Handbook of Chemistry and Physics* with its thousands of specific gravities, melting points, refractive indices, atomic weights, electrical conductivities, etc.

This work appears from time to time in a new edition which is always more rich and comprehensive than its predecessor. The number of primary statements is ever increasing, and new information of permanent value is obtained all the time. Although this happens in a modest and inconspicuous way, it is impossible to deny that this accumulation of primary relations marks a kind of real progress and is a source of wonder and fascination. But here we must distinguish. The numerical values in the Handbook are not fascinating in themselves. I would be just as happy to be told that lead has the specific gravity of 12.7 instead of 11.4. Now a result of this kind is not aesthetically fascinating in the same way as theories may be. It would be strange if it were characterised as "elegant" or "beautiful". The statement that "The moon is illuminated by the sun," may refer to a beautiful phenomenon but is as such neither beautiful nor ugly. It is simply true in the sense that it would be scientifically impossible to replace it by a contradictory statement. Therefore, the real source of fascination is the fact that science does possess a large body of statements which the scientist cannot alter. In contrast to both primary and theoretical statements, we meet here with an open set of propositions which we must always affirm and shall never have reason to forget.

The Role of the Mind

Thus the fascination of science seems to be located in the domain of primary relations and not in that of theoretical speculations, however absorbing the latter may be. At this point one cannot escape a rather painful question. Primary relations and theoretical structures are parts of the same scientific discourse. If scientific theories are nothing more than constructions of the mind, who can say that this is not the case also of the primary relations? This problem has explicitly engaged the attention of philosophers of science, but from time to time scientists have also addressed it, and have come to extremely different answers. We all know the concluding words of Sir Arthur Eddington's *Space, Time and Gravitation* about the explorer on the beach who investigates a set of strange footprints in the sand only to discover that they are his own.[1] Newton had also walked on the beach (at least in his imagination) and was able to liken himself to a small boy who rejoiced in finding a smoother pebble or a prettier shell than others.[2] Thus there are two radically different views. In the first case scientific results are construed as imprints made by the human mind upon nature; in the second they are regarded as something "found by" or "disclosed to" the scientist.

Eddington's attitude may well be founded upon the strange and rather impossible conditions which prevailed around 1920 (when his book appeared), when the new relativistic cosmology was asked to give an account of the universe at large based on practically no numerical data. Even 20 years later all we had to work with were the values of the Hubble constant and the average density of matter, both of them wrong by at least an order of magnitude. This would easily lead to a certain *sub specie aeternitatis* conception of science in accordance with the quotation from *Paradise Lost* placed at the beginning of the book. Here Milton ridiculed theoretical astronomers who were able only

> Perhaps to move
> His laughter at the quaint opinions wide
> Hereafter, when they come to model heaven
> And calculate the stars; how they will wield
> The mighty frame: how build, unbuild, contrive
> To save appearances.[3]

This is as it may be. In any case, Newton seems to have adopted the opposite attitude, thinking that he had "found" his results and feeling a sense of joy during his search. No doubt he also went to an extreme when he thought that he had "found" universal gravitation by inspecting the phenomena of gravity and planetary motion, even if gravitation appeared in this way as a brute fact without support in any known physical theory. Without a complete roll call among scientists over the ages it is impossible to gauge the relative strength of these two camps; but I think that most historians of science share my intuition that the great majority of scientists would come down on the side of Newton. At least the metaphor of "finding" is so widely spread that it would be tedious here to list the many instances in which it has occurred.

The Fundamental Role of Mathematics

There are several reasons why the metaphor of "finding" lends itself so easily to descriptions of how scientific results are produced. It does not imply any particular method of research, but applies equally well to results which seem to appear out of the blue and to those which come after an intense effort guided by preceding theoretical deliberations — even if these were based on a wrong theory as has often happened in the history of science. It also covers the many cases in which the result appears in a way which was completely unexpected regardless of whether the theory was right or wrong. This unexpected character is important since it seems to mark every acquisition of new insight, whether it takes place at the forefront of scientific research or in those more humble situations where it is acquired by students without being new to persons who are already better informed. May I illustrate this by a personal recollection?

As a young science master many years ago, I had to teach specific gravity to a class of eleven year old pupils. Of course, the textbook began with a definition to be learned by heart as a dictate from some unspecified higher authority. Then the teacher was supposed to hand out various small pieces of lead, asking the boys and girls to measure the weights and volumes of their respective bits of metal, and instructing them to perform a prescribed mathematical operation (division of weight by volume) in order to verify that lead has, in fact, the specific gravity of $11.4\,g\,cm^{-3}$, as the book said. Since the results did not always confirm the correct value the pupils felt disappointed, while the teacher tried to dispel the gloomy mood of the class by a boring disquisition on experimental errors.

All this was far from fascinating and a very bad introduction to what really goes on in science. So one year we did it in a different way. The textbook was closed and the pieces of lead handed out right at the beginning. They were weighed and measured and the several pairs of numbers were arranged in two parallel columns on the blackboard. In this way we obtained a great and confused amount of primary statements apparently without rhyme or reason. Asked what they could do with numbers the pupils began to calculate as best they could, applying more and more advanced procedures within their mathematical horizon. The confusion was not removed by adding the two given figures. Subtraction was no better and multiplication made it even worse. The situation became increasingly desperate and in the end they plucked up their courage to try the most advanced mathematical technique they had mastered: they performed a division — and lo and behold! the same result appeared everwhere with only small variations.

Personally I shall never forget the silence which suddenly spread among the children who, contrary to their usual behaviour, sat quite still for minutes wondering at what had happened in this unexpected appearance of order out of confusion. That they were truly fascinated was beyond doubt. Consequently they were motivated for a truly philosophical discussion of both their experiment and their experience. First, it was easy pedagogically to cash in on the discovery and to persuade the young scientists that, since the same figure emerged everywhere, it was probably important and that it ought to be given a name and used as a characteristic of the substance of lead.

Furthermore, the whole procedure illustrated the role of mathematics in science. Only one among several possible mathematical treatments of the primary data was able to produce an interesting result. This meant that mathematics was not just another scientific language to be used or rejected as one pleased. It was much more potent than ordinary language, and the only one which was able to produce a fascinating result. It was not only a vehicle of description but a tool of discovery.

Perhaps the most interesting point was that the children had absolutely no preconceived idea that only one particular mathematical treatment would lead to a result of absorbing interest, whereas other methods were fruitless. This was a fact which no one could have realised by any amount of mental effort. Consequently, the exhilaration was not due to any feeling of being right. It had something to do with the fact that the mind was taught something which, left to itself, it had been unable to discover.

Another point was that this feeling of being taught and thereby made aware of the limitations of one's mind was not accompanied by any depressive sense of intellectual failure. On the contrary, it was followed by an unmistakable feeling of satisfaction, pleasure, and joy. The whole thing had become a personal experience of a new kind, which we might call the fundamental scientific experience. How can it be explained?

The Fundamental Scientific Experience

One possible explanation is that the fundamental scientific experience discloses something new. This is surely what most scientists would claim. After all, it is their task to produce new insights, and it is always pleasant to succeed in one's job. But one must remember that, whereas the scientist may realise this peculiar experience because he has found something which is really new to the scientific community as a whole, the schoolchildren had very much the same kind of experience, even if they knew that their discovery was new only to them. They were perfectly aware that they could have looked it up in their book. This indicates that the fascination of the fundamental scientific experience is not caused by novelty as such, but rather by something which takes place at a deeper level of the mind every time a new insight enters it. Werner Heisenberg clearly expressed this in his autobiography by saying about scientists in general:

> ... one is almost scared by the simplicity and harmony of those connections which nature suddenly spreads out in front of you and for which you were not really prepared. At such a moment one is possessed by a feeling which is completely different from the pleasure one feels when one believes that a piece of physical or intellectual handicraft has been successfully achieved.[4]

However, Heisenberg thinks that the scientist may react in a different way, at least in the beginning:

> When one reduces experimental results to formalized expressions .... thereby reaching a phenomenological description of the event, one has a feeling of having oneself invented these formulae.[5]

This was precisely the feeling expressed by Eddington in *Space, Time and Gravitation*. But Heisenberg continues:

> However, when one stumbles upon these very simple, great connections which are finally fixed into an axiomatic system the whole thing appears in a different light. Then our inner eye is suddenly opened to a connection which has always been there — also without us — and which is quite obviously not created by man.[6]

So, according to Heisenberg, the fundamental scientific experience is unique because it provides a contact with some kind of reality that transcends the human mind. Other scientists have expressed the same idea in even stronger terms. Thus Edwin Hubble, who formulated one of the most important primary relations of modern cosmology, wrote:

> ... sometimes, through a strong, compelling experience of mystical insight, a man knows beyond the shadow of doubt that he has been in touch with a reality that lies behind mere phenomena. He himself is completely convinced, but he cannot communicate the certainty. It is a private revelation. He may be right, but unless we share his ecstasy we cannot know.[7]

If words like "mystical" and "revelation" are used here in a somewhat vague and ambiguous sense, we must admit that both Heisenberg and Hubble shared the same conviction that the fundamental scientific experience is so strong because something from a world beyond the human mind is fed into it in some way which it is difficult to explain and communicate. The mind is certainly full of operations and bustling with activity; but what it really has to work upon is something not provided by itself. In one very pregnant and restrained phrase Pauli expressed the same idea in a nutshell: "The subject matter gives a lot away."[8] Thus the metaphor of "finding" is now connected with the metaphor of "giving".

Finding and Giving

At this point one must raise the question as to whether all this amounts to no more than saying that most scientists are philosophical realists who assume, consciously or in a naive way, that they are dealing with an objective, outer world beyond the mere phenomena. That scientists in general are realists is probably true; but this does not answer the question. In particular it seems difficult to understand why such a realist position should be able to awaken such strong personal responses to the fundamental scientific experience, as we have seen in Heisenberg and Hubble, if this position is held only as a purely philosophical theory of knowledge. Few thinkers have stressed this more clearly than Friedring von Hügel who, unlike his brother, was no scientist but was uncommonly aware that scientific knowledge has personal implications. Writing about biology and archaeology he said that:

> ...everywhere in these newer sciences there is a sense of how much more there is to get, how rich and self-communicative is all reality, to those who are sufficiently detached from their own petty subjectivisms.[9]

Reality is self-communicative, and scientific rationality is not a purely human product. In another passage about mathematics and mechanics he wrote about some of the pioneers in these fields:

> The immensity of their success is an unanswerable proof that this rationality is not imposed, but found there, by man.[10]

But here again we can observe how the metaphor "find" is supplemented by the metaphor of "giving". Referring again to the "newer sciences" von Hügel maintained that:

> A keen yet reverent study of the Given appears here — by a Darwin, be it of but the earth-worm, and by a Wilken, be it but of the scribblings on ancient potsherds. And then the greater Givenness all found in those vast Intelligible orders, ...etc.[11]

According to von Hügel this relationship between mind and reality is apparent also in extra-scientific fields of human experience:

> No true scientist, artist, philosopher, no moral striver, but finds himself, at his best and deepest moment, with the double sense that some abiding, trans-subjective, other-than-human or even more-than-human reality, or force, or law, is manifesting itself in his experiences; and yet these very experiences, and still more his reasoned abstracts of them, give but a very incomplete, ever imperfect, conception of those trans-subjective realities.[12]

Continuing this line of thought one may ask why the metaphor of "giving" is necessary as a complement of "finding" in the account of these deep movements of the mind? Here we may notice that today we have apparatus for "finding" in different fields, i.e. machines which can detect the presence of something outside themselves and also store the retrieved information within themselves, a fact of which von Hügel was presumably unaware.

But if there is a sense in which a machine may be said to "find" something outside itself, it is difficult to see how it might possibly be "given" anything. To receive something as a gift must be the prerogative of personal beings. To recognise a part of the outcome of science as a gift is, therefore, to admit that our relationship with nature has at least one feature in common with personal relationships, and that the knowing subject is more intimately connected with the known object than the metaphor of "finding" is able to convey. Thus von Hügel writes:

> The data of man's actual experience.... are subject and object; the two, and not the one only, are (somehow and to some co-relative extent) included within the single human consciousness.[13]

Signposts in Nature?

Here we have reached a point where it seems difficult to overlook a certain affinity between the "givenness" of the world and the Christian experience of faith. The fundamental scientific experience appears as a disclosure that some of the elements of scientific knowledge stem from a natural order which transcends the human mind. Is that not another way of saying that the world is created? Before this question can be discussed it may be profitable to consider some of the pitfalls which lie in the area of natural theology.

The idea that a road to God is marked by signposts in nature has a long history in Christian thought. It was behind the metaphor of the "Book of Nature," which served, first, to make the study of nature legitimate in the eyes of the theologians, and later to make theological doctrine more palatable to philosophers by appeal to scientific discovery. Since nature was a "book" the attributes of its author could be read on its pages, at least between the lines: he must be intelligent since he had been clever enough to know how to make living organisms so marvellously adapted to their environments. He must also be good, since he took it upon himself to uphold the universe against its inherent tendency to dissolution, an argument already developed by Newton, who thought that gravitational collapse was prevented by what he called a permanent miracle. Thus, the recent tendency to appeal to new theories in cosmology in support of religious faith has venerable precursors, and perhaps no one should be blamed for trying to integrate his or her religious and scientific tenets.

Nevertheless, this kind of apologetics was — and is — unsatisfactory for at least two reasons. On the one hand it is dangerous (as history has shown) to support a permanent article of faith by a scientific theory, which has only a temporary lease on life. On the other hand the natural theology of the Enlightenment disturbed the balance in theology by concentrating its efforts on such prolegomena of Faith as the existence of God, in which Jews and Muslims also believe, whereas the central tenets of the specifically Christian proclamation passed into the background. Already in the 1820's there were strong theological objections to the lack of emphasis on the doctrine of salvation, and soon the idea of evolution in geology and biology began to undermine the scientific presuppositions of natural theology. The resulting separation between natural science and Christian thought is well known and seems to many to be an irrevocable divorce.

Understanding Creation

Nevertheless, I would argue that it is still possible to join von Hügel in construing science, and in particular the fundamental scientific experience, within a Christian perspective in which it is primarily related to the dogma of creation, even if we have to renounce once and for all any attempt to derive divine attributes from our conversation with the phenomena. But this does not mean that we should take recourse in the talk of "mere" phenomena and postulate something "behind" or "beyond" them, even if several philosophically minded scientists seem to be prepared to enter this well-trodden path. What is "beyond" the phenomena is also beyond the grasp of our reason and must remain there. All we have is a growing set of relations between the phenomena, and this insight must therefore be the starting point of the argument.

It is useful to remember the two different modes of logic which have competed for the favour of theologians almost from the very beginning of rational discourse about Christian faith. In one of them positive statements are preferred. "The world is created," appears as such a positive statement. It has a very simple grammatical structure with subject, verb and predicate, and seems to affirm something about the world. Yet what it affirms is extremely obscure. In fact, it has proven impossible to explain what "created" really means in terms of a scientific type of discourse. It does not necessarily mean that the world had a beginning in time since the theological discourse about creation usually maintains that an eternal world might also be "created". Consequently the term "creation" is not equivalent to "beginning in time." It is used in a sense which conflicts with ordinary language, to which its meaning cannot be reduced.

One way of getting around this difficulty is, of course, to describe "creation" as a "mystery" on a par with other Christian mysteries of faith which escape human understanding and which can be approached only through worship and prayer. But there is also the time-honoured way of negative theology, which may here reveal a more penetrating understanding of what "creation" implies. The statement, "The world is created," has a positive grammatical form. It seems to affirm something, although we are unable to describe what this something is. In actual fact it must be regarded as a pseudo-positive statement; in spite of the positive form of grammar, its only function in ordinary human discourse is to negate that which we can describe very well within the horizon of human experience. The negative sense of "creation" is that the world is structured in a way which is not a human construction. Certainly God knows in a positive sense what it means for him that he created the world. To us it can only mean that it was not we who did it. From this point of view the doctrine of creation is not concerned with how the world came into being; it is concerned with how we are related to its fundamental structures. It maintains that we are not responsible for these structures; that we have no influence upon them, can claim no merit for their glory, and can only exercise power if we respect them. Thus the doctrine of creation is not a descriptive statement at all, but a statement which can be understood only if its moral implications are recognized.

It is not difficult to find Biblical support for this view. The Old Testament contains many passages in which God's creative activity is described in very anthropomorphic terms based on a metaphorical use of the word "creation", as if it had something in common with ordinary human "making". The potter working at his wheel is one of these (*Jer.* 18,6). But the notion of creation is usually found in contexts which clearly underline that human beings had nothing to do with establishing the structures of the world. The two different creation stories in *Gen.* 1 and *Gen.* 2 have in common that the human being is described as a late-comer to the universe and not present when it "began". In *Gen.* 1 people appeared on the last day of creation when all the rest of the world had already been given its shape. In *Gen.* 2 human kind was absent when God made heaven and earth. In the *Book of Job* God asks a sort of rhetorical question:

> Where wast thou when I laid the foundation of the earth? Declare if thou hast understanding! Who laid the measures thereof, or who stretched the line upon it? When the morning stars sang together and all the Sons of God shouted for joy! (38,4 ff).

In this joyful commotion God never consulted us: "Who hath measured the waters in the hollow of his hand.... who hath directed the Spirit of the Lord, or being his counsellor hath taught him? (*Jer.* 40,12 f.).

The Old Testament also shows that the absence of humanity at the divine act of creation was not offset by an attempt to make this act accessible to human understanding in the search for what "really happened." The response is not curiosity, but humility: "When I consider the heavens, the work of thy fingers, the moon and the stars, which thou hast ordained: What is a Man that thou art mindful of him?" (Ps 8,3). The answer is worship: "The sea is his and he made it: and his hands formed the dry land. Oh, come, let us worship and bow down, let us kneel before the Lord our maker!" (Ps. 95,5 f.). Ancient Israel was never given to scientific investigation; but the sense of wonder at the strange or beautiful phenomena of nature was always present. Who would not marvel at the wings of the peacock and the wings and the feet of the ostrich (Ps. 40,13), at the geological changes on the surface of the earth (Ps. 107,33 f.), at the treasures of ores and minerals within it (Job 28,1 ff.), at the foetus in the womb (Ps. 139,13 ff.), at the crocodile of the Nile (Job 40-41) and the life of plants and animals (Ps. 104), or the course of the sun and the moon (Ps. 8 and 19). But the contemplation of nature had a moral outcome: "Not unto us, Oh Lord, not unto us! but unto thy name give glory!" (Ps. 115).

Possible Implications

Let us return again to the picture of the fundamental scientific experience and try to compare it with the idea of creation as it was outlined above. The fundamental experience of the scientist conveys a sense of a world imbued with relational structures for the existence of which he is not responsible, although he is in some way able to recognize them. If this is consciously realised, or even just vaguely felt as a determinant behind the conscious mind, the result is a new insight that does not belong to the realm of "mere" scientific results. This insight must be a moving force behind a personal, spiritual, and moral re-orientation, in so far as the mind responds to a call from something which it recognizes as being outside it and at the same time of absorbing interest. This means that our ingrained self-interest and self-absorption is counteracted by the call from that which is not ourselves. It dawns upon us that we are not the only centre of interest in the world; there are other centres of attraction.

In other words, when a scientist realises the implications for one's personal existence of the fundamental scientific experience, he has adopted a relationship towards the world which is essentially the same as that which the believer adopts when expressing belief in creation. Thus, although — unlike the old natural theology — this reflection on scientific practice does not pretend to disclose the attributes of the Creator, it does produce the insight that we are placed in a created world. No single phenomenon and no particular relation between phenomena is behind this conviction, which stems only from the fact that we have to give our assent to the extra-mental character of any relation belonging to that class of statements which science is unable to deny.

At this point one might raise the objection that scientific practice as described above makes us conversant only with the phenomena as such, through the cumulative body of known relationships between them, but without disclosing any "real" world "behind" them. There are two answers to this. First, the whole point of the view presented above is that, even if the phenomena are all we have, they are, nevertheless, able to disclose a growing number of relations which do not change and which we have to acknowledge as "given". Second, the problem of a "real" world behind the appearances is essentially a philosophical question. When all is said and done, it is not materially relevant to the problem of how scientific practice is related to Christian experience. For this experience is, in the last resort, not felt as a solution to an ontological problem. It is much more the recognition of being placed in a correct position in a world whose centre and origin is not the human self, but something else which the believer calls God. This recognition is the essence of religious conversion, and it clearly has much in common with the fundamental scientific experience of a world which has a structure that cannot be derived from the human mind. This seems to explain why natural science cannot have any quarrel with the doctrine of creation, understood in the "negative" sense described above.

The problem remains as to whether this religious and scientific conversion takes us further than to the "natural religion" which the Enlightenment considered to be more fundamental than both Christianity and any other form of monotheistic religion. In other words, is it a conversion to a "deity" who is responsible for the structure of the universe or to the God of the Christian faith who is the Holy Trinity? Here one has to assert unhesitatingly that it is no longer possible to attach any importance to the many attempts to connect the Trinitarian doctrine of God with the many "triads" in nature, where previous ages found vestiges of a triune god. This road seems to be closed forever. If belief in the Trinity has any connection with the scientific discourse on nature, this connection must be established in a different way. Here we again cite von Hügel:

> I believe that not to be aware of the costliness.... of the change from.... self-centredness, from anthropocentrism to theocentrism, means not only a want of awareness to the central demand of religion, but an ignorance or oblivion of the power, the perverse tendencies of the human heart.[14]

Now it was argued above that to admit the personal implications of the scientific conversation with the phenomena is in itself a call to conversion. But this consequence must be accompanied by the admission that the outward movement of the mind can be counteracted, as it were, by inertial forces operating within the mind, turning the movement of conversion back into itself. This is behind the Christian experience of sin and, therefore, behind the Christian doctrine of redemption. This has two facets. One is the conviction that the sinful turning back to self as the only centre is not justified by anything in the world outside the mind. This world is also redeemed by grace and, therefore, worthy of our attention as a safe place for both scientific and other investigations. The other is that the assent to this is experienced by the Christian as a work of the Spirit.

These considerations may seem to shed more darkness than light upon the subject. It may well be the case that a trinitarian "theology of science" has to be hammered out if we shall be able to understand scientific practice fully within the horizon of Christian experience.

NOTES

[1] A. S. Eddington, *Space, Time and Gravitation* (Cambridge, 1920) 201.

[2] D. Brewster, *Memoirs of the Life.... of Sir Isaac Newton*, Vol. II (Edinburgh 1855) 407.

[3] Milton, *Paradise Lost*, VIII, 77-82.

[4] W. Heisenberg, *Der Teil und das Ganze*, Ch. 5.

[5] *Ibid.*, Ch. 8.

[6] *Ibid.*

[7] P. Hubble, *The Nature of Science and other Lectures* (San Marino, California, 1954).

[8] Quoted from Heisenberg, *op. cit.*

[9] F. von Hügel, *Essays and Addresses* I, 1921, 56.

[10] *Ibid.*, 71.

[11] *Ibid.*, 56.

[12] *Ibid.*, 63.

[13] *Ibid.*, 51.

[14] *Ibid.*, 13.

SCIENTIFIC RATIONALITY AND CHRISTIAN LOGOS

MICHAEL HELLER, Pontifical Academy of Theology, Cracow,
and the Vatican Observatory

1. *Introduction*

The empirical sciences constitute a specific type of rationality. For many people it is the only admissible type, or at least a kind of ideal model that should be imitated by other species of rational knowledge as far as it is possible. However, the evolutionary line that has led to this type of rationality has been woven from two threads (strongly interacting with each other), one of which goes back to Greek philosophy, and the other one to the Christian doctrine of creation. Theological reflections on creation, especially in the Middle Ages, stressed the contingency of the world. Since the architecture of the world is entirely dependent on God's will, it cannot be discovered *a priori* by any kind of speculative thinking. This opened the way to the empirical investigation of nature. If modern science and Christian theology have their roots so strongly interacting, the split between them in modern times might seem to be an unexpected surprise. But in fact it was well established by the events and processes of history.

In the present paper I will argue that, in spite of all the differences and conflicts, the deep philosophical affinity between the scientific spirit of rationality and the Christian approach to the created world still exists and still continues to exercise its influence on the very foundations of scientific thinking. Rationality is a value, and the choice of this value (on which all science is based) is a moral one. Without a religious attitude towards the world and towards science (so often emphasized by Einstein) such a choice is reduced to a blind game of purely conventional preferences. From the theological perspective, there is an intimate relationship between the spirit of rationality and the Christian idea of the Logos. Philosophy of science discovers that all science is based on the assumption of rationality and it completes its analyses by elaborating the consequences of that discovery. It is here that a theology of science should take over.

In sec. 2, I will examine the Greek roots of scientific rationality. Sec. 3 elucidates the role of Christian theology in shaping Western ways of looking at nature. The split between the world of science and the world of Christian thought that took place in modern times is briefly considered in Sec. 4. Starting from Sec. 5, I move from the historical perspective to analyze the present situation touching on some key methodological aspects of the problem. My central thesis on the relationship between scientific rationality and the Christian Logos is discussed in Sec. 6.

2. *Faith in Reason*

One of the essential tenets of scientific rationality is the deep conviction that nothing should be accepted without sufficient proof or argument. But what kind of proof or argument should be admitted? This is a secondary, although extremely important, question. In actual scientific practice it is answered by the method of trial and error, rather than by any *a priori* prescription.

However, independently of what one assumes as a sufficient proof or argument, no proof or argument can be given to validate the claim that one should direct one's thinking with the help of *any* proofs or arguments, that is to say, that one should be rational. Any proof or argument in favor of this claim tacitly assumes that one wants to be rational. The decision to be rational is, therefore, a choice.

Rationality is undoubtedly, a value. This becomes manifest as soon as one confronts rationality with its opposite, irrationality. We instinctively treat irrationality as something degrading and almost inhuman. Some philosophers would say that to be rational follows from human nature, from the very fact that we are equipped with the faculty of thinking and choosing. If rationality is a value then the decision to be rational is a *moral* choice.

Freedom is a part of that morality which constitutes the rational attitude of man towards the Universe. The only admissible force is that of proof or argument. Any view imposed by external coercion is irrational, because it is imposed and not inferred from evidence. It seems also that an internal freedom is presupposed by rationality; without being free to choose between possible paths of reasoning, the process of constructing any proof or argument could hardly be imagined.[1]

It was Karl Popper who saw this in a very clear light. He wrote: "The choice before us is not simply an intellectual affair, or a matter of taste. It is a moral decision." Moreover, "it is, in many senses, the most fundamental decision in the ethical field." Since it cannot be argued for or demonstrated, Popper calls it "the faith in reason."[2]

This moral choice gradually matured in the evolution of Western thinking. The adherence to the empirical method of investigating the world may be thought of as a final step in this process. All the successes of this method can be considered as arguments revealing the correctness of that choice. However, this does not change the fact that the empirical method cannot prove itself, and that it still remains a moral choice.

The history of science is nothing else than an attempt to reconstruct, in all its shades and details, the maturation of this choice, its victories and its defeats — in a word, our struggle to develop the astonishing capability of being rational.

There is no doubt that at the beginning of this process lies the discovery of the ancient Greeks that it is worthwhile to ask the world difficult questions and search for answers with no help from outside. The analysis of this event is too well known to be repeated here. I would like only to stress its extraordinary character. We have challenged the reality that surrounds us and that is inside us. We have decided to understand this reality, and presumed that this can be done. With no help, either from the gods, or from some hidden forces, we have stood alone against the silent Universe.

The Greek type of rationality is, first of all, an ethos of thinking. Any thesis has to be clearly stated, and must be argued for. This is a well defined process of thought, and not just any process. It ought to proceed in agreement with the rules of rationality. It is no wonder that the Greeks put a lot of effort and ingenuity into codifying these rules. Greek logic was by no means just a set of technical details appended to Greek philosophy; it was its moral code.

This part of the scheme of rationality should be considered as perhaps the greatest achievement of the Greeks. The geometric system of Euclid, discussed and improved through the centuries, has finally led to modern axiomatic systems that can be thought of as expressing the spirit of rationality, freed of any "matter". It tells us how to deduce one truth (expressed in a set of propositions) from another truth in an absolute and reliable manner. Although, from time to time, some philosophers try to proceed on their own, by discarding the rules of logic, they will certainly be relegated to the marginal areas of the history of philosophy, in spite of a possible short-lived fascination with their originality.

Axiomatic systems are "freed of matter": they are not yet the process of thinking. They constitute a structure that has to be filled with "matter", i.e., with the content of thought. The structure is filled in with this content through assumptions or axioms from which the chain of logical deductions begins. Here the Greeks had serious problems; the more so, as they were not aware of their existence. They highly esteemed thinking itself, and instinctively believed that it was the process of thinking that should establish the starting point of deduction. However, if thinking is to be correct, it has to be governed by the rules of logic. The only way out of this vicious circle was for the Greeks to appeal to self-evidence. But it was the self-evidence of thinking rather than of observation. Usually, our eyes see what thinking orders them to see (or, to be more strict, we try to interpret data in accordance with what the brain has previously registered). Thinking, on the other hand, is usually entangled in the language that is supposed to express it. No wonder that Greek philosophies were based on analyses of common language, and since common language is strictly connected with every day life, some Greek philosophers (Aristotle, for example) believed that their philosophies were based on experiment. It was true at least to a certain extent. We should remember, though, that we cannot impose on the Greeks our concepts of experiment and experimentation. With only a few exceptions (Archimedes was the most eminent one), the difference between scientific experimentation and everyday experience was for the Greeks rather loose and ill-defined.

The ethics of thinking, which the Greeks succeeded in imposing upon all antiquity, was also their aesthetics of thinking. Rules of thinking are simultaneously rules of beauty: the discipline of form, the harmony of deductive movement, the proportion of structure. In Plato's dialogues or

Lucretius' hexameters the literary framework of thought is almost as important as the thought itself; but even Aristotle's coarse phrases are beautiful, since an imposing edifice is being constructed out of them, as if of heavy pieces of granite.

3. *From Theology to Science*

Christianity appeared within the evolutionary chain of Jewish thinking. However, it was not the Old Testament mentality that shaped the intellectual form of the new religion. The Greek type of rationality, not especially caring about experimental details and always hastening to comprehensive syntheses, seemed to provide a suitable background for theological speculations.

Could transcendence be put into syllogisms? Application of the Greek pattern of rationality to theological questions had, sooner or later, to produce Scholasticism. The truth concerning God and his activity could hardly be expected to appear at the end of long chains of distinctions and exclusions, but such a method of doing theology constituted an excellent exercise in applying logic. Owing to medieval theology Europe has learned a great deal of the Greek ethics of thinking.

Logical thinking is more secure if it is defended against extra-logical influences. It is rote manipulation with symbols that replaces the psychology of thinking, and eliminates the possibility of error in the process of inferring logical consequences from their assumptions. Medieval thinkers excelled in inventing subtle formalisms. It seems that they stressed the mnemotechnical aspects of symbols too much. Perhaps they aimed at problems that were too difficult. The fact is that their formalisms were not efficient enough. It was art for art's sake, rather than as a tool for solving real problems.

Mathematics is very close to logic. In both disciplines the rules are practically the same. From the historical point of view, if logical chains began with something which could be expressed in numbers (which could be measured), logic was considered to be mathematics. For a long time mathematics served astronomy well. Without mathematics there would have been no astronomy. The heavens, when contemplated in the light of first principles, remain dark and silent. They speak only when addressed in the language of numbers. As soon as measurement began to enter the philosophy of nature (faint-heartedly in the beginning, but gradually with more and more self-confidence), it was transformed into the modern natural sciences, with the role of logic replaced by that of mathematics. Progress in mathematics soon became an integral aspect of the evolution of science.

However, the role of Christianity in the origin of modern science cannot be reduced to sharpening and transmitting the Greek heritage to our times. Many historians of science agree that Christianity added to this heritage something substantial, something which doubtlessly contributed to the fostering of the empirical method. It is no surprise that this new element was supplied by the Christian teaching on the creation of the world.

Medieval interpretations of the doctrine of creation almost unanimously agreed that the existence of the Universe should be considered as an effect of the free will of the Creator. This free decree of the divine will concerned both the world's existence (the world did not have to be created), and the plan of creation (the world's architecture could have been very different from what we actually observe). The world bears the trait of contingency in both its existence and in its structure. This last point is of great significance. The structure of the world cannot be deduced from self-evident, or otherwise *a priori*, premises. The only way leading one to knowledge of the world is to open one's eyes and to see what can be seen of the world. In other words, one has to experiment with the world.

Does this mean abandoning the Greek conviction about the world's rationality? Not at all. The structure of the universe is the implementation of God's creative plan, and this plan is fundamentally rational. However, its rationality transcends the possibilities of the human mind to such a degree that it cannot be deduced from self-evident axioms. The only realistic strategy is the empirical method. One must start experimenting with nature, and only then try to fit theoretical structures to the results of this experimentation. If the theoretical structures lead to conclusions that turn out to be in agreement with the results of other experiments, there is a good probability that the structures approximate the structure that constituted God's plan of creation.

If Christianity played such an important role in paving the way for the experimental sciences, one might expect that the origin of the sciences would initiate a period of symbiosis between scientific and religious thinking. This was not the case. The succeeding age was marked with conflict, not with harmonious coexistence.

4. *Two Streams of Knowledge*

What was the cause of these dramatic conflicts? There are many studies on this subject. My working hypothesis is that it was the institutionalization of the Church's teaching that was one of the main factors responsible for splitting the way of the Church from the way of science. Traditional structures had reached such a high degree of specialization that they were unable to adapt themselves to new conditions. By institutionalization I mean not only a subordination of philosophy and theology to Church authorites, but also what may be called an "invisible college" (to use a well known expression in a slightly different context), that is, ways of thinking elaborated by long tradition, a balancing of influences among different schools and systems, consolidated methods of collecting and transmitting information, unwritten codes of behaviour for people involved in the ways of knowing. The sciences were born in an entirely new situation, no longer controlled by Church authorities. From the very beginning they started to create their own "invisible college." Conflicts were unavoidable.

The conflicts abounded with drama. Incessant series of successes by the new sciences generated totalitarian tendencies. Church thinkers found themselves on the defensive, and responded by triggering mechanisms of isolation. Two circumstances favoured this process: first, the great inertia of institutionalized structures in collecting and transmitting knowledge (mentioned above); and, second, the extreme specialization of the new sciences. The point is that the understanding of scientific theories (let alone creative work in science) requires protracted studies and great intellectual effort. Proper assessment of scientific theories by an outsider is practically impossible. On the other hand, science is democratic, in the sense that everybody has the right to participate with the condition, however, that a budding scientist would devote enough effort and time to acquire the necessary skills and knowledge. In this way theologians and philosophers of the epoch, busy with their own problems, found themselves somewhat excluded from the possibility of a competent dialogue with the empirical and mathematical sciences, always accelerating in their progress and specialization.

In the long run, this separation and isolation turned out to have important consequences. The stream of knowledge split into two branches. In each of them progress went on independently of the other. Within the empirical sciences it quickly developed into a chain-reaction. Technology, as the natural continuation of the sciences, began to change both social and individual lives. Some totalitarian and positivistic tendencies in the sciences became prominent exit and theology in the eighteenth and nineteenth centuries had is own ups and downs. The exaggerations of Scholasticism contributed to theology's questionable reputation. Neo-Scholasticism and Neo-Thomism should be considered as attempts to exit from this impasse. One must admit that they were partially successful, but only within the Church's stream of knowledge, with negligible effects as far as dialogue with the empirical sciences was concerned. The impact of Neo-Thomism on people engaged in doing science was confined to conversions. These are still occurring, but much less frequently. Some scientists were converted to metaphysics, but it usually had almost no effect on either the sciences themselves or the milieu of a given scientist.

Progress requires a certain continuity, and there is no continuity of knowledge without education. No wonder, then, that both the scientific and ecclesiastical streams of knowledge have developed their own educational systems, surprisingly different, and independent of each other. Many contemporary Catholic universities have excellent departments of mathematics, physics, biology, etc., and, of course, their own philosophy and theology faculties. I know of a very few examples of interaction between them. Usually, two independent streams of knowledge flow through the same university campus.

5. *What Should Be Done?*

Here I stop my analysis. It is not my goal to go into details of the present relations between the Church and the world of science. There are many good accounts of these questions and the interested reader should refer to them. Everything I have said so far was intended as an introduction to the question: what should be done in order to improve these relations?

A possible answer could be: nothing should be done. Both science and the Church benefit from the separation. Philosophy of science has taught scientists to respect the limits of the scientific method. Outside these limits there is ample room for philosophical or even religious belief. On the other hand, the Church has learned not to interfere in the internal affairs of the sciences. The Church is expected, from time in time, to stress the value of science as a human endeavour. If this *savoir vivre* is preserved, there will be no conflicts and perhaps even some mutual appreciation.

I am not quite happy with this solution — if it is a solution at all. An agreement of non-interference proves to be sometimes necessary, and a temporarily efficient means of resolving conflicts. The point is, however, that in the present situation the conflict-frontier cuts through the interior of the human person (especially of the person who believes and also does science). The human personality cannot be split into different zones of influence.

The other extreme would be equally dangerous. Differences in aims, languages, and methods, well established limits of competence, full respect for the different nature of the other side, should always be kept in mind and never trespassed. The answer to the question, "What should be done?" can only be reached under the condition that these methodological differences are strictly respected. There can be no return to the period when theology and the sciences seemed to constitute the same field of human activity. Methodological anarchy solves nothing. The answer should be sought by respecting the individuality and integrity of both the sciences and religion.

6. *The Christian Logos*

Why should we do science rather than engage in some other sort of intuitional creativity? How is an appeal to emotion and intuition worse, as far as our cognitive relations with the world are concerned, than an appeal to reason? In the name of which ideals should we prefer "the awareness of our limitations, the intellectual modesty of those who know how often they err" [3] to a confidence in human nature that simply knows what is good and what is bad?

As we have seen, there are no rational motives compelling us to choose between these two possible options. Rationality is a moral choice. But to choose without any motives whatsoever is heroic, and to be heroic day after day is very hard. No wonder, therefore, that the "spirit of rationality" becomes tired from time to time and gives way to different forms of irrationality. This is what happens nowadays. "...The conflict between rationalism and irrationalism has become the most important intellectual, and perhaps even moral, issue of our time." [4]

The moral choice for rationality could be, after all, based on an illusion. A mortal game in which losers become fools, and winners are declared wise — a struggle for power, in fact — is what science really means in the eyes of many. The degradation of the natural environment and the prospect of atomic annihilation are only external symptoms of a much deeper crisis. If the choice of rationality is not a choice of value but only part of a blind game with nature, it is ultimately an immoral choice. In that case it is fundamentally my own choice which matters. I could have chosen differently. I become a final criterion of my choice; it is up to me to decide how to use the technological achievements of science. With no moral norms besides myself, I may use them to further my own egoistic goals.

It was Einstein who asked the question, "Why is the world comprehensible?" Why? Einstein was not able to answer the question; he could say only, "The eternal mystery of the world is its comprehensibility. ... The fact that it is comprehensible is a miracle."[5] The question ends with astonishment. Philosophy of science can do nothing more. It is the theology of science that has to take over and go deeper in seeking the answer to this question.

In light of Christian theology the choice of the rational method in science is not an unrestricted choice. It is, of course, the doctrine of the creation of the world that is responsible for this constraint. The world is a realization of the rational plan of the Creator; and there is no other way of unravelling the structure of the world except through rational attempts to decipher God's plan. Let us focus on this point for a moment.

The world is for me impregnated with meaning.[6] There are various objects such as a table or a star. I need the table; I can eat or write on it. I also need the star; it can be a source of inspiration for me or an object of intensive study. Both the table and the star constitute values for me.

Only something that has a meaning can be a value, and something can have meaning only for somebody. A table is but a set of physical fields and particles. It becomes the table for somebody who enters into a cognitive nexus with it and identifies it as a table.

The environment of meanings and values is even more important for us than air or food. Without air and food we must die, without an environment of meaning and values we would not even be human.

Once the world, through a long process of evolution, gave birth to human beings, it ceased to be as it had been before. Through human beings the world has been filled with meaning and entered into a complicated fabric of values we have woven out of our ability to think and to will. The process of knowing the world is itself a great value for us. Because it is a value, we want to know what the world would be like without our value-creating presence. To attain this goal we have elaborated an empirical method for investigating the world that consciously prescinds from value and meaning. This strategy is very difficult to implement. Science cannot avoid using human language which by its very nature is full of anthropomorphic meanings and values. To minimize this the empirical sciences adopt, as much as possible, the language of mathematics. Although this language is man-made, it has been created in such a clever manner that its only content is its form. Once the form of this language has been established, we no longer have any power over it.

The world of physics, of astronomy, and of biology is maximally dehumanized. It is true that nowadays we reappear in this world as observers, and through our interpretive activities we influence the investigated object (especially, for instance, in quantum physics). However, we are observers who measure (i.e., translate what we see into numbers), restraining ourselves from any act of evaluation. Silence about values is the price paid for the efficacy of the scientific method. As a result, in the minds of many thinkers, an image of the world devoid of any value has been established.

In the light of the Christian doctrine of creation this is simply not true. Without human beings the world would have no meaning and value, but we are not even the principal creators of meanings and values. The structure of the world is a realization of God's plan, and as such, is totally impregnated with Meaning and Value. Capital letters are intended here to remind us that the words "Meaning" and "Value" are powerless to express the full significance of God's plan. To use Plato's metaphor, meanings and values created by us are only shadows of That Meaning and That Value. "The Word (Logos) was with God. Through Him God made all things; not one thing in all creation was made without Him . . . ."[7] And not only that: "the Logos was made flesh."[8] The doctrine of the incarnation of the Logos certainly has its theological significance as far as the relationship between humankind and rationality is concerned. The Logos made flesh is a profound theological reality, far from being completely explored by theologians, and in the present context opens new vistas for reflection.[9]

All science is based on the spirit of rationality. To follow this spirit was a choice of humanity. Was this a moral choice or just a blind game with values? The answer given by religion to this question should not be seen simply as a service faith can render science. It is immensely more than that. My metaphysical hypothesis is that the spirit of rationality participates in the Christian Logos. In the course of human history the Logos assumed the flesh of scientific rationality. The theological perspective allows us to understand this not only as a literary metaphor. "That Christ is the Logos implies that God's immanence in the world is its rationality."[10]

Only by living in a world of Value and Meaning is it truly worth taking up science.[11]

NOTES

[1] S. W. Hawking and G.F.R. Ellis, speaking of free will, note: "... this is not something which can be dropped lightly since the whole of our philosophy of science is based on the assumption that one is free to perform any experiment." *The Large Scale Structure of Space-Time* (Cambridge: Cambridge University Press, 1973) 189.

[2] K. Popper, *The Open Society and Its Enemies*, vol. II (London: Routledge and Kegan Paul, 1974) 232-233.

[3] *Ibid.*, 227.

[4] *Ibid.*, 224.

[5] A. Einstein, *Ideas and Opinions* (Laurel Edition, 1978) 285.

[6] The following passage is taken from my book: *The Justification of the World* (Cracow: Znak, 1984, in Polish) 94-96.

[7] John **1,** 2b-3.

[8] John **1,** 14.

[9] I am grateful to Nicholas Lash for turning my attention to this aspect of the problem.

[10] This formulation I owe to Olaf Pedersen. It certainly deserves to be more fully elaborated.

[11] M. Heller, *The World and the Word - Between Science and Religion* (Tucson: Pachart, 1986) 175.

CREATION IN THE HEBREW BIBLE

RICHARD J. CLIFFORD, S.J. Weston School of Theology,
Cambridge, Massachusetts.

Introduction

The Hebrew Bible, the Old Testament, is perhaps the most important single source of Western popular images, if not concepts, of creation. The evolutionary theories of Darwin disturbed those images and attendant concepts in the nineteenth century. Post-classical physics is disturbing them again today. The hindsight afforded by the century and a half since Darwin enables us today to see that some of the nineteenth century argument between biblicists and evolutionary theorists was based on a misunderstanding of the biblical accounts of creation.[1] Contemporary dialogue between theologians and scientists may similarly be hindered by false formulations of the biblical material.

This essay interprets the major biblical statements about the creation of the world. The first chapters of Genesis are no doubt the most influential biblical texts in shaping modern images of God's creating activity, but they are not the only biblical creation texts. Creation is an important theme of many psalms, of Second and Third Isaiah, and of the "Wisdom Literature" books of Proverbs and Job. The biblical material is diverse in genres and in dates of composition and cannot be reduced to a single picture.

1. *The Ancient Near Eastern Background*

The biblical texts must be examined against their ancient Near Eastern background. The Bible was written by people who lived in the late second and first millennium and were influenced by the culture of their time. We must look briefly at creation accounts in these surrounding cultures. Of the great cultures of the ancient East amid which Israel lived the most important were those of Egypt, Mesopotamia, and Canaan. The latter two exercised a demonstrable influence upon biblical cosmogonies.[2]

1.1 *Mesopotamia*

Scholars are agreed that Mesopotamian traditions have influenced the creation stories in Genesis 1-11. The material is diverse: theogonies ("genealogies" of gods and elements of the cosmos), allusions to creation in rituals and prayers, and "compendious" cosmogonies such as *Enuma elish* and Atrahasis. No treatise on creation, however, exists; Mesopotamian thought was not much given to generalization. Accounts of creation usually served some other purpose, e.g., to glorify a god, to explain a phenomenon, or to legitimize a value or practice.

A good example of how cosmogonies functioned is an incantation against a toothache. The sufferer goes to a magician who prays to the god Ea to recall the worm back to the function assigned it in the creation order.

> After Anu [had created heaven],
> Heaven had created [the earth],
> The earth had created the rivers,
> The rivers had created the canals,
> (and) the marsh had created the worm —
> The worm went, weeping, before Shamash [the sun god]
> His tears flowing before Ea [one of the three creator gods and god of wisdom and organizing]:
> "What wilt thou give me for my food?
> What will thou give me for my sucking?"
> "I shall give thee the ripe fig,
> (and) the apricot."
> "Of what use are they to me, the ripe fig,
> and the apricot?
> Lift me up among the teeth
> Amid the gums cause me to dwell!
> The blood of the tooth I will suck,
> And of the gum I will gnaw
> Its roots!"
> Fix the pin and seize its foot. [to the dentist]
> Because thou hast said this, o worm,
> May Ea smite thee with the might
> Of his hand.[3]

The magician recalls that the worm was assigned to eat overripe fruit when the world was created, and prays that Ea will make the worm cease attacking the human mouth, a deviation from his original task.[4] The incantation illustrates well how profound was the belief that things were fixed permanently on the day of creation.

Three major Akkadian texts are rightly compared to Genesis texts: *Enuma elish* (sometimes called "The Akkadian Creation Epic,") the Atrahasis epic, and tablet XI of the Gilgamesh epic (the third not a cosmogony). *Enuma elish* was written on seven tablets and is primarily concerned with the exaltation of Marduk (or Asshur in the Assyrian version) to kingship over the gods. In the beginning there is only undifferentiated, unlimited waters; nothing is shaped or living. The waters become differentiated into female and male elements (Apsu and Tiamat), from which cosmic pairs merge — Lahmu and Lahamu, sky and earth, and, after two generations, Marduk. Here are the opening lines:

> *When* on high the heaven had not been named,
> Firm ground below had not been called by name,
> Naught but primordial Apsu [cosmic waters], their begetter,

(and) Mummu-Tiamat [cosmic waters], she who bore them all,
Their waters commingling as a single body;
No reed hut had been matted, no marsh land had appeared,
When no gods whatever had been brought into being,
Uncalled by name, their destinies undetermined —
Then it was that the gods were formed within them.
Lahmu and Lahamu [prob. Atlas figures [5]] were brought forth, by name they were called.
Before they had grown in age and stature.
Anshar [heaven] and Kishar [earth] were formed, surpassing the others.[6]

The "when.... then" construction, italicized in the text, resembles the syntax of Gen 1:1-2 and 2:5-6. As in Genesis 1, at the beginning there is only unlimited water, no firm ground, no humans. Unlike Genesis 1, there is no single creator god standing apart from the primal mass; the gods arise from the mass. Later in the first tablet, an initial revolt of the gods, put down by the wisdom of Ea, is a prelude to the later, more serious conflict among the gods that Ea's favorite, Marduk, will have to resolve by force of arms. Marduk is born two generations after Ea's "settlement," and leads the gods against Tiamat, who has turned hostile to them. After demanding and receiving supreme power from the terrified assembly of the gods, he leads them out to battle and with the weapon of his storm wind, slays Tiamat in single combat. He then makes the universe from her body:

Then the lord paused to view her dead body,
That he might divide the monster and do artful works.
He split her like a dried fish into two parts:
Half of her he set up and ceiled it as sky,
stretched a skin and posted guards.
He bade them to allow not her waters to escape.
He crossed the heavens and surveyed the regions.
He squared Apsu's quarter, and the abode of Nudimmud.
As the lord measured the dimensions of Apsu.
He set up Esharra [the temple in heaven], a counterpart to the Esagila [temple in Babylon].
In Esagila, Esharra which he had built, and the heavens,
He settled in their shrines Anu, Enlil, and Ea. (IV, 135-146.)[7]

The account is highly imaginative. Tiamat, unlimited waters, is skinned and her hide is used to confine part of her waters to the heavens and part to the underground. Marduk then arranges ("measures out") the heavens, and builds the temples of the three great gods. Esagila is the temple at Babylon, which is the counterpart of Esharra, the palace in the heavens, and of Apsu, the palace in the space below the earth. The temple, concrete symbol of the god's (and king's) rule, represents order and fertility. *Enuma elish* was recited annually at the New Year festival, evidence that ancient worshipers felt creation was renewed in the processes of the agricultural year. The text goes on to describe the creation of humans in tablet VI, the gods' imposition of service upon them, and the allegiance of the gods to Marduk, expressed in the fifty names they bestow upon him.

The two cosmogonies in *Enuma elish,* the emergence of cosmic pairs at the beginning and the building of the universe from the body of Tiamat, are both subordinate to the overriding theme of how Marduk became king of gods and humans; they are not told for their own sake. Marduk made the Babylonian's world.

The three-tablet Atrahasis epic is the most important Mesopotamian parallel to Genesis 2-11. Fragmentarily known in the 19th century, the tablets were only published in their proper order in 1969.[8] The plot of Atrahasis resembles that of Genesis 2-9: creation of humans to maintain the universe, their proliferation and "fault," the gods' decision to annihilate the humans in a series of three plagues culminating in a great flood from which only Atrahasis (favored by Ea) is saved; repopulation from Atrahasis is allowed by the gods, at last appreciative of the human labor that maintains them. The repopulation, however, is limited by a new factor: infancy diseases and a class of celibate women. The nature of the human "fault" (lit. "noise") is disputed. Most scholars see it as some kind of moral fault, but E. Rainer and W. L. Moran, correctly in my view, see it as morally neutral — simply the sign of an expanding and exuberant population. The gods did not correctly calculate in their first effort the effect of unlimited population growth.

Comparison with Genesis 2-11 is instructive. Both have the same plot. Both tell of a "fault," though in the Bible it is emphatically moral — acts of injustice in 6:1-8 — and not simply the noise of an expanding population. Both have a flood and a favored survivor. In both, repopulation follows the flood. Atrahasis introduces a mechanism whereby unlimited population growth will no longer be possible — infancy diseases and a class of celibate women. The Bible on the other hand reaffirms the original blessing without any qualification. The purpose of Atrahasis seems to be to show that the brutal god Enlil, inept though he be, ultimately makes the decisions; creation has been by trial and error. Genesis 2-11 follows the plot of Atrahasis and, indeed, seems to be a version of it. The biblical version shows the freedom of the biblical cosmologist to rewrite the tradition, to alter it for a particular purpose. The biblical reinterpretation is through narrative. Narrative was not, as it is generally for us, simply entertainment or illustration, but a means of exploring serious issues like the relation of the gods to the human race and the purpose of human existence.

1.2 *Canaanite (Ugaritic) Cosmogonies*

Since 1929 clay tablets in alphabetic cuneiform have been recovered from the area around Ras Shamra (ancient Ugarit) on the coast of Syria. Six of the tablets, generally described as the "Baal Cycle," tell how the storm god Baal defeats Sea (*yammu*) and Death (*môtu*), builds a temple, and brings fertility and order to the world. After the god's victory, the high god El grants a temple to Baal. In his new temple, Baal hosts a banquet for the gods and celebrates his kingship. Biblical texts such as Psalms 74:12-17, 77:12-21 (cited in the previous section), 89:10-15, and 93, Second Isaiah, and Exodus 15, use vocabulary and traditions similar to those in the Ugaritic texts in describing Yahweh's creating the people Israel. Though scholars agree that the biblical texts drawing on the traditions in the Ugaritic texts are cosmogonies, they are divided whether the Ugaritic texts themselves contain true cosmogonies.[9] If one defines cosmogonies as the bringing to a human community order and life, the Baal cycle contains true cosmogonies. There is, however, no Ugaritic parallel to the creation of the world such as is described in the first tablet of *Enuma elish* or in Genesis 1.

2. *Differences Between Creation in Ancient Near Eastern and Modern Usage*

Enough ancient Near Eastern cosmogonies have been seen to point out several important differences between ancient (including biblical) and modern conceptions of creation. The differences are at least four: the process, the product or the emergent, the description, and the criteria of truth.[10]

2.1 *Process*

Near Eastern texts frequently imagine cosmogony as a conflict of wills in which one party is victorious.[11] Moderns, on the other hand, see creation as the impersonal interaction of physical forces extending over eons, and reject psychologizing of the process. Ancient Near Eastern texts did not make the modern dichotomous distinction between "nature" and human beings, and sometimes offered psychic and social explanations for non-human phenomena.

2.2 *Product or Emergent*

To the ancients, organized human society or some aspect of it was the emergent from the creation process. To moderns, on the other hand, creation usually issues in the physical world, typically the planet amid the solar system. Community and culture do not generally come into consideration. If life is discussed in connection with creation, it is usually life in its most primitive biological sense. The point is worth illustrating.

In *Enuma elish,* tablets VI and VII tell of the organization of Babylonian society under the lordship of the god Marduk paralleling his organizing of the society of gods. Atrahasis tells how there came to be a balance through trial and error between the resources of the land and the population. To anticipate somewhat the discussion of the Bible below, Psalm 77 tells of the "wonders of old," the victory of the storm god over cosmic waters that brought Israel as a people into being.

> [16]The waters saw you, O God,
> the waters saw you, they were convulsed
> Yea, the deep quaked.

[17]The clouds poured forth water,
the clouds thundered forth.
Yea, the lightning bolts shot to and fro.
[18]The crash of your thunder was in the whirlwind,
your lightning lit up the world.
the earth quaked and trembled.
[19]In the sea was your way,
your path through the mighty waters,
your tracks could not be seen.
[20]You led your people like a flock
by the hand of Moses and Aaron.

Yahweh creates a way through his enemy, Sea, destroying with his weapons of thunder and lightning Sea's power to keep the people from their rightful land. What emerges from the conflict and victory is a people installed by Yahweh in his secure land. The psalm ends when Moses and Aaron are installed as the people's leaders; the community is complete.

In language equally "suprahistoric," i.e., with the focus on divine rather than human action, poems like Exodus 15 and sections of poems like Ps 78:41-55 portray the same event: movement from a state of social disorganization because of unrestrained forces to structure and security in Yahweh's land.[12]

2.3 *Manner of Reporting*

Ancients often reported creation as drama; moderns write scientific reports. The difference is a consequence of two essentially different conceptualizations of the process. Modern conceptualizing of the creation is generally evolutionary and "impersonal," and proceeds according to a combination of "laws" and randomness. The ancients saw things differently. Process often meant wills in conflict, hence drama; the result was a story. The mode of reporting corresponds in each case to the underlying conception of the process. Each approach advances the thought and resolves problems in a different fashion. Scientists offer new hypotheses as new data have to be explained. Ancients devised new stories, or wove variations into existing ones, when they wished to explain fresh elements of their world. It is not always easy for moderns, for whom a story typically is either entertainment or illustration, to regard story itself as a carrier of serious meaning.

2.4 *Criterion of Truth*

Moderns expect a creation theory with its empirical reference to be able to explain all the data, to be compatible with other verifiable theories and data. Failure to do so makes the hypothesis suspect. There is a drive toward complete and coherent explanation. The criterion of truth for ancient cosmogonies, on the other hand, is dramatic, the plausibility or usefulness of the story. In one sense it is no less empirical than the scientific account (it draws upon observation), but its verisimilitude is measured differently. Drama selects, omits, concentrates; it need not render a complete account. The story can be about a single aspect and leave others out of consideration. *Enuma elish* is interested in Marduk's rule over the gods and over Babylon; Atrahasis, in the balance of earthly resources and human population; Psalm 89, in the establishment of Davidic kingship in the very creation of the world.

3. *The Bible*

The biblical texts, products also of ancient Near Eastern culture, are marked by the same tendencies — creation issuing in the populated world, preference for narrative, tolerance for versions. Limitations of space allow a review of only the most important texts: some Psalms; the "Wisdom Literature," chiefly Proverbs 8 and Job; Second Isaiah; and of course Genesis 1-11.[13]

Regarding the structure of the created world, all the biblical accounts agree: the divine dwelling is in the heavens, also the storehouse of snow, hail, wind; above the heavens and under the earth were cosmic waters; earth, set on great pillars in the lower cosmic waters and protected from the upper waters by a gigantic disk ("the firmament"), was in the middle. As to how the world came into existence, the biblical accounts offer not a uniform account but a variety of stories. According to some accounts God's mere word arranged the elements to support life and community (e.g., Genesis 1 and Ps 33:6); others tell how God built the firmament like an edifice (e.g., Prov 8); still others narrate how Yahweh defeated the primordial forces, often personified as Cosmic Waters and stygian Night or sterile Wilderness (Psalms 74, 77, 89, and 93; Second Isaiah).

3.1 *The Psalms*

The genre of communal laments is a fair sample of the Psalms. Several (44, 74, 77, 89) contain cosmogonies, which are always accounts of the emergence of the people Israel. In the face of dire threat to the community, the "ancient deed(s)" hymned are the acts by which Israel came into existence. The rhetorical aim of the recital of the ancient deed can be paraphrased: Will you, O Lord, allow your act that brought us into existence to be nullified by the present danger? The community feels itself to be at the brink of extinction as a people in Yahweh's land; it remembers liturgically the originating act in an appeal to Yahweh's honor to renew that act today.

In Psalm 77, Israel's existence is threatened, prompting the psalmist's "Has his steadfast love disappeared forever?" (v. 8). The psalmist then recites "the deeds of Yah.... the deed of old" (v. 11-12). The redemption of Israel is described as a combat in which Yahweh makes a way through the sea, removing the obstacle posed by the waters to entering the land and installing leaders (for the text, see above). The psalm recites the founding event in the face of current threat.

Psalm 74 recalls the primal deed in verses 12-17, again in "suprahistoric" language:

> [12]Yet God my king is from of old,
> working salvation in the midst of the earth.
> [13]You divided the sea by your might;
> you broke the heads of the dragons on the waters.
> [14]You crushed the heads of Leviathan,
> you gave him as food for the creatures of the wilderness.
> [15]You cleft open springs and brooks;
> you dried up ever-flowing streams.
> [16]Yours is the day, yours also the night;
> you have established the luminaries and the sun.
> [17]You fixed all the bounds of the earth;
> you made summer and winter.

The verses depict the victory over Sea and his monstrous allies (vv. 12-14), the channeling of those once chaotic waters into springs and brooks (v. 15), the taming of boundless darkness into the peaceful rhythms of day and night (v. 16), and the establishment of the seasons of the year (v. 17). As often, the coming to be of the world is described as a sequence of cosmic pairs. The speaker recites the cosmogony in the face of an enemy intrusion into the Temple (vv. 1-11).[14]

Outside the genre of communal laments the most striking psalm of creation is Psalm 104, which has been compared to the Egyptian hymn to Aten in the tomb of Ay. The perspective of the psalm resembles Genesis 1. Unlimited Waters and Night, which make human community impossible, are by the action of Yahweh in a storm turned into an environment supportive of human society. In vv. 5-18 the Waters are tamed for human use, and in vv. 19-23, darkness is made into the restful rhythm of day-night. The perspective is persistently that of human community.

3.2 *Second Isaiah*

The author of Isaiah 40-55, who wrote in the 540's B.C. in Babylon, believed that the Jews in Babylonia and other exilic locations had ceased to be Israel, because they no longer dwelt in Canaan but in a land of false gods. They had fallen back into the position of their ancestors in Egypt — oppressed Hebrews needing to be led forth into Canaan (or Zion, his term for the land). He therefore preached a new Exodus-Conquest, a new creation. "In the wilderness clear the way of Yahweh. Make straight in the desert the highway for our God" (40:3). For him Exodus-Conquest and cosmogony were one and the same event. The exodus-conquest (the defeat of Pharaoh and the successful entry into Canaan) and cosmogony (the defeat of Sea, or Desert, interposing itself between the people and its allotted land) accomplish the same purpose — the coming into being of Israel in its land.

Second Isaiah announces that God will repeat the old deed that brought Israel into existence — a new exodus-conquest/cosmogony. The difference is that this time the people will be led through the wilderness instead of being led through the sea in the first exodus. The new act repeating the old, yet with a difference, is expressed with almost mathematical precision in 43:16-21.

> [16]Thus says Yahweh,
> The one who makes a way in the Sea,
> a path in the Mighty Waters,
> [17]the one who musters chariot and horse,
> all the mighty army.
> They lie prostrate, no more to rise,
> they are extinguished, quenched like a wick.
> [18]Recall no more the former things,
> the ancient events bring no longer to mind.
> [19]I am now doing something new,
> now it springs forth, do you not recognize it?
> I am making a way in the wilderness,
> paths (correction from 1QIs[a]) in the desert.
> [20]The wild beasts will honor me,
> jackals and ostriches.
> For I have placed waters in the wilderness,
> rivers in the desert,
> to give drink to my chosen people,
> [21]the people whom I formed for myself,
> to narrate my praiseworthy deeds.

Verses 16-17 narrate the old founding event, the defeat of Pharaoh and the way through the Sea. Verse 18 declares that this story will no longer serve as the national story; it will be replaced by the new story of the way through the wilderness. Both Sea and Wilderness can have "historic" and "suprahistoric" meanings. Sea can mean the Red (or Reed) Sea where the people escaped Pharaoh and his troops and Sea, the primordial enemy of human community. Wilderness can mean the great desert separating Babylon from Zion and impassible Wilderness, the residence of Death.

New creation in Second Isaiah is a renewal of the first act that brought Israel into existence — the Exodus-Conquest, the defeat of Sea. It does not refer, in my opinion, to the act that brought the world of the nations into being; that is the perspective of Genesis, not of Second Isaiah. The people do not properly exist scattered in exile, apart from Yahweh, without land and temple, ritual and officials.

3.3 *Wisdom Literature*

"Wisdom Literature" is a modern designation for various genres of literature, current in learned court circles, for the instruction of young courtiers. Examples are found in Mesopotamia,[15] Egypt,[16] as well as in the Bible.[17] Within the Bible, the books of "Wisdom Literature" — Proverbs, Job, Qoheleth, Sirach, and some Psalms — vary in genre and purpose and are somewhat artificially grouped under the one term.

Prov 8:22-31 is part of chaps. 1-9, a series of poems prefatory to the proverb collections in chaps. 10-31. Wisdom, personified as a gracious woman, invites the inexperienced youth to follow her paths. To ground her claims, she declares that she was created before the world, that her lineaments are to be found in the world.

> [22]The Lord created me at the beginning of his work,
> the first of his acts of old.
> [23]From of old I was fashioned,
> at the beginning, before the origin of the earth,
> [24]when there were no deeps I was brought forth,
> when there were no springs abounding with water.
> [25]Before the mountains had been sunk,
> before the hills, I was brought forth;
> [26]before he had made the earth with its fields,
> or the first of the clay of the world.
> [27]When he established the heavens, I was there,
> when he drew a circle on the face of the deep,
> [28]when he made firm the skies above,
> when he established the fountains of the deep,
> [29]when he assigned to the sea its limit,
> so that the waters might not transgress his command,
> when he marked out the foundations of the earth,
> [30]then I was beside him like a confidant;
> and I was his delight every day,
> rejoicing before him always,
> [31]rejoicing in his inhabited world,
> and delighting in humankind.

God here is architect and artisan, constructing the world as one would build a great edifice, inspired by wisdom (clearly subordinate to him as the first of his creation). The poem draws remotely on a polytheistic picture of a god and his consort creating together. The scholarly suggestion that the Egyptian concept of *maat,* "order," the basis of the world and of human life, often personified as a woman,[18] is possible. More significant, however, is the general ancient Near Eastern assumption of the givenness of reality at creation.[19] At any rate, what is important here is the divine will implanted in creation that can be discovered by the serious seeker. Creation is the statement of God's will.

3.4 *Wisdom Literature: Job*

An important counter to Proverbs' doctrine that the divine will in creation is transparent to the sincere seeker, and indeed to a strong anthropomorphic concept of creation, is the view of creation in Job. The book is designedly provocative. Job is preeminently the just person; God and the reader know this from the outset (cf. chaps. 1-2). Yet he is put through public suffering that in the thinking of the time mark him as a sinner justly punished by God. His comforters only reflect conventional wisdom when they urge him to repent of the sin he must have committed to bring such punishment.

Job speaks three times of creation (9:5-13; 10:8-13; 12:13-25). For him, conscious of being victimized by an uncaring God, creation is only arbitrary power and disregard for human justice.

> 9:[13]God does not restrain his anger,
> under him the allies of Rahab bow low.
> [14]How then can I answer him,
> choose my words with him?
> [15]Though I am innocent, I will not respond,
> From my adversary-at-law I must beg for mercy.

Earlier in the speech, Job mockingly hymned the God who manipulates the elements of the world — the mountains, earth, pillars (9:5-6), the luminaries (9:7), sea and heavens (9:8-9) — with such power as to render true dialogue impossible. How can Job expect an answer from such a creator?

In 10:8-13 Job takes up themes that Psalm 139 uses to express human delight in being transparent before Yahweh; Job's intent, however, is to show how God scrutinizes only to find fault.

> 10:[13]But you hid these things [life and kindness] in your heart,
> I know this is the case with you:
> [14]to watch whether I would sin,
> and not clear me of my guilt.
> [15]If I am guilty, woe is me!
> If I am innocent, I cannot lift my head,
> so sated am I with misery, so filled with shame.

In 12:13-25, Job does not deny God's wisdom and power in creating the world (vv. 13-15) but says that same wisdom and power (v. 16a) is exercised arbitrarily in overturning responsible people (like Job) with no reason given. In the ancient Near East the gods were assumed to support the social system. Bildad, on the other hand, one of Job's comforters, finds no difficulty in praising enthusiastically God's creation (25:1-6 and 26:6-14).

The most important statements about creation in Job, however, are God's two speeches in chaps. 38-41. They are not merely a divine attempt to overwhelm and confuse Job. Job has been insisting on a face-to-face encounter with God, culminating in his speech and oath of chaps. 29-31. Elihu's attempt to play the mediator in chaps. 32-37 has failed. The way is cleared for a theophany and divine responses to Job's questions. According to the analysis of Norman Habel,[20] the first speech of Yahweh refutes Job's charge of capricious governance: "Who is this who obscures design?" (38:2). The second speech refutes Job's charge that Yahweh does not uphold the rights of the poor: "Will you impugn my justice?" (40:8) The two speeches are in parallel. In both speeches Yahweh defends his creation.

In the first divine speech, chaps. 38-39, God asks:

Were you there, or, do you know about:

a. The inanimate physical world (38:4-38):
   — the construction of earth (4-7);

— the hemming in of sea (8-11);
— dawn's role in ridding the earth of sinners (12-15);
— God's dominion over the underworld of death (16-18);
— the placement of light and darkness (19-21);
— the storehouses of earth's weather (22-30);
— the constellations controlling earth's destiny (31-33);
— the thunderstorm fertilizing earth (34-38)

b. The animal and bird kingdoms (38:39-39:30):

— the feeding of the lion (39-40);
— the feeding of the raven (41);
— the ibex and the hind (39:1-4);
— the wild ass (5-8);
— the wild ox (9-12);
— the ostrich (13-18);
— the horse (19-25);
— the hawk and the eagle (26-30).

The speech never loses sight of "design," the divine wisdom in creation, and alludes to Job's earlier accusatory words. For example, Job had accused God in 9:5-6 of a reckless and almost violent attack on the mountains and earth ("He who moves mountains without their knowing it, who overturns them in his anger, who shakes the earth from its place, its pillars totter" 9:5). God asks Job in 38:4-7 if he actually witnessed the foundation of earth, then reveals how like a careful artisan he built with measuring line, utilizing sockets and cornerstone, while a festive chorus sang as at the dedication of a temple (38:4-7). Job had claimed that God does not distinguish between the wicked and the righteous, that the earth has been handed over to the wicked (9:24). The divine response is that dawn exposes the deeds the wicked have done during the night (38:12-15). Job's insistence that as a human being he is to be taken with utmost seriousness is countered by God's questions about the rain upon lands where there are no humans (38:26-7). Job had accused God of hunting him like a lion (10:16); God is rather the one who hunts for the lion (38:39-40). Even the ostrich, proverbial in the culture for stupidity, is so by design (39:13-18). The ostrich is a reminder that the designed universe is not an impersonal machine working smoothly, but that it includes the useful, the bizarre, even the playful; in short, God creates for his own inscrutable pleasure. The creator creates and sustains for himself and not primarily for the world.

Job's answer is: "See, I am small, what can I answer you. I put my hand on my mouth. I have spoken once, and will not reply, twice, I will not do so again." The words are a promise not to speak, an acknowledgement that his words have been pointless.

The second speech, 40:6-41:26, is at first reading extremely strange as a defence of God's justice (40:8). Job had accused God of allowing the wicked to prosper and the righteous, in particular Job, to suffer. God's first questions are those only a god could answer affirmatively: "Have you an arm like God, can you thunder with a voice like his .... Look down upon every proud person and bring him low, and tread upon the wicked where they stand." There then follow descriptions of two great animals: Behemoth (40:15-24) and Leviathan (40:25-42:26). The precise significance of the two beasts, described in such detail, is debated. Are the two animals simply the hippopotamus and the crocodile, which only a god could control? Does Behemoth, a hippopotamus symbolizing the historical enemies of Egypt, and Leviathan, symbolizing the historical enemies in Canaanite culture, stand for Yahweh's control of empires? Do they symbolize chaos, thus illustrating Yahweh's overcoming of chaos?

Any interpretation of the two monsters must respect the narrative line of the whole book, and also keep in mind the parallel speech of 38:1-40:5 about design. The first speech had disposed of the design argument by showing God's care for a world that Job as a human being knows nothing about. The second speech is about "my justice" (40:8) which vv. 9-14 define as the power to bring low the wicked and powerful.

God proposes the two beasts as examples of the exulting proud. Behemoth (40:15-26) is a massive beast. For all his strength, however, he can be taken by the face: "By his eyes he takes him, by hooks he pierces his nose."[21] The second portrait, of Leviathan, is much longer than the first. Leviathan is known from the Bible[22] and from the Ugaritic texts as the great primordial monster who was killed or tamed by God or Baal in a cosmogonic battle. As with Behemoth, God controls him by the mouth (41:1-4). There then follows magnificent poetry: Leviathan is built for no other purpose than to display untrammeled physical might.

What is the purpose of these two beasts in the rhetoric of the speeches? I suggest that they are brutes beyond the power of man to control and, in the second case, beyond his power even to observe. They are allowed to romp within God's universe. They serve no function; they cannot be domesticated nor do they serve humans. But they are under Yahweh's control (40:15, 24; 41:2-4). For reasons not stated he allows them to exist despite their mindlessness or malice. In this respect the beasts are like the Satan, the adversary of chapters 1 and 2. Why should there be an enemy of man within the very heavenly court? No answer is given to this particular problem of evil. But both the adversary and the beasts are under the control of Yahweh. Yahweh says to the Satan in the preface, "So be it. All he [Job] possesses is in your hand, only do not lay a hand on him." (1:12) "So be it. He is in your hand, only watch over his life." No answer is given in the preface about why evil exists, and no answer is given in the speeches; brute force and evil are simply there, ultimately under the easy control of Yahweh.

The anthropocentric perspective of the creation accounts of Genesis 1-11, the Psalms, and Second Isaiah is not in Job. "Here man is incidental — mainly an impotent foil to the God .... Job, representing man, stands outside the picture, displaced from its center to a remote periphery."[23] God, not human beings, is the center of creation. The frequent criticism that the Bible is a charter for humans to exploit the world is refuted definitively in the book of Job.

3.5 *Genesis 1-11*

Examination of Old Testament creation texts have shown that most tell how *a people emerged*; the physical world so important to the modern scientist figures in the ancient texts (except for Job) chiefly as the environment for human society. Apart from Job (and Psalm 104), the texts examined so far all describe the creation of Israel, how Israel came to exist through God's defeat of the forces hostile to it. Genesis 1-11 is noteworthy because it tells how the *goyim,* the nations, emerged; Israel's story begins only afterwards with its ancestors, Abraham and Sarah, in 11:27 ("These are the generations of Terah. Terah was the father of Abram, Nahor, and Haran").

Because Genesis 1-11 has been traditionally read through the lens of Paul's New Adam Christology (Rom 5:12-21 and 1 Cor 15:21-28), which concentrated exclusively on Genesis 2-3 (the sin of Adam) to the neglect of chaps. 4-11,[24] we need to remind ourselves of the coherent plot of chapters 2-11.

According to the scholarly consensus, the Pentateuch was edited by the Priestly redactor (P) from several sources (the most important of which in the early chapters was the tenth century B.C. Yahwist, conventionally J). P wrote Gen 1:1-2:3 most probably in the sixth century B.C. exile as a preface to the whole, organizing and supplementing the venerable J material with material of his own (mostly genealogies and notices, except for the additions to the flood narrative).[25] P used the formula "These are the generations of...." five times in the story of the nations or primeval history (2:4; 5:1; 6:9; 10:1; 11:10) and five times in the story of Israel's ancestors (11:27; 25:12; 25:19; 36:1 [doubled in 36:9]; 37:2). The five-time recurring formula in 2:4-11:26 suggests that this section is a single story. Also suggestive that chaps. 2-11 form a single story is the similar plot of Atrahasis — creation of humankind, fault, flood, survival of a friend of the god, and new beginning. We need, therefore, to look at these chapters with a non-Pauline lens as a single lengthy story, a "compendious" cosmogony.

Before looking at chapters 2-11 as a single story, however, we must look at chapter 1 that prefaces it and highlights important themes. The story is structured as a seven day week.

"Beginning" chaos of Waters and Night (vv. 1-2)

Day 1:	Defeat of darkness by separation of darkness and light into night and day	Day 4:	Luminaries in heaven to regulate day and night
Day 2:	Defeat of waters by separation of waters above and below through "firmament"	Day 5:	Water creatures (fish and birds)
Day 3:	a. waters, dry land b. vegetation	Day 6:	a. earth animals b. man

Day 7: "Completion" God rests

The second set of three days parallels and completes the first set. Conflict is deliberately absent; darkness and night, incompatible with human community, are "defeated" by God's mere word. The climax of creation is not the creation of man but God majestically at rest on the seventh, climactic day. Sexuality in man, "male and female he created them (v. 27)," corresponds to the reproductive power in plants and animals expressed by the phrase, "according to their kinds" (vv. 11-12, 21-22, 24). The "nature" or "essence" of man is given in typical ancient Near Eastern fashion by an expression of will; God's imperatives, "Have dominion over the birds of the air...," (v. 26) "Be fertile and increase," and "fill the earth and subdue it" (v. 28) define what man is.

These imperatives are easily misunderstood. Lynn White has written:

> Man named all the animals, thus establishing his dominance over them. God planned all this explicitly for man's benefit and rule: no item in the physical creation had any purpose save to serve man's purposes. Christianity... not only established a dualism of man and nature but also insisted that it is God's will that man exploit nature for his proper ends.[26]

That imperatives mean something quite different from White's interpretation is clear from the chapters subsequent to chapters 1-3. Having dominion over the animals (1:26) did not include killing them for food (1:29; that permission is given only in chapter 9 as an accommodation to human sinfulness). The meaning of dominion over the animals is given by the actions of the just Noah in chapters 6-9; he sees to the continuance of each species by taking two of each into the ark. Being fertile and increasing (1:28) mandates the continuance of the race and is illustrated chiefly in the genealogies. Filling the earth and subduing it (1:28) means receiving the land God allots to each people and nation. It is illustrated in the nations being assigned their lands and journeying to take them (Gen 10:1-11:9), and Israel's conquest of Canaan.[27] Genesis 1 describes the world before human history and sin. God declares it beautiful seven times. The chapter states God's intent, what will be realized at the end of time.

The prefatory Genesis 1 alerts the reader to what P considers important in chapters 2-11: the continuance of the race through its progeny and their taking of their God-given land. If Genesis 1 described the world as God intended it, these chapters explain how the world humans actually experience came to be. The plot is the key. In chapters 2-4 the man and the woman sin and, thereby, introduce the alienation characterizing human life and culture. The ten-member genealogy in chapter 5 describes an increasingly populated world, not yet differentiated into distinct nations. In 6:1-8 the whole race has become sinful with the exception of Noah and is to be destroyed by the just God. The Hebrew word for God's destruction is the same as that describing what humans have already done to the world; God simply ratifies what humans have already done. The flood purifies the world and establishes the conditions for a new beginning (6:9-8:22). In chapter 9, God reaffirms the original blessing of chapter 1, with the exception of the permission to kill animals for food — an accommodation made in virtue of humans' sinful proclivities (like the clothing of the man and the woman, their removal from the temptation of the tree of life, and the protective mark given to the murderous Cain). Gen 10:1-11:9 is the story of the assignment of a particular land to each people and nation, a task at first refused by the race (by their attempt to remain together in one place and build a city with a tower in the middle of it) but then forced on them by God ("So the Lord scattered them abroad" 11:8-9). The final nine-member genealogy brings the story down to Abraham's father Terah, when the contrasting emergence of Israel will begin.

Chapters 2-11 constitute a "compendious cosmogony" — a collection of many traditions about how the world of men and women came to be what it is — constituting a single story.

4. *The Distinctiveness of Biblical Creation*

The structure of the created universe — earth positioned between upper and lower cosmic waters — is generally the same in the Bible and comparable literatures. So also are the modes of creation — building, divine word, shaping humans out of clay, defeat of chaos.

The biblical accounts of creation differ chiefly from those of their neighbors in attributing creation to a sole transcendent deity, Yahweh, the God of Israel. Whatever does not comport with that belief — creation by several deities or by a consort (sexual generation), the creator as originally within the primal mass, creation by trial and error, creation of humans to maintain the universe in place of unwilling gods — is denied.

Also distinctive is Yahweh's intent, diverse according to the passage but never physical need or caprice. In the communal laments of the Psalter, Yahweh's created people is to reflect his grandeur on earth. In Second Isaiah the return of Israel from Babylonian exile is a renewal of the exodus-conquest and cosmogony that brought Israel into being in the first place. Creation in Proverbs implants God's wisdom in the universe and makes it accessible to the humble seeker. In the Book of Job, God creates for his own pleasure; his plan is beyond human scrutiny. The intent of Genesis 1 is to show how beautifully God has made the world that honors him and the role of humans in giving him honor. Genesis 2-11 explains how the nations fulfilled Yahweh's intent that they continue their species and receive their land, a foil to Israel's carrying out of the same tasks in later chapters. Despite the diversity of Yahweh's intents, the accounts cohere in content and tone. God is serene master in creating and imparting dignity to the human race.

5. *Conclusions*

The biblical authors, like their neighbors, assumed that the world was divinely created (theoretical atheism was virtually unknown). The creator was the unique Yahweh who had made them his people. In expressing Yahweh's creation they made use of the genres and the traditions of ancient Near Eastern reflective literature (hymns, rituals, wisdom literature, epics), preferred narrative to abstraction, and were used to several versions of the same event.

There was no truly scientific writing distinct from the kinds of literature we have been considering. All authors began with a common-sense understanding of reality and assumed the existence of divine powers and an intense relatedness of phenomena that enouraged explanation by analogies. The origin of the world was a privileged moment for them, since the essences of things were definitively established by the gods (or God) at the very first. Development from simple to complex states was generally not part of the ancients' thinking; things were there from the beginning. To explain the "essence" of something they therefore explained its origin. This is the major reason for their interest in cosmogonies.

Granted the diversity between ancient and modern world views, do the biblical accounts of creation still hold meaning for moderns, who distinguish sharply between scientific explanation and religious interpretation? They do. The biblical authors' multiple versions of the creation event show their interest lay not in providing a factual chronicle but in affirming the divine intent of graciousness and in expressing God's majestic power over all things.

NOTES

[1] Shortly before the popularization of Darwin, there was a shift in Christian theology away from accepting the biblical story as the complete account to searching for the history behind the story. The shift has been studied by H. Frei, *The Eclipse of Biblical Narrative: A Study in Eighteenth and Nineteenth Century Hermeneutics* (New Haven: Yale, 1974). In addition, pre-or anti-Kantian epistemology of Scottish Common Sense philosophy deeply influenced American fundamentalism and encouraged the reading of biblical texts as if they were the data of sense observation. The epistemological and political roots of American evangelism have been studied by G. Marsden, *Fundamentalism and American Culture: The Shaping of Twentieth Century Evangelicism 1870-1925* (Oxford: Oxford University, 1980). See also J. Barr, *Fundamentalism* (Philadelphia: Fortress, 1978).

[2] Limitations of this article preclude a discussion of Egyptian cosmogonies. The influence of Egypt upon biblical cosmogonies is in any case limited to Psalm 104 which, according to some scholars, is indebted to the well known Hymn to the Sun of Akhenaten. The basic study of Egyptian creation texts is Serge Sauneron and Jean Yoyotte, "La naissance du monde selon l'Égypte ancienne," in *La naissance du monde* (Sources Orientales 1; Paris: Seuils, 1959) 17-91. For a recent synthesis see B. Menu, "Les cosmogonies de l'ancienne Égypte," *La création dans l'Orient ancien* (Lectio divina 127; Congrés de l'ACFEB, Lille [1985]; Paris: Cerf, 1987), 97-116.

[3] Translation of E. A. Speiser in *Ancient Near Eastern Texts Relating to the Old Testament (ANET)*. (3rd ed.; ed. J. B. Pritchard; Princeton: Princeton University, 1969) 100-01.

[4] J. Bottero has collected twelve such cosmogonies in "Antiquités Assyro-Babyloniennes," *Annuaire École practique des hautes-études*. Sect. hist. et phil. (1978-9) (Paris; Klincksieck, 1982), reprinted in Bottero, *Mésopotamie: L'Écriture, la raison et les dieux* (Bibliotheque des Histoires; Paris: Gallimard, 1987) 279-331. Daniel Arnaud has published an additional cosmogony in "La bibliothèque d'un devin syrien à Meskène (Syrie)," *Comptes rendus des séances — Académie des inscriptions & belles lettres* (1980) 375-87. A good recent synthesis is M.-J. Seux, "La création du monde et de l'homme dans la littérature Suméro-Akkadienne," *La Création dans l'Orient ancien* (Lectio divina 127; Congrès de l'ACFEB, Lille [1985]; Paris: Cerf, 1987) 41-78.

[5] W. G. Lambert, "The pair Lahmu-Lahamu in Cosmology," *Orientalia* **54** (1975) 189-202.

[6] Translation by E. A. Speiser in *ANET*, 60-1, lines 1-11.

[7] Suggestions of W. G. Lambert, "The Cosmology of Sumer and Babylon," in *Ancient Cosmologies*, ed. C. Blacker and M. Loewe (London: Allen and Unwin, 1975) 55, have been incorporated into the translation.

[8] W. G. Lambert and A. R. Millard, *Atra-Hasis: The Babylonian Story of the Flood* (Oxford: Clarendon, 1969). Important commentary has been provided by W. L. Moran's review in *Biblica* **52** (1971) 51-61, and by R. A. Oden, "Divine Aspirations in Atrahasis and Genesis 1-11," *Zeitschrift für alttestamentliche Wissenschaft* **93** (1981) 197-216.

[9] For a summary of the discussion and for arguments that the Ugaritic texts do contain genuine cosmogonies, see my "Cosmogonies in the Ugaritic Texts and in the Bible," *Orientalia* **53** (1984) 183-201.

[10] I draw on my "The Hebrew Scriptures and the Theology of Creation," *Theological Studies* **46** (1985) 507-23.

[11] There are, to be sure, other modes of creating in the ancient Near Eastern and biblical texts — molding humans from clay, divine command, unimpeded construction, and sexual generation (in the Bible only vestigially) — but creation through defeat of forces hostile to human community (chiefly unlimited Waters and Night) is common, especially in the Psalms and Second Isaiah and has not received sufficient emphasis from scholars.

[12] "In Mesopotamia, Ugarit, and Israel, *Chaoskampf* appears not only in cosmological contexts but just as frequently — and this was fundamentally true right from the first — in political contexts. The repulsion and the destruction of the enemy, and thereby the maintenance of political order, always constitutes one of the major dimensions of the battle against chaos. The enemies are none other than a manifestation of chaos which must be driven back." H. H. Schmid, "Creation, Righteousness, and Salvation: 'Creation Theology' as a Broad Horizon of Biblical Theology," in *Creation in the Old Testament: Issues in Religion and Theology*, ed. B. W. Anderson (Philadelphia: Fortress, 1984) 104.

[13] For a good overview from the perspectives of many distinguished scholars, see *Creation in the Old Testament.*

[14] Some of the so called historical psalms also illustrate how creation ends in the formation of a people. Psalms 135 and 136 use creation and redemption language to depict the coming into being of Israel.

[15] W. G. Lambert, *Babylonian Wisdom Literature* (Oxford: Clarendon, 1960) and G. Buccellati, "Wisdom and Not: The Case of Mesopotamia," *Journal of the American Oriental Society* **101** (1981) 35-48.

[16] R. J. Williams, "The Sages of Ancient Egypt in the Light of Recent Scholarship," *Journal of the American Oriental Society* **101** (1981) 1-20.

[17] R. E. Murphy, "Hebrew Wisdom," *Journal of the American Oriental Society* **101** (1981) 21-34.

[18] G. von Rad, *Wisdom in Israel* (Nashville: Abingdon, 1972), cautiously suggests such influence. For a recent study of *maat*, see W. Helck, "*Maat*," *Lexikon für Ägyptologie* (Wiesbaden: Harrassowitz, 1982) III, cols. 1110-19.

[19] Cf. H. H. Schmid, *Gerechtigkeit als Weltordnung: Hintergrund und Geschichte des alttestamentlichen Gerechtigkeitsbegriffes* (Beiträge zur historischen Theologie 40; Tübingen: J. C. B. Mohr) for the general ancient Near Eastern background. Schmid neglects somewhat the relevant Akkadian evidence.

[20] *The Book of Job: Old Testament Library* (Philadelphia: Westminster, 1985). I am much indebted to Habel's analysis of the divine speeches.

[21] The preceding Hebrew verse is overlong and the following is too short, leading many scholars either to add Hebrew *mî hû'*, "Who is there who. . . ." to the beginning of v. 24a or to bring the last two words of v. 23b to begin v. 24a. They vocalize *'il pîhû* as *'ēl*, "God." In any case, the Hebrew text correctly understands the subject to be God.

[22] Pss **104**:25; **74**:14; Isa **27**:1,1; and Job **3**:8.

[23] M. Greenberg, "Job," in *The Literary Guide to the Bible*, ed. R. Alter and F. Kermode (Cambridge, MA: Harvard, 1987) 298.

[24] Claus Westermann, *Creation* (Philadelphia: Fortress, 1974), correctly points out the Pauline influence upon the Christian interpretation of chapters 2-3.

[25] The majority opinion is that Gen 2:4, "These are the generations of the heavens and the earth when they were created," is retrospective; it sums up the story of creation in seven days. Elsewhere, however, the formula is introductory. It seems so here.

[26] "The Historical Roots of Our Ecologic Crisis," *Science* **155** (10 March 1967) 1206.

[27] The Hebrew word for "subdue" is used several times to describe Israel's conquest of Canaan, "The land was subdued before them." For a summary of recent work, see Klaus Koch, "Gestaltet die Erde, doch heget das Leben," in "*Wenn nicht jetzt, wann denn*?", H.-J. Kraus volume, ed. H.-G. Geyer *et al.* (Neukirchen: Neukirchener Verlag, 1983) 23-36.

II.

EPISTEMOLOGY AND METHODOLOGY

KNOWLEDGE AND EXPERIENCE IN SCIENCE AND RELIGION: CAN WE BE REALISTS?

JANET SOSKICE, Ripon College, Oxford.

> We are of the kind to reach the world of intelligence through the world of sense.
>
> (Thomas Aquinas, Summa Theologica, **12**.1,9)

Introduction

Human knowing is knowing through experience. This is no less the case in religious matters than in scientific ones, a point sometimes blurred by the use of specialist terms. Consider *revelation*, the disclosed knowledge of God. Nothing could seem further from the categories acceptable to science, yet both scientific knowing and revelation have a bedrock in experience, observation, and the observing community. In what follows I shall suggest that there are interesting and acceptable parallels at the methodological level between scientific and religious knowing, that models and metaphors are central to this comparison, and that a realist interpretation of our experience is likely to be most fruitful to both.

Let me first set the stage for theology. If some theological notions seem discredited by post-Enlightenment thought, surely none were more so than revelation. The epistemological obstacles to divine self-disclosure, as Hume was not slow to point out, seem great. Increased acquaintance with other great religions has made Christians wary of claiming for their own a distinctive and exclusive illumination. Awareness of culture and change has brought the sometimes grudging acknowledgement that God's self-disclosure must, at least, be greatly influenced by social and historical contexts. In particular, the idea of a propositional revelation, if by that one means "truths written by the finger of God," is increasingly unacceptable to theologians and rank and file believers alike.

Various attempts have been made to circumvent the difficulties. For example, in the nineteenth and twentieth centuries it has become fashionable to claim that it is not truths which are revealed but a person (Jesus Christ) who is revealed. Aligned to this is the tendency to speak, not of individual propositional revelations, but of Revelation in the singular, where what is revealed is God's saving will for humanity. Revelation remains God's self-disclosure, but what is disclosed is a relationship, not a set of propositions. These approaches, while necessary correctives to an overly propositional theory of revelation, are not in themselves sufficient, for even if what is revealed is a person, belief in a person and a relationship with a person involves some, probably many, beliefs about that person. In our relationship with God as in our relationship with a neighbour, belief "in" involves some, probably many, beliefs *that*. It seems that the cognitive element of Christian faith cannot be eliminated, and if it cannot, then the problem of revelation remains, for revelation is an irreducibly cognitive notion. One might even say that Christianity stands or falls on its conviction that its claims concern that which really is the case with God and humanity.

It must not be forgotten, however, that revelation is never simply the disclosure of God but always the disclosure of God *to us*. (Even stone tablets engraved by the finger of God would not be revelation unless some one found and read them.) To put this in an (not uncontroversial) anthropic form: whatever God's revelation is, it must be such that human beings can understand it. Knowing God, even through revelation, is in some sense knowing through experience, whether that experience be ordinary (such as the everyday experience of finite objects which led Aquinas to posit an infinite deity) or extraordinary, as Aquinas would have claimed concerning the appearance of the angel to Mary at the Annunciation.

A further point about the Christian concept of revelation is this: revelation is not characteristically a matter of private illumination. It is of the nature of revelation, properly so called, to be communicable and communicated to others, and this has to do with Christianity as a social, or corporate, religion.

It is for this reason that language becomes important, and in what follows I suggest that God reveals himself to us as creatures of language and as social beings, that language and interpersonal experience are inextricably bound together, and that revelation is not so radically different from other kinds of knowing as it might at first seem. I will suggest that metaphor is a primary means by which God's self-disclosure might be understood without the rigidity which the Enlightenment found so objectionable in earlier propositional theories. It will be argued that our knowledge of both God and the world is at once profoundly social and conditioned by cultural and historical circumstance. Yet at the same time revelation can claim genuinely to be about states and relations which exceed our comprehension of them. Thus a realist case will be made for revelation.

Models and Metaphors in Science and Religion

Philosophers of religion have for some time considered that the use of models in scientific and religious thought makes an interesting comparison. It has also been observed that the metaphors with which theology is replete are linked importantly to the extension of models, so that figurative language becomes not merely decorative but essential to the way in which the theologian struggles to name the unnameable God. Metaphor plays a similar role in extending the models of physical theory.

Virtually all Christians can agree that most of what we say of God is figurative. To say so has no bearing on one's theological conservatism or radicality. There are few true literalists who believe that mention in the Bible of God's "mighty arm" means that God has physical limbs. Certainly Catholic and mainstream Protestant Christianity does not. The God of the Jews, the Christians, and the Muslems, is "He Who Is," the cause of all things who is yet apart from all things.[1] No "name", as Jacob discovered when wrestling with the angel, can capture God. Nor can talk of Jesus Christ be purged of metaphor. Although one might be able to say a good deal of the man, Jesus, in a perfectly literal sense (he was a Jew, etc.), as soon as we speak of his divinity we must speak figuratively. The earliest Christologies we have, going right back to Scripture itself, are ineliminably metaphorical; Jesus is the lamb, the High Priest, the king, the shepherd, he gave his life in "ransom", and so on. This should be uncontroversial once we recognise that metaphor is as satisfactory a linguistic form for making truth claims as is literal speech.

Nevertheless, there is a very real and important debate in modern theology concerning metaphor, and this debate centres not on whether religious language is ineradicably metaphorical, but on what follows if this is so. To put this in an extreme form, some who say that talk of God is metaphorical may be reflecting, in the mode of the prophets, psalmists and mystics, on the inability of human thought or speech to comprehend the deity. Others, however, mean something more like, "Christian language is merely metaphorical, a powerful if somewhat archaic system of images not to be taken as somehow speaking about a world-transcending God in any traditional sense." In this extreme form then, both the mystic and the contemporary "atheologian" (one who wishes to dispense with traditional theism altogether) can agree that talk about God is metaphorical. We might say, though, that their agreement is spurious.

Mere acceptance of the same set of words here glosses over the real issue of whether those speaking are theological realists or theological instrumentalists. By theological realists I mean here those who, while aware of the inability of any theological formulation to catch the divine realities, nonetheless accept that there *are* divine realities that theologians, however ham-fistedly, are trying to catch. By theological instrumentalists I mean those who believe that religious language provides a useful, even uniquely useful, system of symbols which is action-guiding for the believer, but which is not to be taken as making reference to a cosmos-transcending being in the traditional sense.[2] Feuerbach and his latter-day followers would be clear candidates for the second camp, but many others, less obviously radical, put forward ideas whose implications are much the same. Not surprisingly, instrumentalism, in both its theological and non-theological applications, and notably in the philosophy of science, where the debate between realists and instrumentalists has raged for some time, is associated with criticisms of the possibility or necessity of metaphysical explanations.

Realism is attractive because it seems undeniable that Christians and Jews traditionally have been realists of some sort. The difficulty is that, since Locke, Hume and Kant, it has been assumed by many to be philosophically indefensible. One can see why. With traditional metaphysics given short shrift, theologians have judged there to be limited scope in claiming to speak of that which we cannot comprehend.[3]

The perceived weakness of natural theology has added strength to the instrumentalist case. Religious language does not tell us about God, on their account, but evokes our response to God. The difficulty, as always, is response to what? belief in what? Instrumentalism all too easily reduces to a position where religious language is no more than a life-enhancing means of discussing the human condition. For many it is difficult to find any resemblances between that and traditional theism.

To meet instrumentalism, the realist must attempt to say how religious language can claim to be about God at all, given that naive realism in these matters is unthinkable. This task, I suggest, is bound up with giving some good account of how metaphor works in religious language.

The terminology of *realist* and *instrumentalist* is not, of course, native to philosophical theology but borrowed from debates which have taken place in the philosophy of science. As already mentioned, philosophers of religion have made much of the comparison of models in science and religion. They emphasize, for instance, the need for a multiplicity of models, all of which have a tentative descriptive status, and other supposed shared features such as simplicity, elegance, and extensibility. For the most part these comparisons have been inconclusive, amounting at best to a "companions in guilt" argument of the form, "Religion need not be ashamed of its reliance on models if science proceeds in the same way." This does not constitute much of an argument, however, unless the philosopher of religion can demonstrate why these perceived similarities are significant, and here accounts have been weak. Committed, as they often are to a hearty, if uncritical, realism vis-a-vis the role of models in scientific theory (in order to affirm the necessity of models to scientific practice and thus justify their presence in theology), the philosophers of religion have, with some regularity, drifted into non-cognitivist positions when they apply their ideas to theology. Religious models are seen to be challenging, unifying, evocative, and morally valuable. This ghostly gain may be better than no gain at all, but it is far from the promise of cognitive stability which, presumably, was the attraction of the analogy with the philosophy of science in the first place.

Yet we can see why this happens. Scientific realists who place a high value on models do so because they view them as descriptive of states and relations which, while going beyond our powers of direct observation, nonetheless are in important senses independent of the construction we put upon them. The models then, if qualified and limited, are nevertheless held to be descriptive. Yet the very idea that the theologians' model describes God as He is in Himself must be anathema to most philosophers of religion.

Despite these difficulties, or indeed even because of them, the comparison between models in science and religion should continue to interest us, not at the level of individual models (light "waves" and heavenly "fathers") but at the more fundamental level of what constitutes model-based explanation in the two disciplines.

The Argument for Realism

I take it that the theological realist has this much in common with the scientific realist: they both want to preserve their models and the metaphorical terminology to which these give rise, and want to preserve them, not as convenient fictions for the ordering of observables but as terms which somehow provide access to states and relations which exist independently of our theorising about them. So the scientific realist wants to say that speaking of the brain as a computer, of feedback, of programming, and so on, really is talk about brain activity. Similarly the theological realist believes that talk of God as father really is talk about God's relationship to humankind. But neither of these realists wants to claim privileged knowledge of their unobservable subject matters. Indeed, models and the metaphorical terminology to which they give rise are prized in these contexts precisely because of their adaptability; they are always tentative, always qualified. Were this not so they would not be models.[4] But here we come to a problem: how can we claim that these metaphorical terms are in some sense descriptive or, as I prefer to say, reality depicting, prior to and without definitive knowledge of reality?

Would-be scientific and theological realists can seek help from recent studies of reference,[5] particularly those of Saul Kripke and Hilary Putnam. Starting from studies of proper names, Kripke and Putnam have come to challenge traditional theories as to how terms like "cow" and "electricity" refer. For example, traditional theories associated with Bertrand Russell suggested that the reference of a proper name is identified by the application of a definite description. Kripke, on the other hand, argues that reference can take place independently of the possession of a definite description which somehow "qualitatively uniquely" picks out the individual in question. Instead, reference can even be successful where the identifying description associated with the name fails to be true of the individual in question. In one of his examples, a speaker who says Columbus was the man who discovered America and proved the world was round really refers to Columbus, even though Columbus did neither of these things and even if that is all the speaker "knows" about Columbus. And the reason the speaker refers, even though all his particular beliefs about Columbus are incorrect, is because the relevant linguistic competence does not involve unequivocal knowledge but rather depends on the fact that the speaker is a member of a linguistic community who has passed the name from link to link, going back to the man, Columbus, himself.[6]

Kripke's point is in part an amplification of a more modest observation about reference which another writer makes as follows:

> ...successful reference does not depend upon the truth of the description contained in the referring expression. The speaker (and perhaps also the hearer) may mistakenly believe that some person is the postman, when he is in fact the professor of linguisitics, and incorrectly, though successfully, refer to him by means of the expression 'the postman.' It is not even necessary that the speaker should believe that the description is true of the referent. He

> may be ironically employing a description he knows to be false or diplomatically accepting as correct a false description which his hearer believes to be true of the referent; and there are other possibilities.[7]

The point here is that reference depends, in normal speech, as much on context as on content and that reference is an utterance-dependent notion. This, we might note, is what makes metaphor and various other forms of figurative epithet possible; given the right context it will be perfectly clear to your auditor that by "that diamond in the rough" you are referring to your favourite politician.

By extension, and not uncontroversially, Kripke and Putnam argue that the reference of natural kind terms like "gold" and physical magnitude terms like "electricity", need not depend on definitional conventions in the form of lists of attributes, for example, "gold is a malleable, yellow metal." Rather, they argue, reference may be fixed by a kind of "dubbling" or "baptism" such as "gold is whatever this substance is" (pointing), or, "electricity is what caused this needle to jump." We can fix a reference prior to and apart from any knowledge of the essential properties of certain states and relations and yet claim that, when we use the terms, we are referring to the kinds as constituted by those esential properties, whatever those properties might be. Furthermore, if the reference of a term like "electricity" is fixed not by some set of properties but by a "dubbing" or some similar procedure, then the fact that the descripiton associated with the term may change across theories is yet compatible with continuity of reference.

In an interesting article, the philospher of science Richard Boyd develops these comments on reference to support a realist construal of the role of metaphorical theory terms in scientific theory-making.[8] In the past it has been said that metaphors simply lack the precision necessary to science. Over and against this now stands clear evidence that actual theory construction is sometimes heavily dependent on metaphorical terms. Boyd's suggestion is that the old vision of scientific precision is chimerical and, following Kripke, that the "existence of explicit definitions is not characteristic of referring expressions," nor even "a typical accompaniment to sustained epistemic access." If this is so then, he argues, we have the leeway necessary for a realist interpretation of metaphorical theory terms. Indeed, model and metaphor are ideally suited for providing flexible networks of terms which, while not necessarily directly or exhaustively descriptive (their very status as metaphors alerts us to that), can nonetheless claim to be reality depicting.

In the right circumstances, even a substantially false description may put one in the right relationship to a causally significant situation and make genuine epistemic access possible. For example, consider a phenomenon called "rose replant disease." Despite this nomenclature of "disease", I understand that no one is quite sure what it is; what is certain is that roses which are planted in soil where other roses have recently grown fail to flourish. It is not known whether this is because one rose may pass an infection to another through the soil or whether the first rose depletes the resources of the soil in some way that cannot readily be met by top-dressing with fertiliser, or something else entirely. Yet the designation "rose replant disease" successfully refers to the phenomenon, whatever its cause, and the language of "disease" provides the focus by which we may attempt to isolate the cause. It is epistemic access that is important to referring expressions, especially in the sciences. As Boyd puts it, reference is an epistemic notion.

The argument so far has tried to demonstrate how it is that terms may be judged to be reality depicting prior to definitive knowledge. By doing so the argument could vindicate the use of metaphor in theory construction and thereby strengthen the realist's case.

Before attempting to apply these arguments to religious language we should emphasise that the realist program outlined is a cautious one. The realist is not claiming that the particular account of the world which she favours is the only or even the best one. Indeed, models change, theories move on, and descriptive vocabularies accordingly come and go. This descriptive flux, far from debilitating the realist's argument, is exactly why she feels the need to make one. Some explanation must be given for the continuity of access which makes scientific investigation possible. This account goes some way to clarifying how descriptions can change while maintaining that that which is described need not. As Richard Boyd puts it, the world informs our theories, even though our theories never adequately describe the world.

Theological Realism

An important feature of this realism is its significantly social face. This is so much so that the arguments of Kripke, Putnam, and Boyd, at least in the way in which I've made use of them, might best be described as "social" (rather than "causal") accounts of reference and reality depiction. As Putnam insists, it is not words which refer but speakers using words *who* refer. The realism under discussion emphasises rather than conceals contextuality by emphasising that descriptive language, while dealing with immediate experience, will be language embedded in certain traditions of investigation and conviction. For example, the Western geneticist takes it for granted that trait inheritance is not the result of magical spells or configurations of the planets at the time of birth. Rather it is due to some biochemical mechanism which can be explained by the best model available in the tradition in which the investigation stands, that of Western medicine. The descriptive language the gencticist uses is forged in a particular tradition of investigation and a context of agreement on what constitutes evidence, what is a genuine argument. While theories may be reality depicting, they are not free from contextuality, both historical and cultural. This point, that reference is linked to particular contexts of enquiry, is one which any realist should welcome. One important implication is that if reference is an epistemic notion, as Boyd insists, then epistemology is a social enterprise. Knowledge, even scientific knowledge, is a corporate matter.[9]

What of the religious case? Any argument analogous to the ones made in the philosophy of science must involve the claim that we are causally related to God. This seems perfectly acceptable. Indeed it is a basic tenet of most theistic religions that we are so related. But how can this relationship be described? We might propose that God relates to us causally through religious experience; to take a famous example, God is that which on Monday, 23 November, 1654 from about half past ten until half past midnight, Pascal knew as, "Fire. God of Abraham, God of Isaac, God of Jacob, not of philosophers and scholars. Certainty, certainty, heartfelt, joy, peace."

Religious experiences like this one, and also of a more diffuse kind, are of considerable importance to the way in which theists claim to speak of God, a point to which we will return. We should not feel too embarrassed at considering the religious experience of individuals in attempts to ground our talk of God, for even the experiences on which scientific investigations rest are, at some descriptive level, the experiences of *some one*. Nonetheless there is a clear disanalogy here with the scientific case, for religious experiences cannot be repeated under controlled circumstances, and using them to fix a reference involves a commitment to the validity of the experience as reported by the experiencer.

We might then try a designation on which there is general agreement, namely that if it designates anything it designates God. Consider the definition by Anselm, "God is that than which nothing greater can be conceived." This comes near to what we seek, for it is a formula which does not wish to describe so much as to give a designation which, if it designates anything, designates only that which is called God. The wider proof, despite many defenders, is not generally thought to be successful, but this is of no matter here since our object is not to prove that God exists but to provide some designation which, if it designates anything, designates God. The difficulty, however, is that the abstract nature of the formula, "God is that than which nothing greater can be conceived," gives us no suggestion of a causal relation with the world.

If religious experience seems too intimate and Anselm's formula too abstract, we might try the more experiential, "God is the source and cause of all there is." This formula, fundamental to the cosmological argument, retains the kind of epistemic agnosticism we want. God is not described in terms of some set of essential properties, but pointed to as the source of the universe. Now this does not demonstrate that there is such a unified source or that, if there is, it is the God of the Christians. But this possibility of error, even of being radically mistaken about that which is, is a risk the realist takes. This amounts to being willing to admit that the Christian God might not exist, and many Christians admit freely that this is a possibility, however much their own experience leads them to think this is not the case.

Religious Experience and Authoritative Others

Now let's re-examine attempts to ground our speaking of God via religious experience. We sometimes fail to remark that the religious experiences which are significant are not simply one's own — many religious people never have dramatic religious experiences like that of Pascal. As important or even more important to the overall composition of a religion like Christianity is the experience of what one might call "authoritative others." What this means is that, if Pascal has such an experience, and if I'm inclined to trust his judgement about it, then I too can say, "God is that which appeared to Pascal on 23 November, 1654." I use Pascal's experience to ground my reference. But here note that this reliance on "authoritative others" is not unique to religion. What I refer to on the basis of my own immediate experience is a rather small set of things, while what we speak of on the basis of our relationship to others is vast (that is one of the points we can take from Kripke's "Columbus" example.). I have no immediate personal experience of Napoleon, or of the current President of the United States, or of a quasar. I speak about them in virtue of my connectedness, through language and various structures of communication, to others who do have some kind of access to these persons and entities. The astrophysicist is, for me, an authoritative other when I want to speak about quasars or black holes. In religious matters, of course, people come to be seen as authoritative for reasons other than what they say — we may know of their great devotion, their disciplined life of prayer or their concern for the loveless and poor, or we sense a kind of sanctity in what we hear, see, or read of them. Pascal, Dr. Johnson or one's great-aunt might all count for an individual as authoritative others. Ezekiel or St. Paul might count for large groups of people, for whole religious traditions, as authoritative others. But how do we get from the bare experience of individuals to the complex story and formal teachings of, say, a religion like Christianity? We can imagine a situation like this. Such a person has an experience which he or she takes to be "of God" and characteristically describes, often with struggle and hesitancy, by using metaphor. This may be a novel metaphor or one culled from the particular tradition in which the individual stands. "That which appeared to me so is God." Once they have introduced the description, we, or those who regard them as authoritative, may use it to designate "the God Who Is." This is one possible account of what Christians call "revelation."

We are not considering religious experience simply in the restricted sense of one's own personal religious experience or lack of it, but also in the broader sense in which it is also the experience of a community as seen through a particular interpretive tradition. Religion, too, makes claims based on experience — different in kind from that on which scientific judgements are based, but experience nonetheless. And, as in the scientific case, this experience is understood in a context of shared assumptions and shared models, and discussed in terms of a descriptive vocabulary which has been built up by a community over a period of years, or even, in the Judaeo-Christian case, over millenia.

Conclusions: Revelation Revisited

My suggestion is that a good deal of the language which the biblical community calls revelation develops in the way outlined above: metaphors capture someone's experience, for example, Hosea's vision of Israel's relationship to God as like that of wife to husband. Subsequent writers in the tradition then pick up the model and, in recounting their own experience, extend it, as did Ezekiel in an extreme way with the "marriage" model. In the Bible, revelation cannot be separate from tradition (whether literary or devotional or both), for it is pre-eminently within a continuous stream of reflections that models for God's activity have been developed and maintained. (Ian Ramsey's studies of the "wind/spirit" model for divine activity shows this admirably.[10]) It is the claim of the theological realist that the models and the metaphorical terminology, while clearly arising in particular cultures and contexts and modified over time, may nonetheless be reality depicting.

Now if this case for realism is convincing it has a number of implications:

First, it is perfectly respectable to use models and metaphors to speak of a God who "cannot be named."

Second, these models will inevitably be linked to particular historical and social contexts. On my argument this isn't a vice but the very foundation of a realist case: having a shared descriptive vocabulary and a tradition is one's only chance of being able to say anything at all. In theology, science, ethics or any other field of interest and endeavour, a shared and matured descriptive vocabulary gives the possibility of sustained reflection which goes beyond the necessarily limited experiences of each individual.

Third, the word of God comes to us as human word because we are human beings. This seems like a truism, but it is shocking to many students of theology. (One must remember the paradigm of revelation as God writing messages on stones.) On my account, God reveals himself freely, but by means of our shared history, assumptions and common life. This is still a cognitive theory of revelation, but one qualified by the human condition, by the nature of the recipients of divine disclosure.

Fourth, it follows that there is a profoundly corporate aspect to our understanding of God. Although God may extend grace to individuals apart from language (consider the case of the very young or the severely retarded), God *reveals* himself to our understanding in language, not because of what God is, but because of what we are, viz., linguistic beings. Since no one invents language for themselves, to be a linguistic being is to be a corporate being. Thus knowledge of God no less than any other kind of knowledge is a corporate enterprise. For human beings, knowing God depends on knowing other people. An angel might know God immediately, but, to modify Aquinas' argument, we are of the kind to reach the world of intelligence through the world of sense, through language, and with other people.

Fifth, change in the language we use, in models and descriptive vocabulary, is wholly to be expected; nonetheless the realist can still argue for continuity of access. Again we must remember that it is not words which refer, but speakers using words who refer. With natural lanaguage it is not, as the logician might prefer, that individual terms somehow "latch onto" the world but rather whole networks of words, practices and beliefs represent it. "The realist explanation, in a nutshell," says Putnam, "is not that language mirrors the world but that *speakers* mirror the world; i.e. their environment — in the sense of constructing a symbolic representation of that environment."[11] When we consider religious language, we should regard the particular models or, more accurately, sets of counterbalanced models, that a given culture or group find valuable as "housing" something important to the faith, even if no one particular formula is or even could be wholly satisfactory.[12]

And finally, it could be noted that, in defending revelation in this way, one is really reconsidering altogether the much-canvassed polarity between the natural and the revealed. In speaking of the God Who Is and the God of our Lord Jesus Christ, both natural and revealed theology make do with metaphors, approximations and uncertainties. Both must acknowledge their rootedness in groups with shared assumptions, practices and histories, and both must acknowledge that this human and corporate context is not a limitation on God's self-disclosure, but the prerequisite for it — our *praembula fidei*.

NOTES

[1] "He Who Is" is also a name with descriptive content, as feminists point out.

[2] It's important to define the key terms, "realist" and "instrumentalist", for one's own purposes, since in other writings and contexts they have different implications. For a latter-day follower of Feuerbach see Don Cupitt, *Taking Leave of God* (New York: Cromwell Publ., 1981).

[3] For just two influential theologians who have suggested that metaphysical arguments are either impossible or unhelpful for theology see Robert King, "The Task of Systematic Theology," *Christian Theology*, ed. Peter Hodgson and Robert King (Philadelphia: Fortress Press, 1985) and John MacQuarrie in the introduction to *Principles of Christian Theology* (New York: Macmillan, 1977).

[4] The exception, of course, is replica models (homeomorphic models), like model trains. But these, for obvious reasons, are the least interesting kinds of models for theory construction.

[5] This argument is worked out in much more detail in Janet Soskice, *Metaphor and Religious Language* (Oxford: Oxford University Press, 1985) ch. 6-8.

[6] Saul Kripke, "Naming and Necessity," in *Semantics of Natural Language*, ed. Donald Davidson and Gilbert Harman (Dordrecht: D. Reidel Publishing Company, 1972) 295, 301.

[7] John Lyons, *Semantics* (Cambridge: Cambridge University Press, 1977) 181-182.

[8] Richard Boyd, "Metaphor and Theory Change: What is 'Metaphor' a Metaphor for?" in *Metaphor and Thought*, ed. Andrew Ortony (Cambridge: Cambridge University Press, 1979).

[9] See for comparison John D. Barrow and Frank J. Tipler, *The Anthropic Cosmological Principle* (Oxford: Oxford University Press, 1986), 141, concerning the complexity of systems.

[10] See his *Models for Divine Activity* (Oxford: Oxford University Press, 1964).

[11] Hilary Putnam, *Meaning and the Moral Sciences* (London: Routledge and Kegan Paul, 1978) 123.

[12] This is not to say that we have a mysterious pre-linguistic content which, at intervals, we clothe in new terms. Experience and interpretation are inseparable here.

PHYSICS, PHILOSOPHY, AND MYTH

MARY B. HESSE, Department of the History and Philosophy of Science, Cambridge University

Husserl's Objectivized Science

In 1936, two years before his death, Edmund Husserl published two parts of his *Crisis of European Sciences*.[1] The work was a response to an invitation to lecture in Vienna, in a period of unprecedented moral, social, and political crisis in the affairs of Europe, which crisis Husserl traced to a misdirection of reason initiated by the origins of modern science in the 17th century. Briefly his thesis is this.[2] When the Greek ideal of unified rational theory was taken up in the Renaissance, it was still understood as a search for knowledge with, as it were, a human face — an interpretation of life as meaningful in a world pervaded by value, integrating the factual, the practical and the moral. *Scientia* had the connotations of "wisdom" rather than the limited sense that "science" now has in English (though not in other European languages). With the progressive success of natural science over the following centuries, however, knowledge became reduced to positivist facts, "objective" nature became split from the human psyche, and scientifically objectivized nature became "true nature" and lost its concern with the meaningful basis of human life.

This conception of the objectivity of science as factual is so familiar to us that it is difficult seriously to contemplate a conception of reason in which fact and value are inseparable. Husserl defends his thesis by reconstructing the historical situation in which Galileo, as his prototype "scientist", effected this transformation of human consciousness. He traces Galileo's revolution to the Greek geometry of ideal spatial shapes, combined with a new concern with empirical exactitude, encouraged by the requirements of such contemporary technologies as surveying and navigation (although Husserl does not greatly emphasize these social motivations), and made possible by the development of accurate measuring instruments. Galileo conceives nature as "in itself" mathematical, susceptible to ever-increasing quantitative approximation to its real state. Nature, like the realm of geometry, is essentially a unity, its parts being connected by a web of causal laws. Those immediate sense qualities (colour, taste, touch) which do not fall into a sequence of approximations to an ideal are, therefore, to be mathematized indirectly by being reduced to the "primary qualities" of matter-in-motion in space and time.

All this sounds like familiar history of the origins of modern science, but what is different about Husserl's approach, a difference apparently trivial and therefore difficult to grasp, is the new perspective from which it is viewed. Husserl speaks of nature being *objectivized, made objective*, by this process, indeed of being *constructed* by the new methodology of science. Part of Galileo's motivation, as Husserl sees it, is the possibility of identical knowledge being *shared* ever more accurately by all investigators, with the help of their instruments.[3] But what for Galileo was a *discovery* of the "real world" becomes for Husserl a *construction* defined by just that method that proved to be so successful. Mere "facts" are unorganized, subjective experience, until they are objectivized. By objectivized nature Husserl means that aspect that is susceptible to shared, universal, law-like, mathematical description, the function of experiment being to tell us which of the ideal possibilities are actualized. The method has the possibility of successful prediction built into it, because it is a method of successive approximation to the mathematical description. What is not successful is thrown out and replaced by another mathematical hypothesis — objectivized nature *is* just the product of this process which is unfalsified so far.

The concepts of "objectivization" and "construction" must not be misunderstood. Later social constructivists have tried to argue that scientific theory is indistinguishable from any other mythical or metaphysical story, in being imposed *a priori* upon the world rather than discovered within it.[4] This is not Husserl's conception. He expresses his admiration for the immense theoretical and technical successes of the scientific process, and so implicitly recognises that nature, as it were, responds to the method. The method might not have worked this way; that it has done so indicates *something* real that has been discovered about nature, namely that it does contain mathematizable regularities that can be exploited in intersubjective knowledge and prediction. What Husserl does not accept is that this objectivized (and as far as it goes, objective) nature is all there is in "true nature."

I shall not follow his account of what more there is in terms of his phenomenological life-world, but rather take up his challenge in a different way. Speaking of the need to understand and reflect upon the origins of mathematical science, he has some rather sharp things to say about the mathematician and the natural scientist:

> In his actual sphere of inquiry and discovery he does not know at all that everything these reflections must clarify is even in *need* of clarification, and this for the sake of that interest which is decisive for a philosophy or a science, i.e., the interest in true knowledge of the *world itself, nature itself.* And this is precisely what has been lost through a science which is given as a tradition and which has become a *techné*, insofar as this interest played a determining role at all in its primal establishment.[5]

It is this clarification, in terms not just of Galilean but of modern physical science, that should be the contribution of philosophy to this Conference.

It is sometimes held that the mathematical, mechanical world of Galileo and of classical physics in general is no longer with us, and that the split in consciousness that Husserl and others perceive in the origins of science has been overcome in a recognition of the role of the observer in relativity and quantum theory. I believe this is a gross over-simplification.

There are, on the contrary, features of the 20th-century revolution in physics that reinforce Husserl's thesis. Indeed it may be said that Galileo (or at least Husserl's Galileo) initiated a philosophy of science more consonant with modern mathematical physics than the mechanical realism that held sway in physics for most of the intervening centuries. In what follows I shall examine those characteristics that make modern physics an "objectivizing science" in Husserl's sense, and consider briefly what the consequences are for Husserl's problem of the fact/meaning dualism and for the relevance between physics and theology.

Realism and Mathematical Structure

The transition between classical and modern physics, at the epistemological level, can be considered in terms of four problems: realism, causality, reducibility and chance, and the subject/object distinction. I call these "epistemological" problems, because they concern interpretations of the nature of physics, rather than interpretations of the nature of being in the light of physics, which are ontological problems. As we shall see, a generally non-realist solution to these problems will have consequences for any ontology or metaphysics of science. Metaphysics which is derived from other sources may be suggestive for physical theory, and physics may provide useful models for metaphysics and indeed theology, but physical theory will not in itself have any logically necessary implications for metaphysics or theology.

Husserl's account of mathematized science suggests the following theses which define what I will call *structural realism*, in contrast to a stronger *substantial realism*:

(1) Physics presupposes that natural objects have properties and relations that can be specified mathematically with ever-increasing accuracy.

(2) Objects are related by causal laws in an all-embracing network which can be specified with ever-increasing accuracy and universality.

(3) These presuppositions have been notably successful in building up comparatively simple, unified theories which permit accurate application, extrapolation, and prediction, and these predictions are remarkably well confirmed by experiment. It follows that the mathematical programme for "world construction" captures such reality of the world as permits this application to be successful. This is the core of *structural realism*.

(4) The mathematical structure, however, requires interpretation to connect it with empirical observation and measurement. Interpretation ultimately has to take us down to the language of experimental reports, that is, there has to be a set of "correspondence rules," or a "dictionary," in the sense of the deductivist philosophers of science of the mid-20th century. These reports do not have to be expressed in a theory-neutral observation language — it has been abundantly shown that they are "theory-laden" — much less do they have to be in a near-solipsist

sense-data language. But they have to be in a language with a semantics, not just an internally defined network of formal symbols.

(5) The interpretation provides *substance* for the formal network of relations, that is, it provides the material relata which the network relates. The description of substance has to include enough "ordinary language" to permit recognition of the appropriate experiments to test and confirm or falsify particular aspects of theory, but substance is primarily expressed in terms of the theoretical ontology, that is, the set of objects and properties that the theory states to exist in the world. Take a simple example from Maxwell's electrical theory.[6] Pre-Maxwellian theory of electricity has an ontology of charged particles acting on each other across empty space with attractive or repulsive forces. These particles reside at the *inner* surface of conducting matter, and their motion in conductors constitutes electric current and produces heat. The revolutionary Faraday-Maxwell ontology is quite different: a "charge" is not an independent particle, but an epiphenomenon of the cutoff of a line of electric force at the *outer* surface of conducting material, which causes conducting particles (not charges) to repel or attract like- or oppositely-charged conductors. Electric current is the motion of lines of force in the energy field, which causes the transformation of electromagnetic energy into heat in conductors. The furniture of the world is described quite differently; the ontologies are contradictory and there is no "convergence" between them, and yet both at a certain stage of development accurately explained the same elementary electrical experiments, when they were suitably interpreted into experimental language.

(6) *Substantial realism* is the view that successful theoretical ontologies, or models, not only approximately describe real nature, but describe it increasingly accurately as one theory succeeds another in a sort of limiting process towards truth. But as Husserl correctly understood, and the Maxwell example illustrates, a succession of substantial models, differing from each other in essential respects, cannot have the same limiting characteristics as does a sequence of mathematical expressions (such as the successive approximations to a quantitative measure). There is, for example, the possibility of a strict limiting process between special relativity and Newton's laws in a model in which the velocity of light goes to infinity, but there is no limiting process between the concept of charge as particle and charge as a cut-off line of force. In the latter case there is only a variety of different theoretical models whose *structure* may have mathematical limiting properties, but whose *substance* does not.

I shall not argue in detail here for structural as against substantial realism,[7] except to note that the argument is at its strongest in relation to fundamental physics. It is undeniable that mathematical structures become ever more unified and universal with every advance in theory; the structural realm of physics is truly progressive. But the substantial description of what the structures relate changes radically from theory to theory, and indeed seems not to be the major focus of attention when physicists are actually doing physics, as opposed to when they are trying to convey physics to the general public.

To give point to this assertion, we may compare what physical ontology currently says about its "fundamental particles" with what the philosopher Peter Strawson says about the necessary conditions for something to be an "individual", that is a substance.[8] He defines an individual as what is identifiable, reidentifiable, and discriminable from its like. Now it would be easy to say simply that his theory is refuted by successful physical theories in which fundamental particles are not individually identifiable or reidentifiable or discriminable from their like (and in which even worse things happen, such as parts being made up of wholes of which they are the parts). But this would be to miss the significance of Strawson's theory for physics. Its significance is that fundamental particles with these odd properties are *not* the metaphysical or logical individuals of physical theory. What then are these individuals? It is not easy to say, except in the logical sense that every formal or semi-formal theory must at least in principle be formalizable in a logic of some order containing signs for individuals and their properties and relations. The individuals of a field theory may be (necessarily ephemeral) space-time points, the individuals of quantum theory may be state vectors in Hilbert space. What are taken as logical individuals in this sense will vary from theory to theory, and even with different formalizations of a single theory. (Maxwell's elementary theory could be formalized alternatively with particle individuals or space-time individuals.)

These logical individuals are not at all the same as the individuals defined by Strawson as being necessary for the description of experience, that is, for the existence of a shared descriptive language. The individuals in a theoretical ontology do, of course, have properties and causal consequences that enable them to be connected to the Strawsonian objects of everyday experience, so that experimental reports can confirm or falsify the theory. But theoretical individuals in themselves have none of the comfortable concreteness of a substantial answer to Thales traditional question "What is there?" — water, air, fire, earth, atoms, fluids, forces, fields, space-time manifolds...? There is indeed *no* answer to this question derivable from science, but only the ever-changing sequence of theoretical models which will only reach a terminus if science itself comes to an accidental end in history. It follows that no truths about the substance of nature which are relevant to metaphysics or theology can be logically derived from physics.[9] It remains to consider whether the same conclusion can be drawn from the structure as opposed to the substance of physical theory.

Types of Causality

Newton's ontology was not so much an ontology of substances (he vacillated between atoms and ether fields), but rather of *verae causae*, true causes, which he identified with physical forces.[10] With the aid of universal simplifying assumptions (which he called the analogy of nature) he argued that these could be *deduced* from the mathematical expression of matter and its motions. The sole aim of science is to derive all forces in the same rigorous way that the universal law of gravitation was derived from low-level mechanical and planetary laws. Forces then become the surrogate for metaphysical causes: for Newton, their necessity is no longer *a priori* or even synthetic *a priori* as with Kant; rather it is the necessity of empirical grounding plus the analogy of nature and deduction. This necessity is a logical, not a metaphysical, conception: necessity lies in deduction, not in the premises. Newton's God is not under any metaphysical constraints in creation; the physical forces are what he chooses them to be, and the aim of natural philosophy is to discover what he has chosen.

This is why Hume's demolition of the metaphysical concept of causal necessity went practically unnoticed among natural philosophers; what he demolished had not been required in physics for a century. But with a foresight uncommon among empiricists, Hume did notice that some constraints have to be put upon what we habitually come to regard as causes of particular effects. He says they are closely connected with their effects in space and time. In other words, there is no action at a distance across space or time. But Locke had made a similar assumption which he was later to withdraw:

> ... I am since convinced by the judicious Mr. Newton's incomparable book, that it is too bold a presumption to limit God's power, in this point, by my narrow conceptions. The gravitation of matter towards matter by ways inconceivable to me, is not only a demonstration that God can, if he pleases, put into bodies, powers and ways of operation, above what can be derived from our idea of body, or can be explained by what we know of matter, but also an unquestionable and everywhere visible instance, that he has done so.[11]

In the less empiricist climate of the Continent of Europe there was a tendency to ground the contingency of forces in unifying principles which seemed to have metaphysical support: Maupertius' Principle of Least Action (Nature does not do more when she can do less), and the 19th century principles of equality of cause and effect (nothing comes out of nothing), which issued in the energy principles of Mayer and Helmholtz. In Britain these principles were given a more empirical grounding: Joule and Kelvin derive the conservation of energy and the thermodynamic principles as general results not contradicted by any kind of experiment. The Lagrangian and Hamiltonian expressions of mechanics in purely mathematical form became *structural* definitions of the mechanical philosophy, rather than expressions of metaphysical necessity.

The Lagrangian formulation of mechanics can be seen as the precursor and paradigm of all those subsequent mathematical principles which unify the causal structure of physics without determining the nature of the substance of which it is the structure. Causal relations are mathematical relations of contingent concomitance: they are not necessary metaphysical ties. On the other hand they have to be constrained by certain principles to prevent a theoretical explosion of possibilities of connection of anything with anything else that would become quite unmanageable. Such constraints were built into the so-called local causality of late classical and relativistic field theory: cause-effect relations are continuous in space and time, and effects cannot be transmitted faster than the speed of light. At first it was also presupposed that the "same total cause must produce the same effect," but the development of statistical physics and the subsequent denial of this axiom in quantum theory showed that the constraint could be abandoned without rendering causal relations arbitrary. Subsequently the prohibition on action at a distance has also been, in a sense, rejected in non-local interpretations of the Einstein-Podolsky-Rosen thought experiment. This is an apparently mysterious situation in which we seem forced to a holistic view of instantaneous connections across space and time, but one which is quite consistent with the causal nexus of the rest of quantum theory and its experimental grounding. The details of this fascinating episode are not directly relevant here, except as an illustration of the way in which the physical imagination (the "construction of model worlds") can outrun all preconceived metaphysical principles and constraints, while still conforming to the essential aims of science, which are concerned with unified structure and successful prediction and not with the discovery of "true nature" or "true types of cause."[12]

In determining the nature of causality and the types of causal laws, therefore, physics is sovereign in its own domain and not logically beholden to any extra-scientific constraints, although these may always have an heuristic function at different times and places in the history of physics. Conversely, we may ask whether any influence goes the other way, from the character of the laws and causes postulated in physics to significant consequences in religion. Again, there have been contingent heuristic influences in the history of theology, as when 17th-century mechanism and determinism seemed to permit only an absentee Deist God, or when 19th-century biology suggested an evolving creative force. But such influences have been temporary, and highly dependent on contemporary fashions in science, which have themselves undergone radical changes. The consequences for theology are perhaps best expressed as a liberation from constraints upon our knowledge of creation, and hence of our knowledge of God's creative action. It is indeed significant for our models of God that we now believe creation not to be purely mechanical but rather to be a dynamically evolving system. Such beliefs have in any case always entered implicitly into the Biblically-based religions (Judaism, Christianity, Islam), quite independently of physics, and have been maintained by believers without inconsistency in spite of physics, even in the period of the Enlightenment.[13] It would be a mistake now, as it was then, to build the details of such models of causality too firmly into our doctrine of God. They may provide useful analogies for apologetics and a useful liberation from too constrained a notion of God, but they are not essential to central theological beliefs, nor can they logically disprove such beliefs.[14]

Reduction, Explanation, and Chance

More relevant than the nature of causality for our purposes is the introduction into physics of chance as fundamental and irreducible. It is not only the apparent indeterminism of causality in quantum theory that introduces chance, because chance is a pervasive ingredient even in classical theory.[15] In classical statistical mechanics, a complete molecular state description may in principle determine the future course of events uniquely, but the initial conditions are still chance conditions, however far back in the causal chains we go. And there are cogent arguments such as Popper's for the practical unpredictability of all classical systems, even when described by deterministic laws.

The intrusion of chance has seemed to imply the ultimate failure of physics as the basic universal theory to which all other sciences and their subject-matters are in principle reducible. Husserl complained that the psyche and its experiences are necessarily outside objectivized physics, but it has seemed that a successful physics can even overcome this objection by showing that there is nothing in the mind (meanings, intentions motives, beliefs....) that is not in the software of a sufficiently complex physical system (perhaps an artificial intelligence plus a sensitive robot body). I believe that *this* reductive argument regarding mind and body is correct, but, however that may be, it is undercut by an ultimate failure of physical explanation itself. If at some point what happens in a single-case sequence of events has to be ascribed to chance, there is a breakdown, not of physics itself, but of Husserl's constructivist programme for physics in which everything is to be reduced to ever more simple unified laws. I now want to consider three responses to this apparent breakdown from within physics, namely the "many worlds" and "relative state" hypotheses in quantum physics and cosmology, the anthropic principles, and the aspiration, often voiced by cosmologists, for a single universal law of force which will explain the whole of nature.

"Many Worlds" Hypotheses in Quantum Physics

The original source of these hypotheses was the so-called "collapse of the wave-packet" in quantum theory. According to standard quantum theory, Schrödinger's equation describes an entirely deterministic evolution of the wave-function Ψ throughout space-time, but the interpretation of Ψ is wholly probabilistic, that is, $|\Psi|^2$ at any point measures only a probability density at that point. When a particle is detected by means of a measurement or equivalent interaction, the whole energy of the particle, which was previously spread over the wave-front, becomes concentrated at the point of interaction. This particular "collapse" is an undetermined process, subject to probabilistic laws but not further explicable by anything in quantum theory. The appearance of a particle at a particular point is an ultimate chance event, individually uncaused and inexplicable. "Chance" here is *not* the expression of our ignorance of the microscopic determining conditions, as in the case of a coin-toss. It is an ontological indeterminism, since there *are no* law-like and complete determining conditions that would be consistent with quantum theory. A great deal of effort has gone into attempts to supplement quantum theory to avoid this conclusion, but with no success; if anything like current quantum theory is to fit the experimental phenomena, then no set of determining local "hidden variables" can be made consistent with such a theory.[16]

In 1957 Everett proposed a "many worlds" hypothesis in the following radical sense: Schrödinger's equation *does* describe a totally deterministic universe, but one in which a quantum-mechanical multiplicity (perhaps an infinity) of causal sequences "split off" during a measurement or interaction. These will be different from each other in that together they realize all the possibilities permitted by the initial wave function and quantum theory, in the proportions determined by the probability distribution at the interaction point. So, if a particle is observed at A and not at B, although it had a 50% probability of being at A or B, there are now *two* particles, indeed two non-interacting "worlds", differing only in the properties "particle at A" or "particle at B". We can only observe one such world, because we are ourselves physical detecting instruments who are continually splitting into different worlds depending on which way our neurons jump, and "we now" necessarily exist only in one of them, and cannot have memories, or be conscious, of the others. According to Everett's hypothesis every world is as "real" as every other, even though there is obviously a very fast multiplication of separate worlds every time a quantum jump takes place.

In interpreting Everett's hypothesis some writers speak about worlds "splitting" as if there is then a multiplication of "realities". Since we can *ex hypothesi* have no direct evidence of "other realities" this may be objected to as a violation of Occam's razor. But it should be noted that Everett speaks of a "relative state," not a "many worlds" hypothesis. The difference is important. By relative state he means that his theory proposes just one reality, that derived from Schrödinger's equation and its consequences. He shows that its consequences include the possibility of observer-systems with memories as sub-systems of the whole, and that these sub-systems are in a well-defined state (a "perception" of the world), only *relative* to the rest of the composite system of which they are parts. There is no transition from "possible" to "actual" states at a wave-packet collapse: *all* resulting states are actual. But observers, who are "inside" the composite system, are necessarily conscious of only one of its many actual histories. When observers "communicate", the relative world state they each perceive it necessarily the same for all of them. "Many worlds" is in fact a misnomer for one universe containing many different observer (and other) histories; it is better expressed as a "branching" of the universe into many relative states.[17]

The theory is undoubtedly richer than the phenomena accessible to any one set of communicating observers, but it cannot be faulted on that ground alone, since all theories are richer than the phenomena derivable from them. All experiments that support standard quantum theory are derivable from this theory, but the question remains as to whether additional evidence independent of the standard interpretation can also be found to support it. It has been suggested that we are "virtually forced" to the hypothesis, because experimental results indicate that two or more worlds may "come together" again after being separated for some interval of time, as in the two-slit experiment.[18] This argument, however, seems to use the assumptions of the hypothesis to prove itself, because it is open to the one-world theorist to interpret such experiments differently, as a re-establishment of deterministically evolving, but different, wave-fronts after a collapse. Recently Deutsch appears to have overcome this objection in an ingenious argument appealing to the "introspected" states of a consciousness (or sufficiently advanced artificial intelligence) from which he claims a conclusive crucial experiment is in principle possible.[19]

The relative state hypothesis does, however, have the advantage that explanation need not come to an ultimate stop at points where chance is invoked. No explanation is needed as to why a particular chance event occurs and not another, because at least all those with significant probability values will occur. Moreover, the theory appears to be a more unified explanation than standard quantum theory, since there is no need for a separate explanation of why "splitting" takes place at all: this is logically entailed by the Schrödinger formalism and the fact that sub-systems acting as "observers" exist.[20] Observers are essentially intrinsic to the composite system, and partially create their own perceived world by observation and communication within it. Whether, in the absence of crucial experiments, these explanatory advantages provide a strong enough motive for a theory going so far from the phenomena is perhaps a matter of taste rather than of empirical methodology.

My general argument against substantial realism suggests that the relative state theory may not be a permanent part of physics, and that there is no conclusive reason at present to accept it as a serious ontology. There are rival interpretations within physics itself, though perhaps not at present in quantum cosmology. From a theological point of view, taste would surely count against taking it as ontological truth. It is a wholly determinist theory, and like all such theories it makes nonsense of most religious conceptions of freedom and moral responsibility.[21] There is no point in holding oneself responsible for one's actions and one's history if there are other "selves" in every other possible history who have gone through every physically possible chain of events, trivial and significant, good and evil. The conception of a unique meaningful history with serious implications for the relations of God and humankind becomes empty.

Anthropic Principles

The second type of argument for many worlds involves so-called anthropic principles. These arise from the fact that there are many numerical relations and coincidences between the physical constants, the probability of which is exceedingly small, given all currently known physical possibilities and a plausible indifference among detailed values. Moreover, many of these constants turn out to have values, within very tight limits, which are just right for the occurrence of the biological basis of life and hence consciousness. The world appears to have been fine-tuned for the existence of a rational species capable of observing and theorizing about it. Without introducing any concept of design, it is at least uncontroversial to conclude that, if the world had been ever so slightly different, life and consciousness as we know it could not have existed (that is, roughly, what Barrow and Tipler[22] call the Weak Anthropic Principle). If design is left on one side, is this coincidence due to chance or is there a physical explanation yet to be discovered? Introduction of some type of cosmological many-worlds hypothesis provides a middle way between these possibilities. This type of hypothesis assumes that there are many "worlds" in the total real universe, realizing between them all possibilities of the apparently accidental constants and boundary conditions of nature. It is then necessarily true (as a matter of logic, not of metaphysical necessity) that *we* are in a world that has the right values of the natural constants to support life.

The grounds for this hypothesis are weaker than those for the relative state hypothesis, since we have no reason as yet to suppose in this case that a further physical explanation of the constants of this world is impossible. So we are in no sense "forced" to make the hypothesis. It is a more gratuitous hypothesis than that in quantum theory, in that it is difficult to see how evidence can be adduced that would distinguish between the real existence of other worlds inaccessible to us now, and the merely chance existence of this one. But the hypothesis is perhaps less objectionable than that in quantum theory on moral or religious grounds, because it need not postulate splitting histories of each observer (unless of course it is combined in some way with the quantum hypothesis), but only many other worlds with no observers at all.

Another point should be noted about the relation between the quantum many-worlds hypothesis and a particular sort of anthropic principle. I remarked that the existence of "wave-packet collapse" needs no separate explanation in the relative state theory, once the existence of "observer" sub-systems is given. But the theory does not logically *require* that there should be such systems. So there is a sort of anthropic assumption (De Witt's Postulate of Complexity) built even into relative state theory as an unexplained initial condition, and this somewhat reduces the claim of the theory to be a wholly unified one such as is contemplated in the next type of explanation to be considered.

Single Law Explanations

Many writers on the anthropic principles have regarded them as "metaphysical" and have worried that they may be philosophically curious or undesirable.[23,24] It is doutful, however, whether "philosophers" as such have any more to say about their acceptability than do cosmologists. When anthropic principles go beyond the "weak" form, it is a matter of taste whether such explanations can be regarded as properly scientific. We have presented above several examples from the history of physics where taste for styles of ontology and causality have changed under pressure of scientific and other developments. Anthropic principles are not, however, a traditional type of explanation, and where ordinary explanation seems to come to a stop, traditionalists will seek further unified theories to fill the gap, without resorting to many worlds, anthropic principles, or design. The history of physics is replete with stories of the ultimate success of such attempts, which often take the form of incorporation of recalcitrant facts or theories into a wider and more economical synthesis with other successful theories.[25] Is this an indefinitely continuing process, or can we conceive it coming to a natural stop with the discovery of a single all-embracing simple law?

The aspiration towards a single law is expressed in many cosmological writings. Paul Davies, for example, claims that: "For the first time in history we have a rational scientific theory of all existence," and that "all natural phenomena can now be encompassed within a single descriptive scheme."[26] Hawking, more modestly, contemplates a "complete, consistent and unified theory of the physical interactions which would describe all possible observations,"[27] but notes that this would not give detailed predictions of any but the simplest situations (because we cannot solve the equations). He does not, in any case, consider that his unified theory eliminates the need for some anthropic principles.

All this seems to be far removed from Husserl's plea to return from the hubris of such scientific claims to particular, meaningful human life and history. But in another sense it confirms precisely his account of modern science as objectivizing construction. What sense can be made out of the concept of a single explanatory principle encompassing "all phenomena"? As a regulative principle it marks out a constructive programme for science: continue the reductive universalizing sequence of theories that have been largely successful in the past, and incorporate as many particular phenomena as possible within the conceptual schemes thus generated. But as an aspiration towards a final, unique, single theory, the concept seems to be logically flawed in a number of respects.

First, there is the logical point that explanation has to be understood in this context as deduction from universal theoretical premises, plus interpretation into particular experimental results via correspondence rules. No other construal of explanation would make the "single law" aspiration at all meaningful. But it is a logical requirement that from a universal premise no conclusions about particulars can be derived unless there is a particular sentence among the premises. What then can be meant by Hawking's suggestion that the particular initial conditions for the universe may somehow be incorporated non-arbitrarily into the unified theory? There seem to be four possible meanings:

(1) It may be argued that the notion of initial or boundary conditions for the universe does not make sense, because the universe is by definition not one particular system among a number of similar systems discriminated by their boundary conditions.[28] The universe is all that there is. This cogent-sounding argument fails, however, as an objection to the

usual theoretical procedure in dealing with the universe in cosmology. It is usually assumed that there is a set of logically universal physical laws defining physically *possible* worlds, of which the actual universe is one, discriminated from the rest by purely contingent boundary conditions. The population of "universes" is, therefore, well-defined, although they are not all actual. But it is precisely the contingency of the particular features of the actual universe that single-law cosmologists want to explain.

(2) They may seek to do so by showing that the boundary conditions and numerical constants of this universe are unique, or at least overwhelmingly probable relative to other possibilities. Progress may indeed be made in showing that some conditions and constants are derivable from others, but there will still be some fundamental constants yet to be explained. In principle it is possible that those which permit life and consciousness could be shown to be overwhelmingly probable,[29] but in practice it is likely that they will not be derivable without some kind of anthropic principle. In this case the hoped-for unified theory would not supercede the need for anthropic arguments after all.

(3) Another possibility is that the outcome of the single theory is indifferent to boundary conditions, that is, whatever the boundary conditions are, the universe will converge in time to a unique state which is our actual universe. Some progress does seem to have been made in this direction,[30] but again it is very implausible to suppose that such progress can be extrapolated so as the allow *no* alternative physical possibilities or chance events in the detailed history of the entire universe.

(4) Finally, perhaps there are no boundary conditions at all. Hartle and Hawking have developed a theory of "creation" (a veritable *creatio ex nihilo*) in which the concept of "time" is defined only internally and becomes progressively less defined as we approach the vanishingly small universe "at the beginning." Thus, boundary conditions do not arise, because there is no point of time at which the universe "begins." This is the theory discussed in detail in Christopher Isham's paper in this volume.[31]

The second general objection to the claimed reduction to unified theory concerns the necessary particularity and complexity of the consequences of theory. The world is undoubtedly made up of a great number of things of great variety. For the explanation of everything there must in a sense be a conservation of complexity, in other words a trade-off between the simplicity and unity of the theory, and the multiplicity of interpretations of the few general theoretical concepts into many particular objects, properties, and relations. "Mass" is a universal property of Newtonian matter, but to identify various kinds of matter as apples, water molecules, earths, comets, planets, suns, and galaxies requires as many differentiating properties as there are particular discriminable things. Cosmologists and fundamental physicists often refer to these "mere" particulars with dismissive disdain.[32] Nowhere is it clearer that they have a Husserlian constructivist programme: what matters in physics is what can be put into unified, universal, simple models, *not* the many particularities of the world that we in fact inhabit.

Objectivization, Myth, and Value

We have seen how the existence of "observers", that is all sub-systems that register particular eigen states, including conscious perceiving subjects, plays an essential role in the relative state hypothesis and in anthropic arguments. From the beginning of modern physics commentators have been tempted to say that the "human" has been readmitted into the scientific world and Cartesian dualism healed. But what kind of "human" is this that has been readmitted? The "human" subject is only an objectivized observer, who might be replaced by a complex of instruments or a sufficiently intelligent robot. The idea that subject/object duality has been overcome is dependent on the radical reduction which is part of the objectivization process, and no results within physics can show that this reduction is necessary. Nothing that has been said in physics touches Husserl's problem of human values or of meaningful interpretations of the world. So to close, I shall briefly ask whether objectivized science leaves room for any such questions outside itself, and what the status of objectivized science may be within a wider perspective.

Science tells a particular kind of story about the world, issuing in progressive discovery of regularities in nature and hence in the possibility of technical control. The story is constrained by principles that are partly arbitrary, adopted to make possible a shared game of constructing unified universal theories. Part of the game is to eliminate meaning and value from nature, retaining them, if at all, in the aesthetic properties of the theories themselves. No substantial consequences about the world can be drawn from this game except what were put into it, and these were announced from the first to be largely devoid of human meaning. Even if "God" is allowed to emerge from the design-like arguments of the anthropic principle, this god will be simply defined to do what is required: a Deist god who is allowed to choose the right constants at the creation. Nothing follows that is like the God of Abraham, Isaac and Jacob and our Lord Jesus Christ, nor of any of the other traditional religions.

Of course, it is possible that the objectivizing programme of science has, in fact, encompassed the whole of reality, and that the meaning and value that were initially excluded from science are in fact illusory. It may be so, and no argument can refute such a thesis. But nothing in science entails it either, and it will remain open to religious believers to agree with Husserl that the objectivizing programme is a particular historical phenomenon chosen by human beings, and indeed defines its own mathematical values to the exclusion of more important values that are constitutive of human life and social ethos. To reify or ontologize scientific theories is to turn science into myth, and as myth it has no claims to our allegiance different in kind from those of any other myth.

I am not, of course, using "myth" here in the popular sense of "imaginary creation story, such as Genesis, which is false when compared with the results of science." I am using it rather in the sense defined by Northrop Frye in his book *The Great Code*: "Certain stories seem to have a particular significance: they are the stories that tell a society what is important for it to know, whether about its gods, its history, its laws, or its class structure."[33] This is the sense in which social anthropologists and theologians study myths as sacred constituents of a social culture. Scientific "myths" of the creation and destiny of the cosmos clearly have some but not all of the same social functions — that is why they have captured so much of the interest of the popular media and the educational establishment. They are stories with particular significance, as validated by a powerful social institution, namely science; they tell us what a scientific materialist culture thinks it is important for us to know, the latest version of the cosmological "facts"; they tell us about our "gods", that is some kind of sanitized, mathematical, impersonal super-force, or even perhaps the great "external observer" who creates by collapsing all the wave-packets. But they tell us nothing about our history, our laws, our class structure, our social ethos, except implicitly that these things are mere chance perturbations, imposed on the classic beauty of a mathematically structured reality. Even at the level of socially useful myth, this one cannot compete with the traditional religions. Moreover, Husserl is right in concluding that scientific myths are largely a construction of our own making. The stars are, indeed, patient of some of their experimental aspects, but as reified theories they lie in ourselves and they make us moral underlings.[34]

NOTES

[1] Edmund Husserl, *The Crisis of European Sciences and Transcendental Phenomenology* (Evanston: Northwestern University Press, 1970).

[2] Ibid., Parts I and II. See also Jürgen Habermas, *Knowledge and Human Interests*, trans. Jeremy J. Shapiro (London: Heineman, 1972), appendix.

[3] Bacon and Descartes also hold this "democratic" view. But Galileo has, in addition, a lively sense of the way instrumental knowledge reconstructs the common-sense world. For instance, he considered that the Copernican theory of the earth's motion committed violence upon the senses.

[4] Harry M. Collins, *Changing Order: Replication and Induction in Scientific Practice* (London: Sage, 1985).

[5] Husserl, *op. cit.*, 57.

[6] Mary Hesse, *The Structure of Scientific Inference* (London: Macmillan, 1974) chapter 11.

[7] Several contributors in the conference discussions made the point that it is difficult to distinguish the "structure" of a theory sharply from its "substance", and that normal scientific methodology does allow us to infer from observable to unobservable structures, entities, and causes. Both points are quite correct, but the point I want to emphasize is just that structures do not entail substantial conclusions without interpretation, and that interpretations are always ambiguous in their detailed ontologies, and come and go as contending adversaries in the history of science. For more detailed arguments contributing to the current debate of realist versus anti-realist see Mary B. Hesse, *Revolutions and Reconstructions in the Philosophy of Science* (Brighton: Harvester, 1980) chapter 6; and Mary B. Hesse, "Science Beyond Realism and Relativism," in the proceedings of the workshop on *Relativism and Social Science* (Utrecht: University of Utrecht) in press.

[8] Peter Strawson, *Individuals, An Essay in Descriptive Metaphyics* (London: Methuen, 1959) chapter 1.

[9] Attempts to restore realism in terms of hidden variables in quantum theory are concerned with realism in a different sense. See Isham's paper in this volume.

[10] Hesse, 1974, 201 ff.

[11] John Locke, "Second Reply to the Bishop of Worcester," in *Works of John Locke* (London), I, 754.

[12] Current developments in fundamental physics can be described as progressive weakening of extra-empirical constraints, and consequent strengthening of the operational basis. See Bryce S. DeWitt, "Quantum Theory of Gravity I: The Canonical Theory," *Physical Review* **160** (1967) 1113-1148; and Bryce S. DeWitt and Neill Graham, *The Many Worlds Interpretation of Quantum Mechanics* (Princeton: Princeton University Press, 1973), where the quantum theory of gravity is described as "extraordinarily economical" and where the many world hypothesis is said to be "a return to naive realism and the old-fashioned idea that there can be a direct correspondence between formalism and reality."

[13] Even the "dynamic" flavor of modern science may not be irreversible. For example, some accounts of "creation" in cosmological theory have no interpretation of the universe "evolving in time", and hence, if they were to become dominant in physics, static rather than dynamic concepts of the whole creation would result. See Isham in this volume.

[14] I am grateful to John Leslie for helping me to clarify my general thesis about the relation of particular physical theory with religious belief. Some

extrapolations from modern physical theory may undoubtedly constitute threats to central religious beliefs, as we shall see in the example of "relative state" theories. But theologians would be wise to treat such extrapolations with informed but healthy scepticism.

[15] See Alfred Lande, *From Dualism to Unity in Quantum Physics* (Cambridge: Cambridge University Press, 1960) 5; Karl R. Popper, "Indeterminism in Quantum Physics and in Classical Physics," *British Journal for the Philosophy of Science* **1** (1950) 117-133 and 173-195; Ilya Prigogine and I. Stengers, *Order Out of Chaos* (London: Heineman, 1984).

[16] Bernard D'Espagnat, "Use of inequalities...," *Physical Review* **D11** (1975) 1424-1435.

[17] DeWitt and Graham, *op. cit.*, 161.

[18] Brandon Carter, "Large number coincidences and the anthropic principle in cosmology," in *Confrontation of Cosmological Theories with Observational Data*, ed. M.S. Longair (Dordrecht: Reidel) 291-298. DeWitt, 1141, says that "it is possible that Everett's view is not only natural but essential" for quantum gravity. John A. Wheeler, "Assessment of Everett's 'relative state' formulation of quantum theory," *Reviews of Modern Physics* **29** (1957) 463-465, concludes that: "No escape seems possible from this relative state formulation if one wants to have a complete mathematical model for the quantum mechanics that is internal to an isolated system."

[19] David Deutsch, "Quantum theory as a universal physical theory," *International Journal of Theoretical Physics* **24** (1985) 1; and "Three connections between Everett's interpretation and experiment," in *Quantum Concepts in Space and Time*, ed. Roger Penrose and Christopher J. Isham (Oxford: Clarendon, 1986) 215.

[20] Wheeler 1957, *op. cit.*, 463 remarks that the actual decomposition of the state function is not explained in Everett's theory. J.D. Barrow and F.J. Tipler, *The Anthropic Cosmological Principle* (Oxford: Clarendon, 1986) 473, note that it is not obvious *a priori* that the system/apparatus division can be made, but that only that this makes "measurement" meaningful; that is, the universe has to be "sufficiently inhomogeneous" to permit the division. The possibility of system/apparatus distinctions is what DeWitt and Graham, 168, call the postulate of complexity.

[21] This would be disputed by those who put forward so-called "soft determinism." According to this view, physical determinism does not undermine concepts of freedom, guilt, praise, blame, etc., because these are thought of as social reinforcements of negative and positive kinds, having social functions independent of the underlying physical mechanism. I am assuming, however, that such social determinism makes physical determinism doubly unacceptable to most religious believers. The issue of freedom and determinism is one of the few places where physical theory may appear to threaten religious belief, and it is also one of the most intractable problems of philosophy. I have discussed it in Michael A. Arbib and Mary B. Hesse, *The Construction of Reality* (New York and Cambridge: Cambridge University Press, 1986) chapter 5.

[22] Barrow and Tipler, *op. cit.*, 16.

[23] Bernard J. Carr and Martin J. Rees, "The anthropic principle and the structure of the physical world," *Nature* **278** (1979) 605-612.

[24] Carter, *op. cit.*

[25] There are of course historical cases where long-sought reduction fails; for example, the classical programme of reduction to mechanical matter-in-motion, which collapsed with classical electromagnetic field theory.

[26] Paul C.W. Davies, *Superforce* (London: Heinemann, 1984) 5-6.

[27] Stephen W. Hawking, "Is the end in sight for theoretical physics?" in *Beyond the Black Hole*, ed. J. Boslough (London: Collins, 1985).

[28] D. H. Mellor, "God and Probability," *Religious Studies* 5 (1969) 223-234.

[29] Compare the monkey and the typewriter example given by Leslie in this volume.

[30] See Barrow and Tipler, *op. cit.*, 430, 502, on "inflationary universes".

[31] Isham in this volume.

[32] Barrow and Tipler, *op. cit.*, 414, quote Robertson: "Homogeneity is a cosmic undergarment and the frills and furbelows required to express individuality can be readily tacked onto this basic undergarment."

[33] Northrop Frye, *The Great Code* (London: Routledge, Keegan Paul, 1982) 32. I have developed this theme in "Cosmology as Myth," in *Cosmology and Theology, Concilium,* 166, eds. D. Tracy and N. Lash (New York: The Seabury Press, 1983) 49-54. See also Arbib and Hesse, chapter 11.

[34] I am grateful to Jeremy Butterfield for discussion on the subject-matter of this paper, and to all the participants at the Specola Vaticana conference. Particular thanks are due to Michael Heller, John Leslie, Ernan McMullin, John Polkinghorne and Frank Tipler for their patient comments and corrections. All the errors that remain are my own.

OBSERVATION, REVELATION, AND THE POSTERITY OF NOAH

NICHOLAS LASH, Faculty of Divinity, University of Cambridge

Religio Laici

Coined in France a hundred years before, the terms "theism" and "deism" moved into English only at the end of the seventeenth century, remaining interchangeable until well into the eighteenth. As indicating what John Dryden called "the principles of natural worship," [1] these terms stood doubly opposed to "atheism" on the one hand and, on the other, to what was becoming known as "revealed religion."

According to the Oxford English Dictionary, it is in 1682, five years before the publication of Newton's *Principia*, that "deism" makes its first appearance, in the preface to Dryden's *Religio Laici.* Both poem and preface delineate, for political purposes, certain abuses of human reason. At a time of constitutional crisis Dryden, the devout Tory and (at this date) still loyal adherent of the established Church, takes issue with certain forms of dogmatism and sectarian rationalism which seem to him "to threaten the values of human society and to menace the stability of the state." [2]

One of his targets, then, is "deism." The error of the deist is "the belief that nothing of unique value is embedded in tradition or history, that it is possible to wipe the slate clean (as Descartes did) and start all over again, and by the pure exercise of reason to discover 'all ye know, and all ye need to know'." [3]

This is not an historical paper, partly because I lack the competence to produce such a paper on the seventeenth century, and partly because the proposal I wish to offer is systematic rather than directly historical in character. Nevertheless, for reasons which I hope will eventually become clear, this little text of Dryden's may serve (especially if we keep in mind the date of its production) as an engaging parable.[4]

Dryden is not at all disposed to deny the authenticity of deist religion, the reality of communion with God according to "the principles of natural worship." [5] He is, however, convinced that *all* relationship with God, whatever its content or apparent structure, is in response to God's prevenient revealing grace, and not the outcome of unaided human ingenuity.

The historical character and hence the particularity of Jewish and Christian revelation is, of course, an embarrassment to the view that wherever and in whatever form we come into relationship with God, we do so in response to his revealing grace. Dryden has a marvellous conceit for disposing of the difficulty. Noah did, after all, have *three* sons. What seem to be the principles of natural worship, elaborated by unaided reason, are, in fact, "only the faint remnants of dying flames of revealed religion in the posterity of Noah."[6]

With the aid of this device, he is able to correct the rationalism of those "modern philosophers" who have

> too much exalted the faculties of our souls, when they have maintained that by their force mankind has been able to find out that there is one supreme agent or intellectual being which we call God; that praise and prayer are his due worship; and the rest of those deducements, which I am confident are the remote effects of revelation, and unattainable by our discourse.[7]

Or, as he puts it in the poem, addressing the "Deist":

> These truths are not the product of thy mind,
> But dropped from heaven, and of a nobler kind.
> Revealed religion first informed thy sight,
> And reason saw not, till faith sprung the light.
> Hence all thy natural worship takes the source:
> 'Tis revelation what thou think'st discourse.'[8]

Where our knowledge of God is concerned, are we constructors, explorers, or pupils? Dryden in 1682 was clearly concerned to exclude the first two options and keep the third alive. As the following century unfolded, however, the effort and energy and self-assurance required in order to awaken from dogmatic slumbers, finding and fashioning new worlds of knowledge and artefact and social order, rendered intolerable all acknowledgement of pupilage. But if a good part of the eighteenth century remained confident that we were not only explorers but successful explorers, discoverers of God, it seemed increasingly evident to the nineteenth (from Feuerbach to Freud) that we were constructors of all our gods.[9] We cannot now go *back* to Dryden and the dawn of the Englightenment. And yet there may perhaps be appropriate *post*-modern ways of saying. "Tis revelation what thou think'st discourse."

Interaction?

We are invited to consider aspects of "interaction" between physics, philosophy, and theology. Interaction, it seems to me, suggests (as, perhaps, does "dialogue") something approaching parity of reciprocal influence. But is this how scientists regard the relationships between physics and philosophy? Was not Karl Rahner correct in saying, twenty years ago, that the sciences today (and he had in mind the whole range of *Wissenschaften*) "take their decision about their understanding of existence before philosophy is able to have its say. At most it is accepted as reflection on the pluralism of these sciences and their methods"?[10] This being the way things are (and have been, by and large, since Hegel's owl first flew), it is hardly surprising that there are not many learned journals or international conferences devoted to the "dialogue" between philosophy and science.

Why should it be different with theology? Do scientists really expect to have to modify their practices in the light of what they learn from theologians? Not in my experience. And yet, there is much talk of "dialogue" between theology and science.

An agenda paper for our meeting cast the "central issues" which we were to consider in the form of a question: "What are the implications of contemporary physics and cosmology for philosophy (especially metaphysics) and theology?" Notice that we are not asked what the implications of theology are for physics. I make no complaint about this. I merely note, once again, the discrepancy between fact and description where the relationships between science and theology are concerned. The description (whether in terms of "dialogue" or "interaction") suggests a reciprocity or mutuality of influence which the facts belie. My first question, therefore, is: why should this be so?

Theologians have perhaps not sufficiently reflected on the fact that the factors which brought about early modern distinctions of "philosophy" from "science" also helped to generate a quite new sense of what *revelation* might mean. (One way to watch this shift occurring would be to study, with the tools of literary criticism, changing uses of such well-worn metaphors of knowledge and its sources as "light" or the "two books" of scripture and the works of God.)[11]

Consider what happens when "observation" is made the paradigm of learning, and accuracy of representation (rather than, for example, soundness of judgement) becomes the standard of knowledge. Knowledge of nature is arrived at by looking carefully at the world. And knowledge of God? This may come either by imagining what might lie "behind" the world and accounting for its configuration or, according to some people, by the careful study of data which, while constituting items on the list of things that we know, nevertheless do not simply form part of the world in which we come to know them. (And it is, of course, this last proviso which proves increasingly untenable during the eighteenth century.)

Newton, it has been said, "revered [the two books] as separate expressions of the same divine meaning."[12] There have, I think, been few more fateful moments in the history of Christian thought than the early modern subsumption of the whole grammar of revelation and religious belief into that "spectatorial" model of the processes of knowledge which so came to dominate the Western imagination. Before, there had been many ways of reading many texts, but few controls on interpretative ingenuity. Now, there are but two books, and both of them picture-books at that.

My suggestion, then, is that the discrepancy or inconsistency which I mentioned earlier is attributable to the enduring influence, on the imagination of scientists and theologians alike, of seventeenth-century epistemological patterns or structuring metaphors. If there is but *one* way — namely, through disciplined observation — by which we can come to know *anything* (*sive deus, sive natura*, as it were), then we seem stuck with a tale of two sources of truth, two districts in which truths may be "observed". But all such dualisms eventually crumble before the practical acknowledgement of the comprehensiveness of the territory of scientific investigation. Such glimpses as we may have of the unseen, it then seems clear, can only come through observation of the visible described by science. Thus science becomes, in fact, not "partner in dialogue" to theology, but mediator of the latter's truth. On this account, "reductionism" and "scientism" are symptoms rather than disease. The fundamental requirement is to come to grips with the legacy, still tenaciously exercising its influence in both theology *and* science, of early modern "spectatorial" conceptions of human understanding.

"God," said Hegel, "does not offer himself for observation."[13] An eminently sensible remark, but one the significance of which is likely to be missed or misunderstood by someone who supposes that all our knowledge is, directly or indirectly, literally or metaphorically, a matter of observation: that we are simply *spectators* upon our world — and God.

But if not spectators upon our world, then what? Products of that world, undoubtedly, participants in its processes, victims and agents. And one fundamental feature of our agency is the restless quest for freedom and coherence, a quest the centrality of which is not negated by our propensity for producing hideously illiberal and incoherently oppressive caricatures of order and of liberty.

In an essay from which I quoted earlier, Rahner remarked: "In the future, man will not objectify himself in his art and philosophy as the rational and theorising being .... he will appear as practical man in the work of his hands, which changes him in a way he cannot clearly express."[14] And again: "In the future theology's key partner-in-dialogue ....will no longer be philosophy in the traditional sense at all, but the 'unphilosophical' pluralistic sciences and the kind of understanding which they promote either directly or indirectly."[15]

We are a long way here from that vision of individual self-transparency and self-possession which was the ideal of Cartesian individualism. Rahner is suggesting that it was under the spell of the "primacy of pure reason" that Neo-scholastic theology took metaphysics as its dancing-parter. Now, in contrast, it is in irreducible diversity of image and narrative, experiment, labour and technique, and not in any single, overarching description or theory of the world, that such self-understanding as we are capable of finds primary expression. This may seem a very fitful and fragmentary, confusing and dangerous thing to offer in place of that grand order, that perceived simplicity independent of ourselves, which once we thought we had discerned. Still it may at least open up fresh possibilities of fruitful interaction between scientific practice and the labour of Christian discipleship.[16]

On Joining the Conversation

It is nevertheless possible that such fresh forms of interaction will receive rather more philosophical assistance than Rahner expected. Writing in 1967, he could hardly foresee the remarkable extent to which an impressively diverse range of disciplines and traditions of discourse would be brought into a common philosophical conversation (in at least some sense of "philosophical") under the banner of "hermeneutics".[17]

At this point, however, the theologian wishes briefly to register a complaint. In both Europe and the United States theologians are taking an active and often well-informed part in discussion on general hermeneutics. Moreover, some of the major philosophical contributors to these debates (Gadamer and Ricoeur, for example) are sensitive to theological considerations and conversant with the literature. In the English-speaking world, however, it seems largely to be agreed that theology has about as much to contribute as does astrology to what Michael Oakeshott called "the conversation of mankind."[18] Newman, I think, was right when he said that "it is not reason that is against us, but imagination."[19] If anything, however, this merely increases the difficulty of the theologian's task. Be that as it may, what might the theologian have to contribute to what we may call the post-empiricist conversation? On this vast topic I now offer one or two suggestions.

Words and Stories

When Dryden contrasted "discourse" (in the Johnsonian sense of *inference*: the "course" or movement of the mind from premise to conclusion) with "revelation", he was contrasting achievement with gift. And a high estimation of individual achievement becomes the very nerve-centre of the struggle against obscurantism: "have courage to use your own understanding" was Kant's "watchword" for enlightenment.[20] The bourgeois individual, like Prometheus, is a self-made man.

When, however, that contrast was drawn in the way in which it was then drawn, what disappeared from view was language. Or perhaps, to be more exact, we should say, not language but (in a sense now very different from Dryden's) discourse: *parole*, the spoken word; utterance as public, fleshly fact occurring for particular purposes in particular places and times.

In the world of the spectatorial empiricist there are only ("objective") things and ("subjective") thoughts, and endless anxiety about their mediation.[21] The task of language is simply to render private thoughts public and thus depict things thought about or seen. If, in search of a slogan for the turn to hermeneutics in recent decades, I were to speak of "the recovery of conversation," few eyebrows would rise. But if, instead, I spoke of the recovery of the *verbum incarnatum*, I would probably meet with some suspicion. And yet Gadamer himself, in a section of *Truth and Method* entitled "Language and Verbum," says of "the christian idea of incarnation" that it "prevented the forgetfulness of language in Western thought from being complete."[22] It is therefore ironic that it should now have fallen to the philosophers to awaken theologians from forgetfulness of the doctrine that God *is* utterance, *verbum,* Word.

Because we discover this in attending to a Word once spoken in the past, the recovery of this doctrine has, of course, its dangers. A reawakened reverence for given words, inherited meaning, traditioned truth, may (if seen as simply antithetical to enlightenment rationalism) all too easily be used to serve reactionary and most irrational purposes. However, the corrective for this tendency may be sought within that single threefold rule of speech and action which, I shall later suggest, is the Christian doctrine of God: sought, for example, in the insistence that what God "breathes" in his self-utterance is freedom, scope, freshness, inexhaustible possibility; in a word, "spirit". But this is to anticipate.

One area in which the post-Heideggerian insistence on the *eventness* of utterance, the temporality of truth, puts dauntingly difficult questions to both science and theology is that of the status of our narratives. At first sight this may seem surprising, because hermeneutics at least allows us once again to take stories seriously after a long period in which one ideal of positivism had been (we could say) the suppression of story-telling, and especially of autobiography (self-involving narrative), as modes of knowledge.

And yet theologians engaged in the growth industry of "narrative theology" ignore, at their peril, developments which reflect philosophically that declining confidence in the possibility of large-scale, purposive, "plot-linear" narrative unity which has been one of the hallmarks of the story of the novel for nearly a hundred years.

Our world is, in a phrase of Frank Kermode's, "hopelessly plural,"[23] disconnected, disoriented, fragmentary. We work (as Gadamer would say) within "horizons". And though horizons may be expanded, we fool ourselves if we suppose them ever to extend very far.

Cosmologists and theologians, however, not only tell stories but have the impudence to tell stories of the *world.* And even if the cosmologists would claim that their stories are, of set purpose, plotless, it seems to me that both groups could reflect with profit on the problem, not simply of what is meant by claiming that some particular story of the world is *true*, but rather of what *kind* of story a "story of the world" might be. Who could tell it, what would it be announcing, and how would it be told?[24]

Learning and Listening

When scientists go about their work (as Arthur Peacocke has assured us that "the great majority" of them do and will continue to do, whatever the philosophers, sociologists and theologians may say) in a spirit of what he describes as "sceptical and qualified realism,"[25] then they do so as explorers of the world. And, in this world, they represent the only known tribe of agents and utterers, takers of initiative.

Scientists, of course, only exercise their agency effectively, are only fruitful in discovery, in the measure that they are disciplined to that passionate disinterestedness, that energetic stillness of attention, which is the hallmark of objectivity. Nevertheless, the kind of attentiveness, of listening, of contemplativity which is in question here, seems to be qualified by the fact of our sole agency. To put it very simply: there is a difference between listening to a waterfall and listening to another person, and in the natural scientist's world there are only waterfalls.

Human persons, of course, are things, like waterfalls, and we properly treat them as such when we count them, dissect them, and so on. To say that they are not *only* things is at least to say that we deem it improper to treat them only as things. And according to what is probably the most widespread account of the difference between treating things as things (which scientists do) and treating some things as persons, it would seem to follow that scientific attentiveness is quite unlike prayerfulness, when prayerfulness is construed as attentiveness to a personal God. Nor is it only science which is thus deemed properly to be unprayerful, but also theology, insofar as God's existence, attributes, and relations with the world are treated as if "God" were the name of a kind of natural object, a thing beyond the world, to be found, picked up and considered with conceptual tweezers.

But, whether or not natural objects are known in the way in which spectatorial empiricism supposes all objects of knowledge to be known (namely, by constructing mental representations of them), it is certain that whatever is thus known could not be God. God is not a thing, an object over against us, silently lurking in the metaphysical undergrowth, passively awaiting the services of human exploration. (I make no objection to tackling, with utmost rigour and precision, questions concerning the logic and grammar of sentences which contain the word "God". I am simply protesting against the fatuous illusion that we could discover or come across God as a fact about the world.)

This is, in part, the burden of Hegel's remark that God does not offer himself for observation. God, according to Hegel, can only be known *as he is*. That is to say: God can only be known in that eternally still movement of utterance and love which he *is*; known *in* that movement, not by constructing representations of it, whether these be pictorial, narrative or metaphysical (which is not to discount the pedagogic usefulness of such devices). God is known by participating in that movement which he is. And it is this participation which constitutes the reality, the life and history, of everything that is.

My purpose in so absurdly attempting a one paragraph summary of Hegel's philosophy of religion is to ask: what notion of attentiveness is suggested by such an account? The sort of metaphors that come to mind, perhaps, are "being in tune with," or "being on the same wavelength as." And could not such metaphors serve to indicate the character of fruitful attentiveness both to things *and* to persons? And, if so, would we not have begun to erode the sharp contrast between prayerfulness (which, in maturity, requires endless discipline and disinterestedness) and scientific practice? To explore this suggestion would, I think, be to engage in the *kind* of interaction between science and theology which Rahner had in mind. To the Enlightenment, I remarked at the beginning, "pupilage" was no longer an acceptable metaphor for our relation to the world. We had, it was thought, "come of age." But it is, in fact, the hallmark of the adolescent to suppose there to be no further need for teaching. To be an adult is to have discovered, often at great cost, the depth and permanence of the need to set ourselves at school.

Forgetful of language, inattentive to the endless diversity of linguistic practice, we set up sharp dichotomies of fact and thought, experience and idea. The religious counterpart of scientific positivism's "brute fact" was the myth of what we might call "brute revelation" (Fundamentalism is not, as is sometimes supposed, an anachronistically surviving precursor of modern rationalism, but a byproduct of it.) Such science and such religion both work with "a model of truth as something ultimately separable in our minds from the dialectical process of its historical reflection and appropriation."[26] Under the influence of this model we tend to be impatient with "ambivalence, polysemy, paradox. And this is at heart an impatience with learning, and with learning about our learning."[27]

Williams was here reflecting on the notion of revelation, to which, he said, we have recourse in order "to give some ground for the sense in our religious and theological language that the initiative does not ultimately lie with us; before we speak, we are addressed or called."[28] Although it is in theology that this metaphor receives its most sustained elaboration, it is by no means only in religious language that the sense in question is discernible. According to Paul Ricoeur, it is this same sense which makes poetic texts bearers of what he calls "testimony" or "witness", the appropriate response to which requires a certain docility or pupil-stance. But why, Ricoeur asks, "is it so difficult for us to conceive of a dependence without heteronomy? Is it not because we too often and too quickly think of a will that submits and not enough of an imagination that opens itself?"[29]

Are we constructors, explorers, or pupils of the world? I have been trying to give some indication of the extent to which, in a wide variety of disciplines or cultural practices, new possibilities of "pupilage" are opening up in the common conversation of mankind, possibilities that exist, as it were, *on the other side* of the antitheses of modernity. My question to the scientists, therefore, is this: is it of the very nature of research and experiment in the physical sciences that they should seek to stand outside such developments? Or is it possible to imagine the scientists murmuring to each other, without detriment to the rationality and autonomy of their procedures, "Tis revelation what thou think'st discourse"?[30]

Protocols Against Idolatry

How might the Christian doctrine of God be so recast or reread as not only to meet the requirements of the "hermeneutical turn" in philosophy and social theory but also thereby to become, in fact, more faithful to the mainstream of the tradition than modern "theism" could ever hope to be? On this vast subject, I have a proposal to make. Although it is only a proposal, I find it quite difficult to state because of the range of implications and ramifications.[31]

In the world of spectatorial empiricism, God is usually thought of as "a" being, an object or thing standing over against us. The primary task of doctrine or theology is then the construction of conceptual representations of this thing which seek to be, so far as they go, accurate. (And as to how far they go, debate, of course, is endless!)

But suppose we begin, not with whatever may be abstractly considered, thought about, gazed at, but with what we do and say: with human practice. Kant's questions (what can I know? what ought I to do? what may I hope for?) are met, tackled, dealt with, in one form or another, by all normal members of all human societies. And the manner of their handling is often a matter of the patterning of thought and action in story and system, etiology and ethics, constitution, art and enquiry. Such patterns regulate speech and action not (or at least, not necessarily) in the sense of dominating them but, rather, in the sense of providing the ground-rules, the framework, of keeping the show on the road.

Where Christianity is concerned, we have such a pattern in that aspect of public pedagogy which is known as "doctrine". I am now taking this term to refer, not to each and every aspect of that vast diversity of practices — academic and pastoral, liturgical and catechetical — which all, in one way or another, count as "theological" but, much more narrowly and restrictively, to the communal declaration and use of what are acknowledged to be a people's identity-sustaining rules of discourse and behaviour. In a word, the creed.

My first suggestion, then, is that the primary function of Christian doctrine is regulative rather than descriptive. As regulative, its purpose is to protect correct reference: to help us set our hearts on God (and not on some thing which we mistake for God) and make true mention of him.[32]

We require some such pattern for our pedagogy, because we are under continual pressure — from the combined forces of what Martin Buber called "individualism" and "collectivism," and from our own fearfulness and egotism — to seek some grasp on God, to get a "fix" on God, by mistakenly identifying some feature of the world (some tradition, some possession, some dream, or project, or structure, or insight, or ideal) with divinity, with godness, with the "nature" of God. But, as sensible men and women have always known, the nature of God does not lie within our grasp.

My second suggestion, then (or, perhaps better, the second step in my single proposal) is that the Christian doctrine of God, declared in the threefold structure of the single creed, protects the reference to God of Christian action and speech by simultaneously serving as a set of what I have come to call "protocols against idolatry."

The creed performs this single twofold service (the technical correlates of which, in theological grammar, are three "hypostases" and one "nature") by indicating, at each point, where God is truly to be found and then, at each point, by denying that what we find there is simply to be identified with God. Such doctrine leads, at every turn, to simultaneous affirmation and denial; it enables us to make true mention of God and, by denying that the forms of our address (our confession of God as "gift", as "*verbum*", and as "Father", for example) furnish us with some hold upon the "nature" of God, it sustains our recognition of the absolute otherness or non-identity of the world and God.

If all this seems puzzling or somewhat unfamiliar, this is probably because, in the modern world, the tendency has been for the doctrine of God's Trinity either to be misread as the provision of further information supplementary to that contained within theism's description of the nature of God, or simply to be ignored.

Thus, for example, when Walter Kasper makes the striking claim that the history of modern German thought is, at one level, "a history of the many attempts made to reconstruct the doctrine of the Trinity," he at once acknowledges that "the credit for having kept alive the idea of the Trinity belongs less to theology than to philosophy."[33] As Rahner lamented in 1960: "One might almost dare to affirm that if the doctrine of the Trinity were to be erased as false, most religious literature could be preserved almost unchanged throughout the process."[34]

In other words, while the theologians changed the subject, and turned to arguing amongst themselves as to whether the God of modern theism was discoverable by reason or only apprehensible by faith, inferrable from the world or only visible in the light of revelation, the Christian doctrine of God (never, of course, formally denied) did not stay simply dormant, but was active in strange ways and unexpected places, shaping the dialectics of Fichte and Hegel, Feuerbach and Marx.

The next step is to take Kasper's story one stage further by noticing where Gadamer and the "hermeneutical turn" come in. There is, I think, a tendency for much nineteenth and early twentieth-century thought to oscillate between varieties of "realism" and "idealism", absolutism and relativism, "objectivism" and "subjectivism", and so on. The list of labels can be extended but their referents rarely lack resonances of what, theologically, would be known as pantheism and agnosticism. And, of course, if the temper of pantheism is closest to that of the third article of the creed, the doctrine of God's indwelling, life-giving, pervading Spirit, agnosticism develops singlemindedly, undialectically, the insistence of the first article that God is unoriginate, utterly beyond all schemes and patterns of fact and explanation: that it is *ex nihilo* that God creates.

What, then, has been missing since the Enlightenment decision to excise tradition, given meaning, from the calculus of human knowing? Gadamer has already told us: *verbum*, language-as-deed, the territory of the doctrine of God's self-utterance in the world, the subject of the second article of the Christian creed.

Where the interactions between theology and philosophy are concerned, therefore, my proposal amounts to little more than the suggestion that one major school or current in recent philosophy and social theory, a current often thought to be dangerously subversive of theological discourse is, perhaps, only lethal to that "theism" which Newton's world invented and may, in other respects, be just what Christian theology requires to help bring it back into the conversation and put it back in touch with its own proper subject-matter.

What of the interactions between theology and science? This seems to me a much more obscure question which urgently requires a great deal of attention and hard work. The reason for the obscurity (I suspect) lies in the fact that the dialogue between theology and natural science seems so far to have gone most smoothly when both theologians and scientists operate as more or less sophisticated spectatorial empiricists!

I have no conclusion, but only two final thoughts, the first of which follows from my earlier remarks on disciplined attentiveness. My impression is that, in much of the literature, concepts such as "scepticism" and "agnosticism" are too easily used as blunt instruments by people apparently insensitive to the indispensability of modesty, restraint, "unknowingness," reverence for all good conversation — whether simply amongst ourselves, or in consideration of natural objects, or in contemplating the things of God.

Secondly, it may be the residual influence on religious thought of empiricist "exploration", but I sometimes have the impression that some people suppose that, the further we go in our discovery of "grand unified theory" or of what went on in those initial micro-seconds, the nearer we come to the knowledge of God. Here my suspicious nose detects new seeds of gnosticism, for if we read the first article of the creed in the light of the second (or, which comes to the same thing, if we take seriously the Prologue to the Fourth Gospel) — if, in other words, we keep in mind the *singleness* of the Word which God is and utters in his stillness — then we shall be brought to acknowledge, as an implication of the Christian doctrine of God, that we are as close to the heart of the sense of creation in considering and responding to an act of human kindness as in attending to the fundamental physical structures and initial conditions of the world.

NOTES

[1] John Dryden, "*Religio Laici,* Or, A Layman's Faith," in *John Dryden*, ed. Keith Walker (Oxford: Oxford University Press, 1987) 219-239.

[2] Edward N. Hooker, "Dryden and the Atoms of Epicurus," in *Dryden. A Collection of Critical Essays,* ed. Bernard N. Schilling (Englewood Cliffs: Prentice Hall, 1963) 125-135. Reprinted from *English Literary History,* **24** (1957) 177.

[3] Hooker, *Ibid.*, 130.

[4] See Michael J. Buckley, S.J., *At the Origins of Modern Atheism* (New Haven and London: Yale University Press, 1987), in which can be found the historical warrants for several of the claims which I tentatively propose.

[5] Dryden, *op. cit.*, 220.

[6] *Ibid.* Notice the echo of the same story in the last paragraph of Newton's *Opticks*, quoted by Michael J. Buckley, S.J. in this volume.

[7] *Ibid.*, 220-221.

[8] *Ibid.*, 229.

[9] Edward Craig, *The Mind of God and the Works of Man* (Oxford: Clarendon, 1987) 3 and 282, argues, in a way which complements Buckley (see Note 4), that two *Weltbilder* "cover between them a large portion of the philosophy written since the time of Descartes." In the seventeenth century we were explorers, our minds made in the image of God. Today we are constructors or, as he puts it, agents in the void.

[10] Karl Rahner, "Philosophy and Philosophising in Theology," in *Theological Investigations*, Vol. 9, trans. Graham Harrison (London: Darton, Longman and Todd, 1972) 46-63.

[11] On the former the opening lines of *Religio Laici* are most interesting. Richard Rorty, *Philosophy and the Mirror of Nature* (Oxford: Basil Blackwell, 1980), 12-13, is, I believe, mistaken in taking for granted that what he calls "ocular metaphors" of cognition were as dominant before the seventeenth century as they were after it. On the latter Arthur R. Peacocke, *Creation and the World of Science* (Oxford: Clarendon Press, 1979) 3, by beginning his discussion of the uses of this metaphor with Bacon, seems to miss the novelty of seventeenth century practice. On fifteenth century uses see Buckley, *op. cit.*, 69.

[12] Frank E. Manuel, *The Religion of Isaac Newton* (Oxford: Clarendon Press, 1974) 49.

[13] G. W. F. Hegel, *Lectures on the Philosophy of Religion. I. Introduction and the Concept of Religion*, ed. Peter C. Hodgson (Berkeley: The University of California Press, 1984) 258.

[14] Rahner, *op. cit.*, 57.

[15] *Ibid.*, 60.

[16] These comments may set a question-mark against Chris Isham's suggestion that "the meeting ground of physics and theology is philosophy." The implications, however, would seem more disturbing to our theory than our practice: at our sessions at Castel Gandolfo, for example, it did not seem in fact to be primarily (and certainly not simply) upon the ground of philosophy that such meeting as there was between physics and theology occurred. I would tentatively suggest that "interpreter" might be a better metaphor than "meeting place" for the role of philosophy in the dialogue between theology and science.

[17] A conversation which, as Mary Hesse and others have insisted, must include the physical sciences, if only because "the fact that a natural science

requires the existence of a linguistic community of communication as an *a priori* for its own existence cannot be grasped scientifically but must be understood hermeneutically". See Kurt Mueller-Vollmer, *The Hermeneutic Reader. Texts of the German Tradition from the Enlightenment to the Present* (New York: Continuum, 1985) 44; and Karl-Otto Apel, *Towards a Transformation of Philosophy* (London: Routledge and Kegan Paul, 1980). See also the discussion of Hesse's essay, "In Defense of Objectivity," in Richard J. Bernstein, *Beyond Objectivism and Relativism* (Oxford: Basil Blackwell, 1983) 32-34.

[18] Rorty, *op. cit.*, 389.

[19] John Henry Newman, *The Letters and Diaries of John Henry Newman*, Vol. 30, eds. C. S. Dessain and Thomas Gornall (Oxford: Clarendon Press, 1976) 159. The year was 1882, and Newman was discussing the "apparent opposition" between theology and science.

[20] Immanuel Kant, "An Answer to the Question: What is Enlightenment?" in *Kant's Political Writings*, ed. Hans Reiss, trans. H. B. Nisbet (Cambridge: Cambridge University Press, 1970) 54-60.

[21] Bernstein, *op. cit.*, 16-20, gives a masterly account of what he calls "the Cartesian Anxiety".

[22] Hans-Georg Gadamer, *Truth and Method*, trans. Garrett Barden and John Cumming (London: Sheed and Ward, 1975) 387, stresses the importance of the fact that *verbum*, as act and event, cannot be made wholly to coincide with any Greek philosophical account of *logos*: "In developing the idea of the *verbum*, scholastic thought goes beyond the idea that the formation of concepts is simply the reflection of the order of things." The most thorough exploration of Aquinas's contribution to this development is Bernard J. F. Lonergan, *Verbum, Word and Idea in Aquinas*, ed. David B. Burrell (London: Darton, Longman and Todd, 1968).

[23] Frank Kermode, *The Genesis of Secrecy* (Cambridge: Harvard University Press, 1979) 145.

[24] Nicholas Lash, *Theology on the Way to Emmaus* (London: SCM Press, 1986) 62-74, offers some brief reflections on these questions, which, I believe, have something in common with those raised by Isham in this volume.

[25] Peacocke, *op. cit.*, 22. It might be more accurate to describe this spirit as one of skeptical and qualified empiricism.

[26] Rowan Williams, "Trinity and Revelation," *Modern Theology* **2** (1980) 197-212.

[27] *Ibid.*, 198.

[28] *Ibid.*

[29] Paul Ricoeur, "Toward a Hermeneutic of the Idea of Revelation," in *Essays on Biblical Interpretation*, ed. Lewis S. Mudge (London: SPCK, 1981) 117.

[30] Dryden, *op. cit.*, 229. Pedersen's reflections in this volume on the "given" as distint from the merely "found" are particularly pertinent here.

[31] Nicholas Lash, *Easter in Ordinary Reflections on Human Experience and Knowledge of God* (Charlottesville: University Press of Virginia, 1988), provides a fuller version of this.

[32] See Soskice's remarks on "reference" in this volume.

[33] Walter Kasper, *The God of Jesus Christ*, trans. Matthew J. O'Connell (London: SCM Press, 1984) 264.

[34] Karl Rahner, "Remarks on the Dogmatic Treatise *De Trinitate*," in *Theological Investigations*, Vol. 4, trans. Kevin Smyth (London: Darton, Longman and Todd, 1966) 77-102.

III.

CONTEMPORARY PHYSICS AND COSMOLOGY IN PHILOSOPHICAL AND THEOLOGICAL PERSPECTIVE

CONTEMPORARY COSMOLOGY AND ITS IMPLICATIONS FOR THE SCIENCE – RELIGION DIALOGUE

William R. Stoeger, S.J., Vatican Observatory.

1. *Cosmology and Its Present Directions*

Contemporary cosmology embraces a variety of scientific disciplines and addresses a wide range of unresolved fundamental issues concerning the character, structure, origin and even the destiny of the universe as we know it. It is rarely clear in cosmological research and discussion just where strictly scientific analysis ends and philosophical, or metaphysical, reflection begins. It may be even less clear what sort of well-founded conclusions about physical reality cosmology as a science is capable of achieving, and what sort are outside its proper scope — what its actual limits are as a discipline. But many philosophers and theologians will agree that, when properly described and understood, the conclusions cosmology reaches — and perhaps even more so cosmology itself, as a discipline with its own emerging questions, methods, procedures, analyses — have subtle but very important implications for theology and philosophy. Some of these implications are of a positive tenor, specifying a transfer of certain ideas, conclusions or methodological approaches from one to the other. Others, instead, are negative, forbidding such a transfer or dependence of ideas and approaches, which in the popular mind was uncritically thought to be there.

In this paper I wish to describe in broad strokes *some* of the principal features of cosmology, its assumptions and conclusions, which I believe provide important material for such critical interdisciplinary study relative to philosophy and theology. Using these, I shall discuss some key issues which determine the limits of cosmology as a discipline and its relationships with philosophy and with theology.

Throughout this paper I often use the word *philosophy*. In so doing I do not restrict it to metaphysics but insist that it always includes metaphysics as an essential element. I further presuppose a critical realistic position, which I do not attempt to justify here.

1.1 - *Cosmology's Principal Object*

Cosmology focuses very specifically on the observable universe as a single object — upon the origin (or as far back towards the origin as we can get), evolution, and structure — the physics — of observable physical reality as a single object of inquiry, with its own intelligibility. As scientists, cosmologists construct and test physical-mathematical models of this universe, attempting to incorporate in them all the essential large-scale and pervasive features of physical reality in an integral and consistent way. For example, the abundances of the elements are explained quite adequately by stellar evolution, with essential primordial contributions to deuterium, helium and lithium from cosmological nucleo-synthesis about 3 minutes after the Big Bang. The hierarchical structuring of matter into stars, clusters of stars, galaxies, clusters of galaxies, and superclusters is understood as originating from hierarchies of density perturbations which begin to grow after matter decoupled from the radiation field at a redshift of about 1000 to 1500. Before that time, as we know from the near isotropy and homogeneity of the microwave background radiation (MBR) and its blackbody spectrum, the expanding universe was essentially homogeneous — not lumpy at all, as it is today, but very smooth. We know, too, that the observable universe is continuing to expand. We know its present Hubble parameter within a factor of 2 (but not enough yet about how it may vary with direction), and we have some limits on the deceleration of that expansion, and on the present average density of matter in it. Surprisingly enough, however, we are not sure what its dominant components are. About 90% of it appears to be dark or hidden (nonluminous) matter, and we have no certainty at all about the nature of that invisible percentage. Some of it could be baryonic, but not all of it.[1]

As we move back towards the Big Bang — towards higher and higher temperatures and densities, especially as we consider temperatures and densities higher than those at which baryons melt into a quark sea, corresponding in the usual models to a time much less than a second after the initial singularity — the issues cosmology confronts become even more fundamental and basic, one might almost say philosophical. In order to solve the problem of causal connectedness within the universe, for example, as well as several other deep-seated enigmas, most now postulate a very, very early period of exponential (inflationary) expansion, followed by reheating [2,3].

Here, too, arises the whole area of initial underlying symmetries which are broken as these temperatures and energies decline during the intial moments of expansion and which provide the essential frameworks for unifying the four fundamental interactions (see below). How are these four forces related at the very beginning? How do they become differentiated, and when? What are the different families of fundamental particles at different energies which interact through — or mediate — these interactions? Why just these particles, and not others — with these properties and not others? Why the baryon-antibaryon antisymmetry? How does the continuum of space-time arise from quantum unified field theory processes? What determines the arrow of time? Is entropy involved in this definition? These are some of the important questions cosmologists and field theorists grapple with as they move to consider the very early phases of the observable universe.

And then there are a series of related questions, which are even bolder and which many physicists believe will eventually be answered within physics and cosmology themselves. How do the fundamental constants come to possess the values they have? What is the origin of the physical laws, which govern all these fundamental processes? By what mechanism are these realized in nature, while other possible symmetries and physical laws and structures are excluded? Can the initial conditions of the universe be given or determined *within* a cosmological model or theory? Or will they instead always be outside any adequate scientific explanation or account?

These two sets of questions give just a sampling of the large-scale and pervasive fundamental features of physical reality which contemporary cosmology investigates. As we can see, there are some peculiar and unsettling aspects to a number of them which distinguish them from those of other areas of physics and the natural sciences. We shall discuss some of these below in section 2.

At this point it is important to ask what the warrants are for treating the observable universe as a single object. Can we justify so doing on scientific grounds? Does the observable universe have intelligibility as a single object of study?

One such warrant for pursuing cosmology as a science is simply that we *can* construct meaningful models of the observable universe as a whole. To some extent these are testable, in principle at least, and account for a large number of its general characteristics and for the various phases of its history. From a more philosophical point of view, we tend to assume that the sum total of what we observe and will observe (what is observable) must have a common intelligible origin, or at least belong to, or constitute, a common intelligible whole, and that it must have a common intelligible destiny — that all that makes it up is interrelated prior to our observing it.

Several specific features of the observable universe strongly support our consideration of it as a single object: (1) The cosmic microwave background, particularly its near isotropy and homogeneity, assures us that all we presently see *was* causally interrelated prior to a redshift of 1000-1500 — so causally interrelated that all the antecedents of presently observed objects had the same temperature and density at that epoch. Only much later were there individually distinguishable objects and lumpiness; (2) The second such feature is the rough homogeneity of texture throughout the universe at the present time, along with the fact that the same physical laws and the same values of the physical constants seem to hold everywhere, so far as we can tell; (3) And finally, we observe systematic redshifting of distant galaxies, which indicates that all we observe is undergoing systematic large-scale expansion, that the whole of space-time is undergoing expansion. Together, these three features strongly force us to consider the universe as a single object.[4]

1.2 - *The Evolving Universe*

Another general characteristic of cosmology is that it describes the universe as *evolving*. This is clear from what we have already discussed in Sec. 1.1 above. Evolution is certainly obvious on small and intermediate astronomical scales — stars, clusters of stars, galaxies, and clusters of galaxies. But the microwave background and the primordial element abundances, among other indicators, show us that there was a succession of hotter denser phases going back at least to a temperature of 10^{12} K, and that there was a time when the universe as a whole was homogeneous on relatively small scales. As expansion and cooling proceeded, complexification on many different levels occurred — to give us the hierarchical structuring we observe today.

1.3 - *The Origin and the End Point of the Universe*

From this evident evolution, we quickly come to consider that the universe must have both a physical origin or starting point, and an end point or future, with certain definite, though at present uncertain, characteristics. Depending upon the amount of matter it contains, it will either expand forever — eventually evanescing in the process — or it will recollapse.

Let's reflect briefly on the origin or starting point of the observable universe. It is difficult to specify exactly what "origin" should mean in this context, for a number of philosophical and scientific reasons. We cannot specify an absolutely initial state in physical terms, susceptible to scientific analysis and critique. In fact, generally speaking, as is well known, there is no *a priori* reason why there needs to be a starting point — a universe which has always existed cannot be ruled out on either strictly scientific or philosophical grounds.[5,6] We can, however, specify an antecedent state to each known state the universe has occupied — and work our way back towards what we might call the beginning, the limit towards which such a series of states tends, if it exists. In the equations which we usually use to model the universe, this succession of states is reflected in the time parameter t, and $t = 0$ is in many models a singularity (at which certain parameters, like density or temperature, go infinite). We refer to it as "the initial singularity," or the Big Bang — and in some sense we can call it "the beginning" of the universe. In fact, under very general conditions, the Penrose-Hawking singularity theorems[7] assure us that such a singularity will occur in practically all of our acceptable cosmological models. However, because it is a singularity, it is not clear that it represents, or can represent, what *actually* occurred — the model of which it is a part probably breaks down before the singularity in reached. This is also obvious from the point of view of quantum field theory. Before the singularity is reached, the effects of quantum gravity will dominate and determine the space-time structure, and for technical reasons we are not at a point at which such quantum effects can be properly included in our theories. Furthermore, even if the singularity did represent what actually occurs, it is not an origin or a beginning in an absolute sense — but only within the context of the model — and any possible state prior to the singularity is inaccessible to us. Long before reaching it, in fact, we gradually lose our ability to make observations or perform experiments which directly test these extreme epochs. We run into a variety of serious limitations to our verification of these models in practice, if not in principle. People speak of the "accelerator barrier," for example, in this context.[3]

1.4 - *Cosmology and Unification*

Connected with our approach to the initial singularity, or Big Bang, is the central question of unification, the attempt to unify the four fundamental interactions (electromagnetism, gravitation, the weak and the strong nuclear forces) in such a way that all of physical reality as we know it is comprehended in a single detailed model. We do not yet have an acceptable superunification model, but remarkable progress has been made. There is the Weinberg-Salam model, which unifies the electromagnetic and weak interactions. And then there are the grand unification theories (GUTS), supergravity theories and superstring theories, which show definite promise of being developed eventually into acceptable models.[8,9] A key component in all these attempts is symmetry. A very large symmetry group provides the basis of unification at very high energies. Then the differentiation of forces and particles occurs as this symmetry is broken down into appropriate smaller symmetry groups as the temperature (or energy) falls to lower values.

Considerations such as these lead to an even more fundamental investigations concerning the origins of physical laws and the determination of the values of the fundamental constants. In some cases, these issues can be resolved within the framework of the unified field theories; in other cases, it may be that an even more fundamental investigation will be needed. What are the limits of a scientific explanation of these origins? Will it ever be possible to have a "theory of everything"? Some researchers at present believe so.[10,11] Such a theory would have to contain within itself, presumably, not only the explanation for all physical processes, particles, and the laws governing their interactions and transformations, but also the explanation for itself. It would also have to account in a non-arbitrary manner for boundary conditions, and especially for initial conditions, which in almost all other theories must be specified separately. Is it ever possible for a model to account for itself and for its concretization in reality? I would maintain that in science — certainly in physics and in cosmology — there is no such current case, nor is there likely to be in the future. There is nothing in a physical or cosmological model which *a priori* necessitates its realization in the concrete world. We adopt it, instead, on the basis of its correspondence to reality — through experiment and observation we gradually determine that it more or less fits the *de facto* physical world. But there is no absolute necessity involved. There is nothing within the model which specifies why this model is to be realized rather than another one. And it is precisely this which must be specified, among other things, if a given theory is to explain itself, if it is to be a model of everything.

Setting aside this deeper philosophical question, it is still not clear whether it is possible in principle to construct an adequate theory which, while not accounting for itself, would nevertheless comprehend all physical laws, all manifestations of mass-energy and their mutual interactions, the values of all the fundamental constants, and at the same time specify the necessary boundary conditions and initial conditions in a non-arbitrary manner. Many are aiming for this, certainly hoping for this. And there are some indications that the hope might be partially realized.[12] But, it seems to me to be an open question whether it can be done in a comprehensive way. To my knowledge, it has never been done even for a very specific and narrow range of phenomena. Again, we are dealing with a question of limits and horizons, which are important for us to recognize, both for the healthy development of a given discipline itself, as well as for fruitful interdisciplinary dialog and research.

1.5 - *Some Key Assumptions in Cosmology*

In describing contemporary physical cosmology, it is important to recognize that some key assumptions are usually employed. There is always at least some common-sense justification for these; but often such justification is not as scientifically rigorous or compelling as we would wish. Here I mention several of these crucial working assumptions.

The first and most commonly recognized assumption is what is known as "the cosmological principle", that our space-time position in the universe — the position from which all observations are made and experiments are performed — is not in any significant sense a privileged position, but is like any other point of space-time. We shall discuss what this can mean later. Obviously, there are many different ways in which our space-time point might be privileged. Would it be sensible to set aside all privileged features? Or while, setting aside some possible privileged features, to maintain others which seem reasonably well justified? It turns out that the cosmological principle, strictly adherred to, implies that the universe is isotropic and homogeneous, on some very large length-scale. In fact, the most frequently employed cosmological models — the Friedmann-Robertson-Walker models — are spatially homogeneous and isotropic.

And certainly, given the homogeneity and isotropy of the cosmic background radiation, the more or less uniform texturing of the universe on very large scales, and the systematic indications of universal expansion in every direction, it seems reasonable to assume — to a zeroth-order approximation at least — that the universe is isotropic and spatially homogeneous. However, it is clear at the same time that the universe is *not exactly* isotropic and homogeneous. It is very lumpy and anisotropic on small and intermediate length scales[13] — with some rough indications of *almost homogeneity and isotropy* on scales above 200 Megaparsecs. It thus seems crucial to find ways of formulating and applying our assumptions very precisely in weakened "almost homogeneity," "almost isotropy," "almost..." forms, which better reflect the actual situation. This is just beginning to be done by some workers in the field.[14]

A second working assumption in cosmology is the manifold-metric model of space-time. The universe is always identified with a 4-dimensional Lorentz manifold which constitutes its space-time — and a metric, or precise "distance function," is always specified.[15] This is a continuum model. And, as such, one imagines a 3-dimensional membrane — like a 3-D balloon — which as such is already an intelligible whole — and which can expand and contract in all directions. The 4th dimension, of course, is time.

A third assumption is the universal validity of physical laws and constants; this seems to be well justified on the basis of experience. But it is not a completely justifiable assumption in the scientific sense. There are just a number of indications that it may be true; and if we cannot assume it, we are unable to move any further.

Finally, there are two "intermediate level" assumptions which are always employed in cosmology. They are, strictly speaking, unjustifiable — in practice anyhow — on cosmological scales; but they are very testable on smaller scales. The first is a theory of gravity; the second is the fluid approximation for the matter in the universe.

A theory of gravity is an essential component of a cosmological model. It specifies the dynamics of the universe — how it evolves as a whole. Usually, Einstein's general relativity is assumed as the theory of gravity in cosmological models. In weak field, slow motion situations, general relativity is well-tested on solar system-scales. One of the difficulties is that it is not at all well tested — nor is any gravitational theory — on intermediate and large scales.[16] Furthermore, its effective formulation on large scales in a lumpy universe involves some rather tricky smoothing — or averaging — over smaller scales of the stress-energy tensor (the right-hand-side of the field equations), the Einstein tensor (the left-hand-side of the field equations), and the metric tensor itself.[17] Then, too, it is clear from what we said above, that at very high energies — very near to the Big Bang — we will not be able to use general relativity — or any other classical continuum theory of gravity. We will have to use, instead, a suitable quantum theory of gravity, of which general relativity or some other continuum theory is the low-energy limit.

The fluid approximation is used to characterize the overall distribution of matter in the universe in a simple way. With the fluid approximation, we can derive an equation of state, which is a necessary component for solving the equations which describe the model.[18] It is not clear that we are strictly justified in using the fluid approximation — in order to do so, we must have homogeneity on the largest scales. If we do not, and we have, instead of spatial homogeneity, an unending hierarchy of clustering, then we will have to turn to a kinetic theory description of matter, which will be much more complicated.[19]

2. *Some Reflections on Cosmology and on Its Conclusions*

2.1 - *Cosmology as a Discipline*

I should first like to reflect on cosmology as a discipline. I mentioned in Section 1 that there are a number of features peculiar to cosmology which distinguish it from other areas of physics and the natural sciences, and from philosophy. What are some of these?

Cosmology — and the areas of physics it embraces — does deal with many of the fundamental characteristics of physical reality in general, space and time, matter (mass-energy) and its transformations, causality and its physical roots, in a way which sometimes reminds us of philosophy. But it does so by probing the detailed processes and structures — and the evolutionary history they determine — of the pervasive features of the universe and by gradually fitting them together into a coherent explanatory whole. So, like other scientific disciplines, it examines particulars and their relationships and interactions in a dialectic of theory and experiment/observation. It is just that in this case these particulars pertain to some of the most general and universal features of physical reality; the focus is on these structures and particulars with the aim of uncovering their significance and relevance for the larger whole, the observable universe. Science may be analytic. But it is also cumulative — and with a vengeance!

Philosophy, instead, looks at some of the same realities from the standpoint of how they are given to us as knowers and of the role they play in the general structures of knowing and of being as we know it — and not so much from the point of view of the detailed particulars which give rise to them and explain their coming to be in a physical way. Philosophy concentrates on the intelligible wholes which are given in physical reality in general — including those given by the sciences — independent of our knowledge and understanding of the processes which explain them. This pervasive and general structuring to which philosophy is attentive is fundamentally pre-scientific, and is not subject to the same sort of rigorous theory/experiment dialectic which is the hallmark of the sciences — though it is based on experience. It is pervasive in our experience as knowers, and enters naturally into our language.[20]

As I have said above, the subject of cosmology is also pervasive — but pervasive, so to speak, in its object, not in our experience or knowledge of the object. In philosophy, both the object and our experience of it are pervasive or general. Thus, philosophy attends to the intelligible wholes and structures, and to their interrelations, which must be assumed or presupposed by the sciences, including cosmology. Cosmology and philosophy deal with the same realities — but each with a different focus, and relying on different evidential grounds.[21] Philosophy may attend to the details, when invited to do so by the sciences, but only as instances of the general; the details are not its focus. Cosmology may attend to what is pervasive or general in our experience — in order to orient itself — but that is not its focus. It is on the particular.

There is and should be a critical interaction here, of course. The fruits of research in cosmology and field theory may force us to alter our understanding of the details which underlie our general experiences of known physical reality, and to modify or even abandon certain general assumptions. The effect of science on philosophy can be to purify its objects — to provide better articulated generalities on which it can focus its attention. On its side, philosophical reflection can help science to articulate its assumptions and presuppositions, reformulate them in an improved way, and to indicate how they might be justified.

As an example of this interaction, we may recall that some philosphers — Kant especially — concluded that space is Euclidean on the basis of philosophical reflection on our sense experience.[22] We now know from our more precise physical analysis of space-time through experimentally verified theoretical description that, though a space section can always be approximated at a point by a Euclidean manifold on a macroscopic level, it is not necessarily globally Euclidean.

In our analysis of the relationships between physics/cosmology and philosophy — and of the philosophical questions flowing from physics/cosmology — we are aware that analogous investigations of other natural sciences would yield some results which would be similar. However, some of our key conclusions will have no parallel for the other sciences. This is because physics and cosmology do not presuppose the conclusions of other disciplines — as does biology, relying on chemistry and physics, and chemistry, relying on physics. When we step back from physics and cosmology to justify the assumptions and presuppositions we employ in pursuing them, we have nowhere to go, except to some sort of philosophical reflection. Thus, physics and cosmology, and the disciplines which are directly related to them, have much closer ties to philosophical considerations and tend to pose fundamental philosophical questions much more insistently than the other natural sciences — except perhaps some of the areas of biological science, and there in a rather different way.

Oftentimes, in discussing the methods of science, people accentuate the two processes of deduction and induction. But, as Ernan McMullin has stressed, there is a much more important and powerful general method which is commonly employed in science. That is what has been called retroduction,[23] in which one argues from an observed effect — or an observed congeries of effects — to their cause. That cause may be observable or unobservable in practice, observable or unobservable in principle. If it is known to be observable in practice, it falls within the competency of the sciences, strictly speaking. If it is, instead, unobservable in practice and in principle, it does not. If it is unobservable in practice but observable in principle, it may be considered to fall within the realm of the sciences. In the second case, the consideration belongs, rather, to philosophy, though not necessarily so. It may, as a matter of fact, fall outside the competency of every discipline. But this is not the place to discuss that possibility.

In fact, it may not be obvious whether some effect is unobservable — or unverifiable — in principle, or just in practice. The "practice", or even the "principle", with respect to which an effect, or a prediction, is unobservable, or untestable, may not be well-defined, or clear — and, most importantly, it may not be static. What was unverifiable in practice or in principle in the past, relative to the practice and the principles as understood then, may not be so now. And what is seemingly out of reach now, relative to our present grasp of the possibilities, may not be so in the future. This is just a reflection of our own ignorance of what is possible — or practically possible — in an absolute sense.

When I speak of philosophy and of metaphysics in this connection, it is worth emphasizing that I am not talking about some timeless, independent system, as we may have understood it in the past. There are not *independent* grounds of evidence for philosophy. But there *is* a different way of abstracting from evidence, compared with the way that is characteristic of the sciences, which leads to philosophical conclusions. In my discussions here I have tried to describe some of the key characteristics which distinguish that way of abstracting. So, as Ernan McMullin has argued, metaphysics is not — and cannot be — an Aristotelian philosophy of nature, where ordinary experience of the middle range is sufficient for philosophical generalization. Nor can it be a Kantian *synthetic a priori* process. Instead, it must take into account all that science and other disciplines reveal about reality — the broadest possible range of human experience — in arriving at its conclusions and generalizations. The philosophy and metaphysics I mean is what Polkinghorne and Isham, for instance, have done in their contributions to this volume, and what Russell has done in the first part of his paper.

Returning to cosmology, others of its peculiarities are: (1) The object of its study — the universe — is unique. Thus we have no other examples with which to compare it, and statistical analyses in the usual sense, are out of question. It is, however, fashionable to examine the ensemble, or space, of all possible model universes and to see how ours might be picked out in some natural way; (2) Because of its uniqueness — and also because its evolution spans all time, as we know it, so that none of its phases or transitions is directly reproducible — cosmology is, in a way, very much more like history than like science. Paleontology, or geology, are similar, but in these two sciences, unlike in cosmology, there are at least different examples of similar things; (3) In cosmology we cannot see or examine the entire object of our study. We are immersed in it — a very small part of it. And we have extremely limited access to it, both spatially and temporally. As George Ellis has suggested, doing cosmology is like trying to do geography from a single fixed position on the earth. In a very definitive way, we are studying an object — the observable universe — whose temporal and spatial boundaries and limits we cannot comprehend. We do not have any direct way of examining its beginning — whatever that may have been — and we have no access to its final state. Furthermore, even for distant regions to which we have some access, adequate precise data is very hard to come by. All of these problems lead to the questions, which I shall briefly discuss next in Section 2.2: What are the limits of verification in cosmology? What is verifiable in principle? What is verifiable in practice?[24]

From a philosophical point of view, there is also the question pointed out to me by Michael Heller: The need for what one might call intermediate levels of metaphysics which pertain to the special categories of reality the different disciplines present us with.[25] This is particularly true of both cosmology, and quantum field theory, and their daughter specializations. The point here is that, before we can rise to a general metaphysics with any confidence, we must first analyze knowledge of different aspects of reality gained through their appropriate special disciplines, and the structures of known reality manifested in those special areas of knowledge. This is particularly true of disciplines in which we are investigating the limits of physical reality — and where, therefore, fundamental and limiting characteristics both of our knowledge and its scope, and of the realities known and their essential structure, are much more apparent. These intermediate and specialized levels of metaphysics must then be comprehended, or subsumed, either positively or negatively by a more general or universal metaphysics. (Many may baulk at my use of the term metaphysics in these two ways — the terminology itself is not important, but the idea of taking seriously the total import of what is revealed in all areas of special knowledge about the structure and dynamics, the limits and essential characteristics, of our knowledge and the contents of our knowledge — what is known — certainly *is* important.) Otherwise, we end up establishing a particular range of our knowledge, and a particular level at which we know about the outside world — or about ourselves — as a privilieged one, or the canonical one, against which all others are measured. For example, as often happens, we can end up taking ordinary experience as privileged; or we can end up taking the special experience of the physicist or biologist as normative. In either case, we find ourselves in difficulty.

2.2 - *The Limits of Verification in Cosmology*

A key initial question here is, "How large is the observable universe?" If it is as large as we have good reason to believe it is, then this in itself places severe limits on what we can know concerning its structure and its history, and on our ability to gradually approach and approximate with precision a unique description of it which is testable, or falsifiable.[26]

To begin with, there is overwhelming evidence that the universe is so large that we cannot move through any significant part of it from our space-time point over a period of hundreds of thousands of years. Even if we could travel at the speed of light, it would take 100,000 years to go from one side of our galaxy, the Milky Way, to the other — and our galaxy is essentially a single point in space; by travelling from one side of it to the other, we do not significantly change our position relative to cosmological scales. Similarly, if we allow a very long time to elapse, say 100,000 years, we do not move significantly forward in time — along our world line — relative to cosmological time scales. Thus, the perspective we have of the universe — the point of view from which we can examine it, study it and make observations which reveal its structure, history and evolution — is essentially fixed. We cannot travel around the universe and gather data — nor examine it from different space and time perspectives. Our situation is represented by our past light cone — we are at the vertex, and we must stay there, effectively; we cannot move measurably forward in time (vertically), nor can we travel any significantly measureable distance in space away from it.

Practically all the information we can obtain about the observable universe comes to us through electromagnetic signals travelling from the past along our past light cone. The only exceptions are the bits of information hidden in the material which makes up our earth and our solar system, to which we have more direct access, and the cosmic rays which bombard us. In fact, there is a vast quantity and a great variety of information which is communicated by these electromagnetic signals from the past, if we analyze them properly. Nevertheless, this dominant channel of relevant data about our universe is constrained by some very important limits.

First, this data by itself — along with an assumed theory of gravity — will, at very best, allow us to determine the structure of space-time only in a relatively small region near our world line, unless, as is usually the case, very strong *a priori* assumptions (such as isotropy and spatial homogeneity) are made. Such assumptions are impossible to justify scientifically in their exact form, and it is very unclear what their theoretical import is — and how one works with them — in their approximate or "almost" form. Secondly, in practice, it seems impossible to acquire enough precise data of the sort that is needed, in order to determine uniquely a cosmological model in these limited regions, unless the strong assumptions mentioned above are imposed. Thirdly, it seems unlikely that the astrophysical information contained in these signals will ever by able to be separated cleanly from the relevant cosmological information (the so-called evolution problem). Finally, even given that we overcome these problems or successfully circumvent them, it is not clear that our data sets would determine stable cosmological solutions to the field equations. If this turns out to be the case, we would know very little concerning the universe in the distant past or in the distant future, even if we possess adequate data to determine its "present" structure in our cosmological neighborhood. In fact, just fitting a Friedmann-Robertson-Walker model to an available set of data on our past light cone in a rigorous and meaningful way involves a number of ambiguities which have not yet been resolved.[27]

It is still just possible, of course, that the universe is really not as big as we think it is, and that the large number of galaxy images we see at intermediate and large distances are really duplicate images of objects we are seeing over and over again at different points in their history.[28] We may be seeing many times around a small universe. We suspect that this in *not* the case, but no one has definitively ruled out this intriguing possibility.

2.3 - *The Ontological Status of Cosmology's Principal Object*

We have already indicated above the strong reasons for treating the observed universe as a single object. But we must also reflect on the consequences of doing so. What is the minimum import treating it as a single entity implies, philosophically speaking — from the point of view of epistemology, and from the point of view of an intermediate specialized metaphysics? For instance, in modelling the universe as a 4-D, or a 3-D manifold which evolves (viz. expands), because of the energy dumped into it at a certain time, and because of its mass-energy content, what sort of thing are we implying the universe to be? We certainly seem to imply that it is to some extent — at some level — causally self-connected, with some common origin. But then what does this imply about the character of space and time? How do we get away from the implication that they are containers in some definite sense — and have some absolute or independent existence — apart from the mass-energy they contain? The manifold model is a model, which works reasonably well — the question is: Which of its characteristics are to be taken seriously in describing space and time on a philosophical level? And, if there are certain characteristics we do not take seriously on a philosophical level, because they are merely irrelevant bi-products of the model, can we continue to take those aspects seriously on the level of science? Apparently not! And what criteria are we to use in distinguishing the important aspects of the model — those which are really supposed to approximate reality — from those which are secondary and discardable? Are these criteria to be primarily scientific (cosmological)? Or might they also be philosophical? If the latter, why?

2.4 - *Linguistic Problems in Relating Cosmology to Philosophy and to Theology*

From our discussions above, it is obvious that certain key words have different meanings in cosmology, philosophy and theology. Words like "universe," "time," "space," "cause," are cases in point. So are "nothing," which must be distinguished from "vacuum," and "matter," which is different from "mass." Rarely do people discussing a particular issue within an interdisciplinary context adequately make these crucial distinctions. Failure to do so, at the very least, prevents proper precision from being achieved and often leads to real confusion.

It is true, of course, that these words usually have closely related or overlapping meanings in, say, physics and in philosophy. But they usually differ significantly both in their comprehension and in their extension, and in the way they function within the discourse of each of the two disciplines. Therefore, the precise meaning of a word like "cause," or "time," will change a great deal from one to the other.

A particularly relevant and important example of these linguistic difficulties is provided by Michael Buckley's contribution to this volume.[29] In describing "The Newtonian Settlement" and its role in the birth of modern atheism, he gives an analysis of the different but related meanings "mechanics" has had for various physicists and philosophers, and the misunderstandings this has sown:

> One does significant violence to the history and the achievement of ideas, if one takes "mechanics" as a word possessing a single meaning and designating one obvious subject-matter. Isaac Newton endows mechanics with a comprehension of meaning and an extension of subject that neither Galileo nor Descartes before him nor many in contemporary science would

admit. For Newton, it was mechanics which both provided the foundations for geometry and also established the existence of God.[30]

2.5 - *The Gaps Between Cosmology and Philosophy*

In doing cosmology or any science, as I have pointed out, there are always assumptions which the science itself cannot justify — or at least not completely justify. Some of the justification may come from other scientific disciplines, and some from what might be called philosophical or common-sense considerations. For cosmology, and other fundamental areas of physics, practically all the assumptions fall in this latter category. However, working cosmologists are acutely aware that there is a large category of assumptions and presuppositions which are unverifiable and unjustifiable not only by cosmology and physics themselves but also seemingly by philosophy. They fall between the two disciplines, and seem to require special consideration — whether by some intermediate specialized metaphysics, or by some other methodological considerations. Examples of this in cosmology would be the cosmological principle itself and the manifold-metric model of space-time.

Another type of issue along this line involves those features of the universe which are physical but which fall outside the limits of verification (in principle) for cosmology. They, therefore, fall outside science, strictly speaking. But they cannot be justified or resolved by philosophical considerations either. We can specify what sort of data would determine those characteristics, but we discover that in principle such data will never be attainable. Certain aspects of the global structure of space-time, and of the very early universe *may* fall into this category. Examples are the large-scale homogeneity of the universe and the verifiably operative theory of gravity on cosmological scales.

3. *The Focus and the Experiential Grounds of a Discipline*

In interdisciplinary discussion, those who not only utilize diverse methodologies and techniques but also have very different foci of inquiry and experiential grounds to which they appeal, attempt to understand and appreciate one another's points of view. They hope to discover in disciplines other than their own — and in the often broad, hazy interfaces between disciplines — clues, indications and new pieces of evidence which will help them push forward the limits of their own disciplines and hone their procedures and techniques. Most of all, perhaps, they look for new heuristic structures, while moving more surely towards a critical synthesis and creative communication with other disciplines. However, the different foci and experiential grounds involved — as well as the diverse methodologies — give rise to different languages and contexts for interpretation and understanding, which make interdisciplinary communication and understanding on any deep level difficult and uncertain, even in areas of study which many outside the two disciplines in question would believe to be very close — for instance, pure differential geometry as a field of mathematics, and gravitational theory as a field of theoretical physics, which employs differential geometry as a tool. Thus, it is vital, in interdisciplinary study and discussion to be aware of — and to specify and clarify, as we go along — not only these differences of language and method but also, even more, the radical differences in epistemological focus and in evidential grounds of which they are but surface manifestations. *

All of this is relatively clear and straightforward when dealing with two natural sciences — for instance chemistry and physics — or two sciences in general, experimental psychology and biology, or psychology and cultural anthropology. Here there can be some very important and delicate questions, to be sure, but the fundamental differences in foci and in evidential grounds are usually transparent and well-recognized. With respect to the relationships between the natural sciences, philosophy and theology, it is also obvious that there are fundamental differences in focus, in experiential grounds, in method. But it is not at all obvious how those differences are best characterized or described, and how we can most correctly specify these disciplines in relation to one another. That in large part is because many people do not have a clear idea concerning the essential foci and experiential grounds of philosophy or of theology. That there might be some confusion here can be easily seen if we look at the history of the sciences and of philosophy. Before the diversification of disciplines, there was "natural philosophy," which embraces much of what now belongs to very distinct disciplines — physics, chemistry, biology, astronomy, meteorology — along with questions which still form an integral part of philosophy. The usual account of this separation of disciplines stresses the role of the experimental method and the appeal to experimental and observational data, and of the theoretical modelling correlative to it. Both of these aspects, in turn, developed an essential reliance on mathematical and quantitative considerations in physics, chemistry, and astronomy. But, it is more fundamental to characterize the diversification in terms of differing foci and experiential grounds.

The reason I believe a discussion of foci and experiential grounds can be so fruitful is because it is really here that all the differences between the disciplines as ways of knowing and methods of inquiry are founded — and not so much in their material objects, except in so far as they help constitute the focus or provide the experiential grounds. I use the word "focus" advisedly — to be distinguished from "locus". The *focus* of a discipline indicates the primary aspect or part of experienced reality to which it gives attention, and provides its primary point of reference. It may, from that focus — in virtue of, or from the perspective of that focus — turn to consider a large number of other aspects of reality. All these constitute — together with the focus — the locus of the discipline.

The result of doing this may be either positive or negative. That is, in moving our consideration outward from a discipline's focus, we may discover that certain positive content is added to our knowledge. At the same time, we may become aware that we have moved up against certain limits — from the perspective of a given focus, and using a certain method consonant with it, we may discover that we are able to say nothing or very little about another aspect. From another focus, and using another method, we may be able to unravel those same phenomena in all their richness. Such negative components of a discipline's "locus" are extremely important — both for that discipline itself and for knowledge of the object or aspect of reality under consideration. Correlative to the negative content, there is usually a profound and simple positive content, which can only be revealed against the background of that negative experience of "coming up with empty hands." And for the discipline itself, that negative experience helps to purify it and make it aware of itself — its range of competencies and its limitations — so that it does not exhaust or compromise itself pretending to do what it cannot do.

Once we have specified the focus of a discipline, we can move on to consider its experiential grounds. These are the type of data, of phenomena, or of experience to which the discipline appeals, which it analyzes, and on which it reflects, in arriving at and justifying its conclusions, and in testing and modifying its models. Again, a discipline, in the course of its application by a practitioner, may consider an object or an aspect of reality which is also an object of another discipline. However, the experiential grounds to which it appeals in examining that object will be quite different from those of another field. In the natural sciences the focus is on detailed, reproducible behavior, on patterns of structure and behavior of physical, chemical and biological systems, as given by systematic and controlled observation and experiment, and by precise measurement. This also specifies the experiential and evidential grounds of the sciences.

One can, from this point of view — with reference to this focus — and employing the familiar methods characteristic of it, turn one's attention to other rather different aspects of our world. For instance, one can look at certain pervasive features of physical reality, within which or in terms of which the usual objects of the natural sciences are investigated. What is time? What is space — and place? What is motion? Life? What is the origin of physical laws — and of the values of the physical constants? Why are they as they are, and not different than they are? Why is there something, rather than nothing? With such considerations, we become aware that the focus and the experiential grounds of the natural sciences are not wholly adequate. We begin to run into limits, and find ourselves in the negative sectors of the loci of the sciences.

At the same time, however, the natural sciences *do have* something to contribute to the exploration of these more general questions. On the basis of what we know from contemporary physics and cosmology, we can bring important information to bear on questions concerning the nature of space, time and matter — and even on the origin and maintainance of physical constants and laws. But we begin to get the strong impression that their focus along with the experiential grounds and the methods to which we appeal in the natural sciences — helpful, indicative and penetrating as they are in many cases — fall short of grasping the fundamental reality involved in these questions. A new focus, new experiential grounds, new or different heuristic structures, complementary to those provided by the natural sciences themselves, are needed. Many would question whether anything in terms of focus, experiential grounds, and heuristic structures outside of the sciences can take us any further in answering the above questions. That is a possible point of view to be discussed and examined. But, at least, we must recognize that a certain experience of essential methodological limits manifests itself — whether or not those limits can be transcended by other disciplines, or by other methods. Others, however, would immediately turn to the focus and experiential grounds proper to philosophy, and specifically to certain types of philosophical analysis, like phenomenology or the various brands of critical realism.

Another, rather different, description of this discovery of horizons or limits beyond which the focus proper to the natural sciences and the investigative methods which articulate them cannot take us, is in terms of justifying the presuppositions or assumptions upon which the natural sciences rest. All of them begin with certain assumptions, which are considered obvious or at least eminently reasonable and well-substantiated on the basis of our common and specialized experiences. No clear instances to the contrary are available. On these bases we go forward to construct our different sciences — with great success, and with the promise of even greater success in the future. Indirectly, that strongly confirms the initial assumptions we have made, whether for reasons of providing essential order (as in the assumption that physical laws and the values of the physical constants are the same in different places and at different times in cosmology) or for reasons of simplicity (as that the Universe is isotropic and spatially homogeneous).[31]

What happens, however, when we turn to examine these assumptions themselves — not only our rigorous justification for them, but also our explanation of them — our accounting for them in terms of cause and effect? We are almost always forced to assume a new focus, to appeal to new evidential grounds, to develop new and rather different methods of description, analysis and inquiry.

There are at least three levels here. First, there is the level of properly articulating the assumption we are examining and what it means, or does *not* mean. Secondly, there is the level of finding a way, or ways, of justifying it, either by using the methods and experiential data provided by the natural sciences themselves, by the science in question itself, or by another more fundamental discipline (e.g., justifying a principle in chemistry by appeal to what is known in physics), or by appealing to other justifying considerations, which may lie outside the realm of the natural sciences. If the latter, then these justifying considerations have their own different focus and experiential grounds, and they must be articulated in terms of these, examined carefully according to the canons and methods proper to that focus and the type of experiential grounds which are appropriate to it. An example would be, of course, the various types of transcendental philosophical reflection.

Thirdly, there is the level of accounting for or explaining the content of the assumption. Why is the world that particular way? And not some other way? How did it get that way? Why and how does it stay that way? Was it always that way? Does it always have to be that way? If so, why? If not, what other way could it be? And, why and how was this particular alternative picked out? If it is initial conditions, or boundary conditions, how do we account for them; how do we, or how can we, explain their being set in this or that particular way?[32]

These are the three basic levels on which we are called to analyze and justify the assumptions we make in a group of disciplines like the natural sciences. It is clear that oftentimes foci, experiential grounds and methods of inquiry rather different from those of the sciences themselves are needed to accomplish this task. Even if such methods and such evidential grounds are not forthcoming or available, it is crucial to recognize the frequent inability of the sciences to do this adequately for themselves.

By examining the horizons and limits of the natural sciences — and their presuppositions — we have already, of course, moved into philosophy. We now describe metaphysical knowledge more explicitly and more precisely. What is its fundamental intentional focus, and what are the experiential grounds to which it appeals? We have discussed very briefly how we sometimes are pushed to examine very carefully, in order to justify and account for, the assumptions and presuppositions we make in the sciences, and how the sciences themselves are often not capable of adequately doing this. So, we might say perhaps that philosophy deals with the underlying assumptions upon which the sciences are built. It *does* that, but not just in virtue of the fact that they are assumptions unjustified by the sciences. It does that in virtue of the fact that these assumptions often involve the pervasive, deep and fundamental structures which underlie reality as we know it — reflecting the reality itself in some definite way, and our knowledge of it. Those structures involve space and time, place, identity and distinctiveness, causality in its different forms, and the laws governing them, etc. They provide the horizon against which, and the matrix within which, all other phenomena, situations and events occur, and are experienced. That does not mean, as I have already stressed, that certain phenomena studied by the physical sciences, for instance, do not reveal some of the fundamental details of this deeper more pervasive structuring. But that does not happen regularly, and even when it does, the sciences themselves do not have the competency to situate properly what they sometimes reveal about these deep structures within a critical and comprehensive description of them.

The focus of philosophy essentially is on the knower, a reflection on the experience of knowing, and on the structure of what is known, or can be known, that is, being as such. And the experiential grounds of philosophy are just the experiences of knowing, or more broadly, the experiences of being, knowing and acting: *All* such experiences, not just those pertinent to a given class of phenomena, but not excluding any of those either (e.g., the specialized experiences of the theoretical or the experimental fundamental particle physicist). In so far as these are experiences of knowing, they reveal aspects of those deep, pervasive structures we have been speaking of. And so one develops methods of reflection upon, and analysis of, our experience of knowing and of what is given in knowledge, of its essential organization or structure, both static and dynamic, in order to articulate, describe, explain and account for it in a suitable fashion. Usually this involves prescinding or abstracting from those features which are transient, are not pervasive — those features which specify what is known and what is and what becomes in its marvelous variety and diversity, but not in its radical giveness as being known, existing, becoming. When one does philosophy, the focus is on oneself as knower, and on that experience of knowing. One moves through transcendental methods of reflection and analysis, through an analysis of the structure of knowing and what is given in knowing, and its conditions of possibility, until one arrives at what is known, at what is.

In some way, all this can be developed in terms of the ancient philosophical maxim: "Know thyself." In coming to know ourselves as both knowers and agents we come to know our strengths and our limitations in understanding, relating to, and affecting the world around us; and we come to know what is given and what is not given — but only hinted at or intimated — as a result of our intentional reach outwards and inwards. In critically coming to know that, we come to know at least something of what is, and of its structure and internal constitutive relationships. It is clear in this case, how the focus and experiential grounds of philosophical knowledge are different from those of scientific knowledge, although there is, I believe, an area of overlap — or at least of nearness — particularly when in contemporary physics the role of the observer as participant becomes essential. Quantum phenomena are given, never in themselves but only in terms of a measurement made by an observer. What is given is never the thing in itself, but always the thing in relationship, in interaction with, the observer or the detector. (That, however, is not necessarily the conscious and self-reflective knower!) In this the issues of limits and limitations are obvious. The observer, the detector, the knower are always limited — in perspective, in range of sensitivity, in distance from the object, in being in some sense separate from it, outside it. To the observer and to the knower, the object reveals itself in some of its features, and hides itself in others, inviting the knower to assume new and different perspectives, new and different methods which will reveal heretofore hidden properties. But the philosopher generally does not occupy him or herself with these transient or nonpervasive properties. That is left to the scientist. The philosopher probes, rather, the essential properties without which it would not be known at all, without which it could not be described and could not be said to be, without which it could not be accounted for in terms of other things — if it is not self-explanatory.

We might also characterize philosophy — or least a certain large part of it — as focusing on ultimates. But such a description, I believe, is not as fundamental as that which I have given above in terms of reflection on the experience of knowing, and what is given in knowing. Concern with ultimate issues brings about a poignant consciousness of the limits. Then we strive and struggle to transcend those limits by scouring our radical experience as knowers to find a way of knowing something of what lies beyond them — of rendering the horizons revelatory!

It is at this point that philosophy begins to open itself to the possibility of theology — or, rather, the foundations of theology, religious experience and revelation. The focus and experiential grounds which characterize philosophy begin to disclose the possibility of the focus and the experiential grounds which characterize theology. As one focuses on ultimates, on the limits of our experience, our living and our doing, we still have a philosophical focus. We acknowledge these limits beyond which we cannot go by appealing to our usual experiences and using our ordinary ways of acquiring knowledge. However, by assuming the philosophical focus, and by appealing to the experiential grounds upon which it relies along with its methods of analysis, we may begin to transcend these limits — at least coming to some knowledge that there is reality *beyond* them and that it must possess certain characteristics — either on the basis of what we already know about being in general, or on the basis of what the qualities and limitations of the reality we know tell us concerning the realities which lie beyond these horizons.

The focus then becomes theological — as do the experiential grounds — when we attend primarily to the experienced presence of the Other which comes from beyond these limits — these ultimates — and reveals, or discloses what lies beyond. Insofar as the experience of the Other lies on this side of the limits, we are still in the area of science or philosophy; but insofar as it and its primary aspect transcend the limits, we are in the realm of theology. Obviously, the communication of this revelation will always be in terms of what lies on this side of our horizons — but its content, significance and focus may be deeply revelatory of what is beyond them. When we reflect on the experiences of genuine love, of faith, of permanent commitment, and on the realities they intend, then we attend to the revelation taking place in our lives.

This sketch of the focus and evidential grounds of the natural sciences, of philosophy, and of theology has emphasized the philosophical or metaphysical, and then the distinctions between it and the scientific, on one hand, and the theological on the other. Much more could be said about both the focus and evidential grounds of the natural sciences and those of theology. But they are better recognized and much less problematic for the dialogue between science and religion.

4. *The Implications for the Dialogue between Religion and Science*

We can indicate the various avenues through which the sciences in general impact religion, and influence the interaction between science and religion. The first avenue is that through which the findings of the sciences, or the methods and procedures proper to them, directly confront theology, altering its own conclusions, the ways in which they are reached, or the terms in which they are expressed. The second is through philosophy; the sciences modify the metaphysics which is employed in theological reflection and articulation. And finally, the sciences influence religion through the new images, perspectives, symbols, and stories with which they enrich the common cultural field. On all these levels, we see how the physical and the biological sciences, the psychological sciences, and the sociological and anthropological disciplines have affected religion and theology.

Within this framework we could begin to describe *the impact* cosmology is having on theology, and on the dialogue between religion and science. But our immediate concern with *its implications* for theology, and for the dialogue, is somewhat narrower. The impact cosmology, or any other science, has on religion is inevitable and, to a large extent, unarticulated and perhaps unrecognized. The implications, though they certainly flow from our awareness of this impact, are, in contrast, mediated by those who have perceived them. They are the articulated consequences of cosmology which are consciously and critically brought to bear on theological questions, or taken into consideration in pursuing rapport between religion and the sciences. If they were not adverted to, theology would still be done, the dialogue would continue, and science would have its impact on both; but there would be less control and enlightened research, and undoubtedly, more isolation and confusion.

On the basis of what we have already discussed, what are the implications of cosmology for the religion-science dialogue?

Considering the first avenue mentioned above, we must take very seriously the general conclusions and findings of contemporary cosmology — that the universe is as large and as old as it is, that it is evolving, that all that is within it has had a common physical origin in time, and that all it contains is explicable in immediate terms and at the level of contingent being by the natural sciences. (Obviously, in ultimate terms, the natural sciences are incapable of providing an explanation.) At the same time, we must take seriously, too, the limits one encounters in doing cosmology — limits of verification, in principle and in practice, and limits of competency. There will be certain things about the universe we shall never know for sure. And there are certain questions about physical reality which cosmology will never be able to address. We discussed some of these in Sections 1 and 2. Both these positive and negative aspects of cosmology have further important implications.

One which flows from what it has contributed to our knowledge of physical reality is that *any* "God of the gaps" form of explanation is to be avoided. There are many such explanatory gaps in the scientific account of the evolutionary development of the universe, for instance, and more will become evident. But, wherever they occur within a causal chain in a scientific context, they should be left for the scientific disciplines themselves to fill. Whether they occur between different phases of the universe, or different stages in the evolution or the development of a biological entity, they inevitably end up being bridged by science itself. *At its own level*, science is capable of providing adequate and complete explanations — at levels other than those which deal with "why there is something rather than nothing," with the experience of knowledge, with the personal, with value. Science should not be helped from the outside to answer the questions it has raised and which are proper to its focus.

In this regard, even establishing a rough parallel, or consonance, between "the beginning of time" in the Big Bang and "the beginning of time" in the doctrine of creation (I insist on distinguishing this latter concept from the radical meaning of *creatio ex nihilo*) is very questionable. It seems highly unlikely that cosmology, or any physical science, will ever be able to unveil a point of *absolute* beginning — before which *nothing* existed, before which time of any sort was not — which would require the direct influence of God. That does not mean that such an event did not occur. It does mean that cosmology is *not* able to discover it and reveal it as the "Ur-event", the event needing other than secondary causes for its immediate explanation. Nor does it mean that miracles do not occur — just that science is incapable of revealing them as such, or of providing positive evidence of their occurrence.

Another implication of contemporary cosmology for the dialogue between science and religion is, as I have already said, that the limitations of science — and of theology, too, of course — must find practical acknowledgement in interdisciplinary interaction. Even better, we must explore the significance of these limitations — for cosmology and the physical sciences themselves, and for their relationships with other disciplines. The different limits and horizons we encounter in science, in philosophy and in theology are essential for the growth of each in itself, for the purification of each, and for promoting fruitful interdisciplinary collaboration. The more self-awareness a given subject attains of its radical limitations, of the areas in which it cannot move with confidence and which it must cede to other disciplines, the more its own specific role and competencies can be discovered, focussed and developed. There is a deeply purifying role to be played by the interactions between the disciplines, which is just a reflection of the limitations we ourselves constantly experience as both knowers and doers.

A third implication, still referring to the first avenue of influence, is that it is important for the theologian to take into consideration what cosmology, and science in general, reveals to us of the universe and our place in it. As we have seen, material reality is on every level more vast, more intricate in its structure and development, more amazing in its evolution, in its variety flowing from fundamental levels of unity, and in its balance of functions, than we could have imagined without the contributions of the sciences. Certainly, at least in some way, such a perspective and such understanding enriches theological reflection, and provides *some* of the detailed experiential points of reference from which we consider who God is, and who He is not, and who we are in relation to Him, to one another, and to our world. Much more could be said with regard to this. Here, also, we begin to slide over in the third avenue of impact I mentioned above — the way the sciences influence religion by modifying the common cultural field.

We now move to the second avenue of influence, cosmology affecting theology, or religion, through philosophy. The major implication is that, in the dialogue between religion and science, the intermediate dialogue between philosophy and science must always be considered. Without it, that between relgion and science will be partially blind. Our articulation of our faith commitment, and of the community's faith experience, always involves implicit philosophical stances. At the same time, the possible relevance or challenge of a particular scientific conclusion or perspective to theology can only be judged by a critique which relies heavily on those same stances or presuppositions.

Without a metaphysics developed with critical attention to the conclusions and perspectives of cosmology, and of the other sciences, the implicit and explicit philosophy used in mediating between science and religion will be unpurified, so to speak, and, at the very least, will be dominated by unrefined generalizations. These probably will rely too heavily on "the middle range" of experience. What "cause" is, and is not, what "time," "space," "matter," etc. are, must be formulated with an understanding of what contemporary physics and cosmology reveal about these concepts. Otherwise they will be inadequate to the task imposed upon them — facilitating the dialogue between two radically different and fundamental disciplines. This is not to say that philosophy should simply adopt these concepts from the going cosmology or science. That would be irresponsible. It must, as I indicated in Section 2, develop them for itself, taking the relevant contributions and viewpoints of the sciences into serious critical consideration.

Along with this there must be a sensitivity to the different ways the same word may be used in two different disciplines, or within a single discipline at two different periods of its history, as I mentioned in Section 2.4. Linguistic and historical studies bridging science, philosophy and theology are essential complements in nourishing the dialogue between science and religion, and in insuring its success.[33]

Finally, in discussing the mediation of philosophy in such a dialogue, we should stress once again that it is important to distinguish among the different foci and evidential grounds of the three disciplines (see Section 3). A great deal of misunderstanding is eliminated if these distinctions are maintained. And often the linguistic confusions just referred to are thereby quickly resolved.

When we come to the third avenue of scientific impact on religion and on the religion-science dialogue, the main implication is that in interdisciplinary research and discussion this important, multivalent channel of influence must be recognized and taken into account. The images, symbols, and perspectives generated by science and by its applications enter our culture in very fundamental ways, helping to mold our values and attitudes, shaping our ways of looking at ourselves and our universe, and dictating the stances we take towards reality itself. Not infrequently they coalesce into powerful myths which threaten, modify and even replace more traditional ones.[34] These images, perspectives and new myths can either enhance and purify religion when properly relativized and integrated, or they can distort or even destroy it, when they are not acknowledged and properly engaged by religion in its dialogue with science and with our scientific culture.

One often speaks of the compatibility, consonance or coherence between science and religion in their perspectives upon and their conclusions regarding ourselves, our world, our universe.[35] What kind of compatibility, consonance or coherence do our discussions imply? At the very least they imply an absence of essential conflict or contradiction. But they stop far short of anything that smacks of disciplinary unity or integration. Certainly, from what we have already seen, one of the principal bases for such a nuanced compatibility — or for a more general "coherence of world views" as McMullin[36] conceives it — is found in effectively recognizing and honoring the radical limitations and the strengths of each of the disciplines involved. When we have grounds for suspecting that the representatives of a field — be it one of the sciences, philosophy or theology — are systematically overstepping its limits, exaggerating its presumed competencies, or reaching conclusions which are in apparent conflict with the basic findings of another discipline, then the situation must be honestly studied from both sides and the flawed positions must be reformulated in light of the more precisely revealed horizons and limits and of the legitimate challenges received from others. Examples of such cases abound. Buckley's study of "the Newtonian settlement" is a well-known one.[37]

We have so far characterized the compatibility of science and religion in negative terms. Can we also discern important positive elements? I believe so.

One such element is related to the first avenue of influence we discussed above. To put it in the baldest of terms, it is simply that religion and theology must maintain a radical openness to, and a critical acceptance of, the range, evolution and structure of physical, biological, psychological and cultural reality which the sciences reveal to us. Indeed, they must acknowledge whatever experience genuinely gives us. Implicit in this contemplative stance is a deep reverence for what they reveal, and a conviction that relishing the world as it really is — in all its richness, variety, and fragility, and sometimes in its harshness, hostility and absurdity — is more consonant with true religion than any other defensive, reluctant or controlling stance we could have taken. The validity of this approach is confirmed again and again by those who have taken it. For the theologian — and more importantly for the believer — God is somehow working in and through *all* that is — manifesting Himself both in obvious and in very subtle ways.

Is such a disciplinary openness and critical acceptance reciprocal? Should scientists *as scientists* maintain an explicit receptivity to the conclusions and findings of philosophy and theology? Obviously not. The status each discipline employs within the framework of the other is by *no* means reciprocal. One can do impeccable scientific research without adverting to philosophical or theological findings or principles. In fact, such external referencing can lead to serious distortions and errors in scientific work. Science is complete and self-sufficient at its own level. Of course, as I have emphasized again and again, the sciences must cultivate an awareness of their own limitations — of the questions they cannot answer, and those which it seems they will never be able to answer. In a sense, this recognition constitutes a *partial* reciprocity with theology and philosophy, an acknowledgement that certain avenues are outside their realm of expertise. But the sciences, as sciences, cannot go further and accept the positive content philosophy and theology provide in those areas. As disciplines the sciences are just not equipped to receive and integrate such conclusions.

perhaps they should be

In contrast, philosophy and theology *are* essentially interdisciplinary. The philosopher or theologian can be seriously faulted for not taking into consideration the findings and the perspectives of the sciences, or of another discipline, insofar as they have direct or indirect bearing on a given philosophical or theological issue. Not infrequently, the detailed content of another discipline will be found to be irrelevant, but the general conclusions and perspectives — the overall context — it has uncovered will at the same time provide important and even crucial data for philosophical and theological reflection. Both of these disciplines, each with its own methods and foci, must be able to draw from the full range of experience. This will always be done in a critical way, of course, and to the extent the question at hand demands it. But both philosophy and theology must jealously maintain their abilities to do this. One touchstone of their responsiveness would be the seriousness with which they have developed workable, detailed criteria for judging a given datum or perspective as irrelevant.

Another positive component in the emerging consonance between religion and science consists in what we might call common or similar underlying thematic concepts,[38] and at a deeper level in their common operative philosophical presuppositions. Both of these provide similar centers of integration and new understanding for each discipline, with a similar dynamic intention. Themata like evolution or development, unfication, diversification, relationship, complementarity, symmetry, etc. often possess a common core of meaning in religion, in philosophy and in science, despite their different contextual or concrete meanings or referents. The deep influences these have on two different disciplines induces in them certain similarities of structure and growth, of perspective, of openness and approach, and assumed value which are essential for their quests, even though their methods and competencies, their foci and evidential grounds are quite different and fail to overlap in any significant way. Such controlling themata are also the basis of rich metaphors and analogies which can be transferred, always with care and certain important disclaimers, from one field to the other. They also undoubtedly have a great deal to do with the fascination many of the images and ideas generated in the sciences have for the general public — with the reasons why the third aveue of impact functions with such powerful inevitability.

In light of these reflections, I tentatively propose several principles of interdisciplinary interaction. One is simply that, in order to decide which discipline takes precedence on a given question, the discipline whose focus includes that question is to be preferred. Another follows. If this is unclear, and we examine the possible experiential grounds which can be brought to bear on the question, then those of the lower disciplines (the natural sciences) are always to be preferred to those of the higher disciplines (philosophy and theology), when they are pertinent and when they involve detailed phenomena which are subject to verifiable experiment, or particular evidence which is susceptible to critical examination. Only questions whose experiential grounds fall outside these categories should be the focus of philosophy or of theology. When such questions are at issue — or are strongly suspected of being at issue — then philosophical approaches and in their turn theological approaches must be taken seriously, but never uncritically. On certain issues they, too, may encounter limitations which are every bit as severe as those the natural sciences experience.

These are some of the major implications contemporary cosmology and its allied fields have for the dialogue between science and religion. As we delve more carefully and thoughtfully into these ideas, other significant ones will undoubtedly emerge.

I wish to thank all those who helped me in writing this paper, particularly Chris Isham and Mary Hesse who commented upon it officially at the Conference, Bob Russell who made a number of important suggestions in the final stages of revision, and George Ellis, my collaborator in cosmological research — with whom I have shared the fascinations and frustrations of probing the universe.

NOTES

[1] L. W. Krauss, "Dark Matter in the Universe," in *Sci. Amer.* **255** (Dec. 1986) 50.

[2] A. H. Guth and P. J. Steinhardt, "The Inflationary Universe," *Sci. Amer.* **250**, No. 5 (May 1984) 90. Such inflationary expansion would be occurring at a rate equivalent to many orders of magnitude greater than the speed of light. There is no problem here, in the standard account, because this is the rate at which space-time itself is expanding, not the rate at which material objects are moving in space-time.

[3] A. D. Linde, "Particle Physics and Inflationary Cosmology," *Physics Today* **40**, No. 9 (Sept. 1987) 61.

[4] Other features, less well established, could also be cited, such as what appears to be a universal primordial abundance of certain light elements (deuterium, helium, and lithium), which we mentioned above.

[5] St. Thomas Aquinas, *Summa Contra Gentiles,* I, 44.

[6] See also E. McMullin, "How Should Cosmology Relate to Theology?" *The Sciences and Theology in the Twentieth Century,* ed. Arthur Peacocke (Notre Dame: University of Notre Dame Press, 1981) 39.

[7] See S. W. Hawking and G.F.R. Ellis, *The Large Scale Structure of Space-Time* (Cambridge; Cambridge University Press, 1973) 256-275.

[8] See for instance, D. Z. Freedman and P. Nieuwenhuizen, "Supergravity and the Unification of the Laws of Physics," *Sci. Amer.* **238**, No. 2 (Feb. 1978) 126; J. Ellis, "Hope Grows for Supersymmetry," *Nature,* **313** (1985) 626; H. E. Haber and G. L. Kane, "Is Nature Supersymmetric?" *Sci. Amer.* **254,** No. 6 (June 1986) 42.

[9] See M. B. Green, "Unification of Forces and Particles in Superstring Theories," *Nature* **314** (1985) 409; A. De Rújula, "Superstrings and Supersymmetry," *Nature* **320** (1986) 678; M. B. Green, "Superstrings," *Sci. Amer.* **255**, No. 3 (Sept. 1986) 44.

[10] See the contributions by C. J. Isham and F. J. Tipler in this volume.

[11] See S. W. Hawking, "The Boundary Conditions of the Universe," in *Astrophysical Cosmology,* ed. H. A. Bruck, G. V. Coyne, and M. S. Longair (Vatican City State: Pontificiae Academiae Scientiarum Scripta Varia, 1982) 563-574; and J. B. Hartle and S. W. Hawking "Wave Function of the Universe," *Phys. Rev.* **D28** (1983) 2960.

[12] See C. J. Isham in this volume.

[13] See for instance J. O. Burns, "Very Large Structures in the Universe," *Sci. Amer.* **255**, No. 1 (July 1986) 30; A. Dressler, "The Large Scale Streaming of Galaxies," *Sci. Amer.* **257**, No. 3 (Sept. 1987) 38.

[14] See some of the contributions in *Theory and Observational Limits in Cosmology,* ed. W. R. Stoeger (Vatican City State: Vatican Observatory, 1987) particularly those of Ellis (pp. 43-72), MacCallum (pp. 121-141), Partridge (pp. 173- 195), Rowan-Robinson (pp. 203-210), Tyson (pp. 211-230), Geller (pp. 231-254), Nel (pp. 255-274), Stoeger (pp. 275-305), and Dyer (pp. 325-343); henceforth Stoeger (TOLC).

[15] See standard reference works in gravitational physics and cosmology, such as C. W. Misner, K. S. Thorne, and J. A. Wheeler, *Gravitation* (San Francisco: W. H. Freeman and Co., 1973); S. Weinberg, *Gravitation and Cosmology* (New York: John Wiley and Sons, 1972); S. W. Hawking and G. F. R. Ellis, reference 7 above.

[16] W. R. Stoeger, in Stoeger (TOLC), pp. 489-495, and references therein.

[17] G. F.R. Ellis, "Relativistic Cosmology: Its Nature, Aims and Problems," *General Relativity and Gravitation,* eds B. Bertotti, F. de Felice, and A. Pascolini, *Proc. of GR 10* (D. Reidel, Dordrecht, 1984) 215-288; and G. F. R. Ellis in Stoeger (TOLC) 43-72.

[18] See for instance G. F. R. Ellis, S. D. Nel, R. Maartens, W. R. Stoeger, and A. P. Whitman, "Ideal Observational Cosmology," *Physics Reports* **124** (1985) 315-417, particularly 317-321; G. F. R. Ellis, in *General Relativity and Cosmology,* ed. R. K. Sachs (London: Academic Press, 1971) 106-182.

[19] See Note 17 above and G. De Vaucoleurs, *Science* **167** (1970) 1203.

[20] See for instance, W. R. Stoeger, "The Evolving Interaction between Philosophy and the Sciences: Towards a Self-Critical Philosophy," *Philosophy in Science,* Vol. I (Tucson: Pachart Publishing House, 1983) 21-43, and the references therein.

[21] For a description of what I mean by the focus and the evidential grounds of a discipline, see Section 3. In speaking about the different evidential grounds to which science and philosophy each appeal, I do *not* intend to say that they appeal to different ranges of experience. Both the sciences and philosophy appeal to the same range of experience. But each selects somewhat different aspects of that experience as crucial for its own analyses; each abstracts from that common range of experience in a different way, as is clear from what I have said above.

[22] See the detailed discussion of Kant's philosophy of space and time by Peter Mittelstaedt, *Philosophical Problems of Modern Physics,* Boston Studies in the Philosophy of Science, ed. R. S. Cohen and M. W. Wartofsky, Vol. XVIII (Dordrecht: Reidel, 1976) especially 25-29 and 41-50.

[23] I am using the word "retroduction" here in the same way as E. McMullin has used it in his remarks – in terms of moving from effects to causes ("Models of Scientific Inference," *CTNS Bulletin* Vol. 8.2, 1988). "Retroduction," or "abduction," is a term originally used by C. S. Peirce to denote a key process in arriving at scientific knowledge – the construction of hypotheses – to be complemented by deduction and induction. In this sense it also implies the movement from effects to causes which McMullin stresses, citing Pierce to this effect:

> Every inquiry whatsoever takes its rise in the observation .... of some surprising phenomenon, some experience.... The inquiry begins with pondering these phenomena in all their aspects, in search of some point of view when the wonder shall be resolved. At length a conjecture arises that furnishes a possible Explanation, by which I mean a syllogism exhibiting the surprising fact as necessarily consequent upon the circumstances of its occurrence together with the truth of the incredible conjecture, as premises. On account of this Explanation, the inquirer is led to regard his conjecture, or hypothesis, with favor.... Its characteristic formula (that of the "First Stage of Inquiry") of reasoning I term Retroduction, that is reasoning from consequent to antecedent.

C. S. Peirce, *Scientific Metaphysics,* Book II, Ch. 3, paragraph 3, in *Collected Papers of Charles Sanders Peirce* ed. by Charles Hartshorne and Paul Weiss, Vol. VI, paragraph 469, pp. 320-321 (Cambridge, Massachusetts: Belknap Press of Harvard University Press, 1960) and references contained therein.

[24] See particularly G. F. R. Ellis, *Ann. N.Y. Acad. of Sciences* **336** (1980) 130; G. F. R. Ellis, *South African J. of Science* **76** (1980) 540; the references listed in note 16, and first reference listed in note 17.

[25] M. Heller, "On the Cosmological Problem," *Acta Cosmolgica* 14 (1986) 57-72.

[26] There is an outside chance that the universe is not as large as we think it is – that we are seeing around it many times already, and therefore, observing many

widely separated images of each distant object, but are unable to identify those sets of images. On small universes see Dyer, 467-473, and Ellis, 475-486, in Stoeger (TOLC).

[27] G. F. R. Ellis and W. R. Stoeger, "The 'Fitting Problem' in Cosmology," *Class. and Quant. Grav.* **4** (1987) 1627-1729.

[28] See references in Note 26 above.

[29] M. Buckley, this volume.

[30] *Ibid.* 86.

[31] The motive for introducing the assumption of isotropy and spatial homogeneity, of course, is not only simplicity. It is also essentially equivalent to the cosmological principle, which is formulated in terms of the lack of privileged locations in the universe.

[32] In the case of initial conditions, or boundary conditions, we are not concerned with what we normally mean by *fundamental* assumptions or presuppositions. But they *are* assumptions relative to the theory: they are not specified by the theory, but can be specfied by the data, or in some other justifiable way. In many situations it appears that initial conditions can be set arbitrarily, when they are not tied to the data. But in a given case, how far outside the theory do you have to go to find out how these conditions were chosen, picked out from all possible ones? In some cases it is clear that they are given and explained by other complementary, or more fundamental, theories. But in other cases, e.g., cosmology, dynamical systems, etc., it is not at all clear. Is it sometimes just a matter of chance? Can philosophical considerations suitably justify some initial, or boundary, conditions rather than others? Are there examples of this?

[33] Cf., as an example, Michael Heller, "Adventures of the Concepts of Mass and Matter," in *Philosophy in Science*, Vol. III, eds. M. Heller, W. R. Stoeger, J. Życiński (Tucson: Pachart Publishing House, 1987).

[34] See, for instance, Mary Hesse, this volume.

[35] See for instance the development of this theme in Ian Barbour, this volume; Ernan McMullin, "How Should Cosmology Relate to Theology?" in *The Sciences and Theology in the Twentieth Century,* ed. Arthur Peacocke (Notre Dame: University of Notre Dame Press, 1981) 17-51; Ernan McMullin, this volume; Ted Peters, this volume. In speaking of the compatibility, consonance or coherence of science and religion, different authors mean rather different things.

[36] McMullin, 1981, 52. I substantially agree with McMullin's characterization of the desired consonance or coherence. Here I have attempted to elaborate what its basis might be.

[37] Michael Buckley, this volume.

[38] This notion of controlling themata I have derived from Gerald Holton's important works, *Thematic Origins of Scientific Thought* (Cambridge, Mass: Harvard University Press, 1973) 21-29 and 53-68; hereafter cited as *TO*; and *The Scientific Imagination: Case Studies* (Cambridge: Cambridge University Press, 1978) vii-xv and 3-23; hereafter cited as *SI*. Here I am using this concept in a significantly broader way than Holton does to apply to disciplines outside the sciences, like philosophy and theology. I doubt that Holton would agree to such an extension, but I believe it to be within the spirit of his general approach. "It appears therefore that the work of mapping and classifying themata can lay bare basic commonalities between scientific and humanistic concerns that are not equally likely to become evident through other means." (SI). Holton is careful to distinguish his themata from Platonic or Jungian archetypes, myths, synthetic *a priori* knowledge, intuitive apprehensions, etc.. But he does not deny that they may be rooted in, or related to, such psychological structures. *TO*, 60, *SI*, 23.

MODELS OF GOD FOR AN ECOLOGICAL, EVOLUTIONARY ERA: GOD AS MOTHER OF THE UNIVERSE[1]

SALLIE MCFAGUE, Department of Religion, Vanderbilt University

Introduction

David Tracy and Nicholas Lash have called recently for a "collaborative" relationship between science and theology in order to "help establish plausible 'mutually critical correlations' not only to interpret the world but to help change it."[2] They note that relations between science and theology are not only those posed by a recognition of analogies between the two areas on methodological issues, but, more pressingly, by a common concern with the cosmos. Thus, a focus on the cosmos with the intent both to understand it better, and to orient our praxis within it more appropriately, is one collaborative effort for science and theology in our time.

As we near the close of the twentieth century, we have become increasingly conscious of the fragility of our world. We have also become aware that the anthropocentrism that characterizes much of the Judaeo-Christian tradition has often fed a sensibility insensitive to our proper place in the universe.[3] The ecological crisis, epitomized in the possibility of a nuclear holocaust, has brought home to many the need for a new mode of consciousness on the part of human beings, for what Rosemary Ruether calls a "conversion" to the earth, a cosmocentric sensibility.[4]

One collaborative task, therefore, for scientists and theologians is cosmology. While cosmology may mean several different things, the theologian's contribution is concerned with "accounts of the world as God's creation," and, within that broad compass, one specific enterprise especially needed in our time involves "imaginative perceptions of how the world seems and where we stand in it."[5] In other words, I propose that one theological task is an experimental one with metaphors and models for the relationship between God and the world that will help bring about a cosmocentric in place of an anthropocentric sensibility.

This kind of theology, by no means the only kind, could be called heuristic theology, and in analogy with some similar activities in the sciences, it "plays" with possibilities in order to find out, to discover, new fruitful ways to interpret the universe.[6] In the case of an heuristic theology focussed on cosmology, the discovery would be oriented toward "remythologizing" creation as dependent upon God. More specifically, I propose as a modest contribution to the contemporary understanding of a theological cosmology for our time an elaboration of the model of God the creator as mother who gives birth to the universe.[7]

This essay, therefore, will be a "case study" with a theological model for re-envisioning the relationship between God and the universe. Before turning to this study, however, we will make some preliminary comments on the method employed in this kind of theology as well as on metaphors and models, their character and status.

Heuristic Theology

Christian faith is, it seems to me, most basically a claim that the universe is neither indifferent nor malevolent but that there is a power (and a personal power at that) which is on the side of life and its fulfillment. Moreover, the Christian believes that we have some clues for fleshing out this claim in the life, death, and appearances of Jesus of Nazareth. Nevertheless, each generation must venture, through an analysis of what fulfillment could and must mean for its own time, the best way to express that claim. A critical dimension of this expression is the imaginative picture, the metaphors and models, that underlie the conceptual systems of theology. One cannot hope to interpret Christian faith for one's own time if one remains indifferent to the basic images that are the lifeblood of interpretation and that greatly influence people's perceptions and behavior.[8]

Many of the major models for the relationship between God and the world in the Judaeo-Christian tradition are ones that emphasize the transcendence of God and the distance between God and the world: God as king with the world as his realm, God as potter who creates the cosmos by molding it, God as speaker who with a word brings the world to be out of nothing. One has to ask whether these models are adequate ones for our time, our ecological nuclear age, in which the radical interdependence and interrelationship of all forms of life must be underscored.[9] Quite apart from that crisis, however, responsible theology ought to be done in the context of contemporary science, and, were it to take that context seriously, models underscoring the closeness, not the distance, of God and the world would emerge. A. R. Peacocke makes this point well when he says, "there is increasing awareness not only among Christian theologians, but even more among ordinary believers that, if God is in fact the all-encompassing Reality that Christian faith proclaims, then that Reality is to be experienced in and through our actual lives as biological organisms who are persons, part of nature and living in society."[10]

For a number of reasons, therefore, experimentation with models underscoring the intimacy of God and creation may be in order and it is this task, with one model, that I will undertake. I have characterized the theological method operative here as heuristic and concerned with metaphors and models. Let us look briefly at these matters. Heuristic theology is distinct from theology as hermeneutics or as construction.[11] The *Shorter Oxford English Dictionary* defines "heuristic" adjectivally as "serving to find out" and, when employed as a noun related to learning, as "a system of education under which pupils are trained to find out for themselves." Thus heuristic theology will be one that experiments and tests, that thinks in an as-if fashion, that imagines possibilities that are novel, that dares to think differently. It will not accept solely on the basis of authority, but will search for what it finds convincing and persuasive; it will not, however, be fantasy or mere play but will assume that there is something to find out and that if some imagined possibilites fail, others may succeed. The mention of failure and success, and of the persuasive and the convincing, indicates that although I wish to distinguish heuristic theology from both hermeneutical and constructive theology, it bears similarities to both.

If the characteristic mark of hermeneutical theology is its interpretive stance, especially in regard to texts, both the classic text of the Judaeo-Christian tradition (the Hebrew Scriptures and the New Testament) and the exemplary theologies that build on the classic text, then heuristic theology is also interpretive, for it claims that its successful unconventional metaphors are not only in continuity with the paradigmatic events and their significance expressed in this classic text but are also appropriate expressions of these matters for the present time. Heuristic theology, though not bound to the images and concepts in Scripure, is constrained to show that its proposed models are an appropriate, persuasive expression of Christian faith for our time. Hence, while heuristic theology is not limited to interpreting texts, it is concerned with the same "matter" as the classic texts, namely, the salvific power of God.[12]

If, on the other hand, the distinctive mark of constructive theology is that it does not rely principally on classical sources but attempts its articulation of the concepts of God, world, and human being with the help of a variety of sources, including material from the natural, physical, and social sciences as well as from philosophy, literature, and the arts, then heuristic theology is also constructive in that it claims that a valid understanding of God and world for a particular time is an imaginative construal built up from a variety of sources, many of them outside religious traditions. Like theology as construction, theology as heuristics supports the assertion that our concept of God is precisely that—*our concept* of God—and not God. Yet, while heuristic theology has some similarities to constructive theology, it has a distinctive emphasis: it will be more experimental, imagistic, and pluralistic.

Its experimental character means it is a kind of theology well suited for times of uncertainty and change, when systematic, comprehensive construction seems inappropriate if not impossible. It could be called "free theology,"[13] for it must be willing to play with possibilities and, as a consequence, not take itself too seriously, accepting its tentative, relative, partial, and hypothetical character.

Its imagistic character means it stands as a corrective to the bias of much constructive theology toward conceptual clarity, often at the price of imagistic richness.[14] Although it would be insufficient to rest in new images and to refuse to spell out conceptually their implications in as comprehensive a way as possible, the more critical task is to propose what Dennis Nineham calls a "lively imaginative picture" of the way God and the world as we know it are related.[15] It is no coincidence that most religious traditions turn to personal and public human relationships to serve as metaphors and models of the relationship between God and the world: God as father, mother, lover, friend, king, lord, governor.[16] These metaphors give a precision and persuasive power to the construct of God which concepts alone cannot. Because religions, including Christianity, are not incidentally imagistic but centrally and necessarily so, theology must also be an affair of the imagination.

To say that heuristic theology is pluralistic is to insist that, since no metaphor or model refers properly or directly to God, many are necessary. All are inappropriate, partial, and inadequate; the most that can be said is that some aspect or aspects of the God-world relationship are illuminated by this or that model in a fashion relevant to a particular time and place. Models of God are not definitions of God but likely accounts of experiences of relating to God with the help of relationships we know and understand. If one accepts that metaphors (and all language about God) are principally adverbial, having to do with how we relate to God rather than defining the nature of God, then no metaphors or models can be reified, petrified, or expanded so as to exclude all others. One can, for instance, include many possibilities: we can envision relating to God as to a father and a mother, to a healer and a liberator, to the sun and a mountain. As definitions of God, these possibilities are mutually exclusive; as models expressing experiences of relating to God, they are mutually enriching.

In summary, the theology I am proposing is a kind of heuristic construction that, in focussing on the imaginative construal of the God-world relationship, attempts to remythologize Christian faith through metaphors and models appropriate for our time.

Metaphor

What, however, is the character and status of the metaphors and models which are the central concern of heuristic theology? A metaphor is a word or phrase used *in*appropriately.[17] It belongs in one context but is being used in another: the arm of the chair, war as a chess game, God the father. From Aristotle until recently, metaphor was seen mainly as a poetic device to embellish or decorate. Increasingly, however, the idea of metaphor as unsubstitutable is winning acceptance: what a metaphor expresses cannot be said directly or apart from it, for if it could, one would have said it directly. Here, metaphor is a strategy of desperation, not decoration; it is an attempt to say something about the unfamiliar in terms of the familiar, an attempt to speak about what we do not know in terms of what we do know.

Metaphor always has the character of "is" and "is not": an assertion is made but as a likely account rather than a definition.[18] For instance, to say "God is mother" is not to define God as mother, not to assert identity between the terms "God" and "mother", but to suggest that we consider what we do not know how to talk about — relating to God—through the metaphor of mother. The point that metaphor underscores is that in certain matters there can be no direct description. It used to be the case that poetry and religion were thought to be distinctive in their reliance on metaphor, but more recently the use of metaphors and models in the natural and social sciences has widened the scope of metaphorical thinking considerably and linked science and theology methodologically in ways inconceivable twenty years ago.[19]

The difference between a metaphor and a model can be expressed in a number of ways, but most simply, a model is a metaphor with "staying power," that is, a model is a metaphor that has gained sufficient stability and scope so as to present a pattern for relatively comprehensive and coherent explanation.[20] The metaphor of God the father is an excellent example of this. In becoming a model, it has engendered a wide-ranging interpretation of the relationship between God and human beings: if God is seen as father, human beings become children, sin can be seen as rebellious behavior, and redemption can be thought of as restoration to the status of favored offspring.

It should be evident that a theology that describes itself as metaphorical is a theology "at risk." Jacques Derrida, in defining metaphor, writes, "if metaphor, which is *mimesis* trying its chance, *mimesis* at risk, may always fail to attain truth, this is because it has to reckon with a definite absence."[21] As Derrida puts it, metaphor lies somewhere between "nonsense" and "truth", and a theology based on metaphor will be open to the charge that it is closer to the first than the second. This is, I believe, a risk that theology in our time must be willing to run. Theology has usually had a high stake in truth, so high that it has refused all play of the imagination: through creedal control and the formulations of orthodoxy, it has refused all attempts at new metaphors "trying their chance." But a heuristic theology insists that new metaphors and models be given a chance, be tried out as likely accounts of the God-world relationship, be allowed to make a case for themselves. A heuristic theology is, therefore, destabilizing: since no language about God is adequate and all of it is improper, new metaphors are not necessarily less inadequate or more improper than old ones. All are in the same situation and no authority — not scriptural status, liturgical longevity, nor ecclesiastical fiat—can decree that some types of language, or some images, refer literally to God while others do not. None do. Hence, the criteria for preferring some to others must be other than authority, however defined.

Language about God

We come, then, finally to the issue of the status of language about God. R. W. Hepburn has posed it directly: "The question which should be of the greatest concern to the theologian is.... whether or not the circle of myth, metaphor, and symbol is a closed one: and if closed then in what way propositions about God manage to refer."[22] The "truth" of a construal of the God-world relationship is a mixture of belief (Ricoeur calls it a "wager"), pragmatic criteria, and what Philip Wheelwright terms a "shy ontological claim," or, as in Mary Hesse's striking remark, "God is more like gravitation than embarrassment."[23] Belief in God is not taken to be purely a social construct. At least this is what a critical realist would claim: thus, metaphors and models of God are understood to be "discovered" as well as "created", to relate to God's reality not in the sense of being literally in correspondence with it, but as versions or hypotheses of it that the community (in this case, the Church) accepts as relatively adequate.[24] Hence, models of God are not simply heuristic fictions; the critical realist does not accept the Feuerbachian critique that language about God is nothing but human projection. On the other hand, any particular metaphor or model is not the only, appropriate, true one.

The "wager" of this essay is the belief that to be a Christian is to be persuaded that there is a personal, gracious power who is on the side of life and its fulfillment, a power whom the paradigmatic figure Jesus of Nazareth expresses and illuminates. But when we try to say something more, we turn, necessarily to the "loves" we know in our deepest personal experiences — the loves, for instance, of parents, lovers, and friends. Can we say that these loves, the love, for instance, of a mother, is descriptive of God as God *is*? We do not *know* whether the inner being can be described with this model: at most we wager that it can and, more significantly, we live *within* the model, testing our wager by its consequences. These consequences are both theoretical and practical. An adequate model will be illuminating, fruitful, have relatively comprehensive explanatory ability, be relatively consistent, able to deal with anomalies, etc. Of equal importance, an adequate model will, given the "wager" that God is on the side of life and its fulfillment, support that wager.

This is largely (though not totally) a functional, pragmatic view of truth, with heavy stress on the implications of certain models for the quality of human and nonhuman life (since the initial assumption is that God is on the side of life). The principal points I would stress are two: a model of God is verified mainly by its consequences, and this verification takes place within the community, the Church. That is to say, a novel metaphorical construal of the God-world relationship is tested principally by functional rather than metaphysical criteria and that testing must win the acceptance of the Church. On the first point, a praxis orientation does not deny the possibility of the "shy ontological claim," but it does acknowledge both the mystery of God and the importance of truth as practical wisdom. Thus, it acknowledges with the apophatic tradition that we really do not *know* the inner being of divine reality: the hints and clues we have of the way things are, whether we call them religious experiences, revelation, or whatever, are too fragile, too little (and often too negative) for heavy metaphysical claims. Rather, in the tradition of Aristotle, truth means constructing the good life for the *polis,* though for our time, this must mean for the cosmos. A "true" model of God will be one that is a powerful, persuasive construal of God as being on the side of life and its fulfillment in our time.[25] On the second point, the decision concerning acceptable models rests with the community, the Church, which in its wisdom must judge whether novel models are in continuity with the deepest beliefs that characterize the Christian faith.

God as Mother of the Universe

"Father-Mother God, loving me, guard me while I sleep, guide my little feet up to thee." This prayer, which theologian Herbert Richardson reports reciting as a child, impressed upon his young mind that if God is both father and mother, then God is not like anything else he knew.[26] The point is worth emphasizing, for as we begin our experiment with the model of God as mother, we recall that metaphors for God, far from reducing God to what we understand, underscore by their multiplicity and lack of fit the unknowability of God. This crucial characteristic of metaphorical language for God is lost, however, when only one important personal relationship, that of father and child, is allowed to serve as a framework for speaking of the God-human relationship. In fact, by excluding other relationships as metaphors, the model of father becomes idolatrous, for it comes to be viewed as a description of God.[27]

In this essay the model I have employed has sometimes been "God as mother" and sometimes "God as parent;" the emphasis will be on the former, but the latter will have a role as well. Our tradition has thoroughly analyzed the paternal metaphors, albeit mainly in a patriarchal context. The goal of the present reflections will be to investigate the potential of the maternal model but to do so in a fashion that will provide an alternative interpretive context for the paternal model—a parental one.[28]

God as the giver of life, as the power of being in all being, can be imaged through the metaphor of mother—and of father. Parental love is the most powerful and intimate experience we have of giving love whose return is not calculated (though a return is appreciated): it is the gift of *life as such* to others. Parental love wills life and when it comes, exclaims, "It is good that you exist."[29] Moreover, in addition to being the gift of life, parental love nurtures what it has brought into existence, wanting growth and fulfillment for all.[30] This agapic love is revolutionary, for it loves the weak and the vulnerable as well as the strong and beautiful. No human love can, of course, be perfectly just and impartial, but parental love is the best metaphor we have for imaging the creative love of God.[31]

A caveat is necessary at this point. We have characterized this love as agape, but that designation needs qualification since the usual understanding of agape sees it as totally unmotivated, disinterested love.[32] The discussions on the nature of divine love, especially in Protestant circles and principally motivated by the desire to expunge any trace of interest or need on the part of God toward creation, paint a picture of God as isolated from creation and in no way dependent on it. Reflections about agape as definitive of divine love have, unfortunately, usually focussed on redemption, not creation, and as a result have stressed the distinterested character of God's love, which can overlook the sin in the sinners and love them anyway. But if it is God's creative love that we characterize as agapic, then it is a statement to created beings: "It is good that you exist!" Agape has been characterized as the love that gives, and as such it belongs with the gift of creation.

Let us now consider our model in more detail: the model of parental love for God's agapic, creative love. Why is this a powerful, attractive model for expressing the Christian understanding of creation in our time? If the heart of Christian faith for an ecological, and nuclear-threatened, age must be a profound awareness of the preciousness and vulnerability of life as a gift we receive and pass on, with appreciation for its value and desire for its fulfillment, it is difficult to think of any metaphor more apt than the parental one. There are three features basic to the parental model which will give flesh to this statement: it brings us closest to the beginnings of life, to the nurture of life, and to the impartial fulfillment of life. Much of the power of the parental model is its immediate connection with the mystery of new life. Becoming a biological parent is the closest experience most people have to an experience of creation, that is, of bringing into existence. No matter how knowledgeable one is biologically, no matter how aware that human beings by becoming parents are simply doing what all animals do in passing life along, becoming a biological parent is for most people an awesome experience inspiring feelings of having glimpsed the heart of things. We are, after all, the only creatures who can think about the wonder of existence, the sheer fact that "things are," that the incredible richness and complexity of life in all its forms has existed for millions of years, and that as part of the vast, unfathomable network of life, we both receive it from others and pass it along. At the time of the birth of new life from our bodies, we feel a sense of being co-creators, participating at least passively in the great chain of being. No matter how trite and hackneyed the phrases have become — "the miracle of birth," "the wonder of existence," and so on — on becoming a parent one repeats them again and joins the millions of others who marvel at their role in passing life along.[33]

The physical act of giving birth is the base from which this model derives its power, for here it joins the reservoir of the great symbols of life and of life's continuity: blood, water, breath, sex, and food. In the acts of conception, gestation, and birth all are involved, and it is therefore no surprise that these symbols became the center of most religions, including Christianity, for they have the power to express the renewal and transformation of life — the "second birth" — because they are the basis of our "first birth." And yet, at least in Christianity, our first birth has been strangely neglected; another way of saying this is that creation, the birth of the universe and all its beings, has not been permitted the imagery that this tradition uses so freely for the transformation and fulfillment of creation. Why is this the case?

One reason is surely that Christianity, alienated as it always has been from female sexuality, has been willing to image the second, "spiritual", renewal of existence in the birth metaphor, but not the first, "physical", coming into existence.[34] In fact, as we shall see, in the Judaeo-Christian tradition, creation has been imaginatively pictured as an intellectual, aesthetic "act" of God, accomplished through God's word and wrought by God's "hands", much as a painting is created by an artist or a form by a sculptor. But the model of God as mother suggests a very different kind of creation, one which underscores the radical dependence of all things on God, but in an internal rather than an external fashion. Thus, if we wish to understand the world as in some fashion "in" God rather than God as in some fashion "in" the world, it is clearly the parent *as mother* that is the stronger candidate for an understanding of creation as bodied forth from the divine being. For it is the imagery of gestation, giving birth, and lactation that creates an imaginative picture of creation as profoundly dependent on and cared for by divine life.[35] There simply is no other imagery available to us that has this power for expressing the interdependence and interrelatedness of all life with its ground. All of us, female and male, have the womb as our first home, all of us are born from the bodies of our mothers, most of us are fed by our mother. What better imagery could there be for expressing the most basic reality of existence: that we live and move and have our being in God?[36]

If the symbol of birth were allowed openly and centrally into the tradition, would this involve a radical theological change? Would it mean a different understanding of God's relation to the world? We will be dealing with that issue soon in more detail, but the simple answer is yes, the view associated with birth symbolism would be different from the distant, anthropocentric view of the tradition's monarchical model in which God relates to the world as a king to subjects in his realm. It would not, however, identify God and the world. By analogy, mothers, at least good ones, encourage the independence of their offspring, and even though children are products of their parents' bodies, they are often radically different from them.[37]

The power of the parental model for God's creative, agapic love only begins with the birth imagery. Of equal importance is the ability of the model to express the nurturing of life and, to a lesser extent, its impartial fulfullment. It is at these levels that the more complex theological and ethical issues arise, for the divine agapic love that nurtures all creatures epitomizes justice at the most basic level of the fair distribution of the necessities of life, and divine agapic love impartially fulfilling all of creation epitomizes *inclusive* justice. The parental model of God is especially pertinent as a way of talking about God's "just" love, the love that attends to the most basic needs of all creatures. It is important to look more closely at the way the model expresses the nurture and inclusion of all of life.

Parents feed the young. This is, across the entire range of life, the most basic responsibility of parents, often of fathers as well as of mothers. Among most animals it is instinctual and is often accomplished only at the cost of the health or life of the parent. It is not principally from altruistic motives that parents feed the young but from a base close to the one that brought new life into existence, the source that participates in passing life along. With human parents, the same love that says, "It is good that you exist!" desires that existence to continue, and for many parents in much of the world that is a daily and often horrendous struggle. There is, perhaps, no picture more powerful to express "giving" love than that of parents wanting, but not having the food, to feed their starving children.

The Christian tradition has paid a lot of attention to food and eating imagery. In fact, one could say that such imagery is probably at the center of the tradition's symbolic power: not only does the New Testament portrait of Jesus of Nazareth paint him as constantly feeding people, and eating with outcasts, but the Church has as its central ritual a eucharistic meal reminiscent of the passion and death of Jesus and suggestive of the eschatological banquet yet to come. The power of food imagery, however, as with other basic symbols of life, depends upon acknowledging its physical connection, for the use of food as a symbol of the renewal of life must be grounded in food's basic role as the maintainer of life. Unfortunately, however, in Christianity, the practical truth that food is basic to all life has often been neglected. A tradition that uses food as a symbol of spiritual renewal has often forgotten what parents know so well: that the young must be fed.

A theology that sees God as the parent who feeds the young and, by extension, the weak and vulnerable, understands God as caring about the most basic needs of life in its struggle to continue. One can extend nurture to include much more than attention to physical needs, but one ought not to move too quickly, for the concern about life and its continuation that is a basic ingredient in the sensibility needed in our time has often been neglected by Christianity in its interest in "spiritual" well-being. An evolutionary, ecological sensibility makes no clear distinction between matter and spirit or between body and mind, for life is a continuum and cannot flourish at the so-called higher levels unless supported at all levels. God as the parent loves agapically in giving with no thought of return the sustenance needed for life to continue. This is creative love, for it provides the conditions minimally necessary for life to go on.[38]

Finally, God as parent wants *all* to flourish. Divine agapic love is inclusive and hence epitomizes impartial justice. Parental love can model the impartiality of divine love in only a highly qualified way; yet it is central to the essence of agapic love to stress that it is impartial , or better, inclusive. This is a more desirable way to express what is at stake than to call the love disinterested, which suggests that God's love is detached, unconcerned, or perfunctory. In fact, the opposite is intended, for agapic love functions in spite of obstacles and in this way can be love of *all,* whatever the barriers may be. God as mother is parent to *all* species and wishes all to flourish.[39] We can reflect this inclusiveness in the model of parent only in partial and distorted fashion, for as parents we tend to focus on our own species and on particular individuals within that species. To be sure, when we extend the model beyond its physical base to include our parental inclinations toward human children not our own, as well as toward life forms not our own, a measure of impartiality, of inclusiveness, emerges, but only as a faint intimation of divine agape.[40]

The kind of theological statement issuing from the model of God as mother is, of course, the doctrine of creation. The doctrine of creation, so basic to the Judaeo-Christian tradition, has in the past three hundred years undergone various revisions as scientific knowledge has questioned the received view of many centuries. The received view consisted of a nest of shared beliefs, but the two most important for our concern are that God created *ex nihilo,* from "nothing", and that God created hierarchically, with the physical subordinated to the spiritual.[41] Both of these notions support dualism: the absolute distinction of God from the world, and the inferiority of matter to spirit, body to mind.[42] Quite apart from the scientific difficulties of the traditional view of creation, the imaginative picture it paints is of a God fashioning the world, either intellectually by word (a creation of the mind) or aesthetically by craft (a creation of the hands), but in either case out of what is totally different from God, and in a manner that places humanity above nature, spirit above body. The principal elements of the artistic model of creation are evident in the Genesis stories: the earth that was "without form and void;" the "words" of God that bring into existence light and earth, sky and water, plants and animals; the special creation of man, sculpted by God from the earth; the superiority of human beings to nature, which they are to "subdue;" and the superiority of man over woman, who is formed from his side. The two versions of the story differ, but the picture that endured and fed into the tradition's consensus was of a creation totally different from God and structured hierarchically, descending through angels (all spirit), to man (mainly spirit), to woman (mainly body), and on down the line. Although this picture has been discredited scientifically and has certainly faded considerably in the popular mind as well, its principal force hangs on, bespeaking distance and difference between God and the world, and the superiority of spirit to body, humanity to nature. It hangs on in part because, in spite of impressive philosophical and theological attempts at revision, ranging from deism and idealism to process thought, no new imaginative picture has replaced the old one.

But just such an alternative imaginative picture does emerge from the model of God as mother. The kind of creation that fits with this model is creation not as an intellectual act but as a physical event: the universe is bodied forth from God, it is expressive of God's very being: it could, therefore, be seen as God's "body".[43] It is not something alien to God but is from the "womb" of God, formed through "gestation". There are some implications of this picture we need to follow out, but first we must remind ourselves once again that this is a picture — but then so is the artistic model. We are not claiming that God creates by giving birth to the world as her body; what we are suggesting is that the birth metaphor is both closer to Christian faith and to a contemporary evolutionary, ecological context than the alternative craftsman model.

The first implication of this model is that the universe and God are neither totally distant nor totally different. Is this not going against the heart of Christian faith, which proclaims the utter majesty and sovereignty of God, the transcendence of God over all reality, the absolute difference between the infinite and the finite? The rendering of Christianity implied in this question derives from the monarchical model (God as king and the world as his realm), aided as it was by Aristotelian and Platonic notions of the distance of God from the world. It does not come from the Hebraic roots of Christianity, where even the high, holy One was in intimate, covenantal relationship with his chosen people, nor from Christian beginnings, for however one interprets the incarnation, it implies that God has "come near." To say that the universe and God are neither distant nor different implies that they are close and similar, in a way, for instance, that a mother and her child have a sense of affinity and kinship. What is critical in our model of creation from God as mother is not whether this makes God and the world identical, for obviously it does not. What is critical is that the model underscores, as the artistic model does not, God's closeness to us in the world in which we live, rather than portraying God as a being who miraculously intervenes in our lives or public affairs.

Is this creation, then, God's child or God's body? The model of creation as the birth of the universe from God wavers at this point, and its nonsense side emerges. For when we give birth it is not to our bodies but to children of our bodies. We are not creators, we are only those who pass life along, and though at the time of giving birth we may feel like co-creators, we are mainly the passive conduits of the life growing in and passing through our bodies. But God is creator, the source of life, of all forms of life: that is the critical theological statement, and the theological way to imagine that statement for our time must be commensurate with the holistic, evolutionary sensibility. The picture of the universe as the visible creation coming from God's reality and expressive of God — the picture of God giving birth to her "body", that is, to life, even as we give birth to children — provides a model of kinship, concern, and affinity markedly different from the distance and difference of the artistic model. The dualism of God and the world is undercut.[44]

The other implication of our model is that it also overturns the dualisms to body and mind, flesh and spirit, nature and humanity. God's body, that which supports all life, is not matter or spirit but the matrix out of which everything evolves. In this picture, God is not spirit over against a universe of matter, with human beings dangling in between, chained to their bodies but eager to escape to the world of spirit. The universe, from God's being, is properly body (as well as spirit) because in some sense God is physical (as well as beyond the physical). This shocking idea — that God is physical — is one of the most important implications of the model of creation by God the mother. It is an explicit rejection of Christianity's long, oppressive, and dangerous alliance with spirit against body, an alliance out of step with a holistic, evolutionary sensibility as well as with Christianity's Hebraic background.

To say that God is physical is not, however, to reverse the hierarchy and to proclaim a new gospel celebrating nature and the body. Christianity is not a nature religion or a fertility cult. But that should not be taken to mean, as it has often been, that Christianity is antinature or antibody. Rather, it suggests a special kind of relationship between God and the world: the universe as God's "other". If the universe is God's "other", if God's body is the entire organic complex of which we are a part — that is, if something similar to but not identical with God is God's other — then, God is physical as well as spiritual. God will therefore need the world, want the world, not simply as dependent inferior (flesh subordinated to Spirit) but as offspring, beloved companion.

But how can God relate to the universe as her own body? Is this not a monistic or narcissistic relationship? And if God *does* relate to the universe as body, how can this be a significant relationship with a personal "other"? Are there not tensions here between the model of the world as God's body and the personal model of God as mother? The central issue in this mix of questions concerns how divine relationality is perceived in the overall picture of God as mother of her body, the world or universe. One's initial response to that picture may be that the identification of God and God's "other," the universe, is too close, for the universe is, after all, God's body. But the tradition has always struggled with monistic,indeed, narcissistic tendencies when attempting to speak of divine relationality. Consider orthodox Trinitarianism: here God's "other" is God's own self, for the relationality of God is seen in terms of the relationships among the persons of the immanent Trinity. This solipsistic view is epitomized in C. S. Lewis's statement that God is "at home in the land of the Trinity," and, entirely self-sufficient and needing nothing, "loves into existence totally superfluous creatures."[45] In our model, God's "other" is the universe, which, to be sure, comes from God, but is not identical with God. Is not this understanding of divine relationality less monistic and narcissistic than the traditional one?

Let us look briefly at two issues of divine relationality: God as the mother of her own body, or the question of the source of that body, and God as interacting with it as mother.

To say that God is the mother of her own body, we must first recall that this particular "body" is nothing less than all that is — the universe or universes that cosmologists speak of. The body of God, then, is creation, understood as God's self-expression; it is formed in God's own reality, bodied forth in the eons of evolutionary time, and supplied with the means to nurture and sustain billions of different forms of life. And what could that body be except God's *own* creation? Could some other creator have made it? If so, then *that* creator would be God. *We* give life only to others of our own species, but God gives life to *all* that is, all species of life and all forms of matter. In a monotheistic, panentheistic theology, if one is to understand God in some sense as physical and not just spiritual, then the entire "body" of the universe is "in" God and is God's visible self-expression. This body, albeit a strange one if we take ours as the model, belongs to God. God is mother of all reality; God is the source of all that is. As Julian of Norwich writes of God as mother: "We owe our being to him and this is the essence of motherhood."[46] The seeming incoherence here, I think, comes from the fact that our bodies are given to us, as are all other aspects of our existence. But as the creator of all that is, God is necessarily the source, the mother, of her own body.

But how does God as mother interact with a body as her "other"? Does not the personal model of God as mother demand a personal "other"? We need to recall again the nature of God's body: it is nothing less than all that exists, which includes creatures with various levels of spirit and mind. In the model of God relating to the universe as mother, there would be personal counterparts, though they need not all be human ones. The body that God relates to in this model includes various levels of responsiveness — as not only the process theologians have pointed out, but as a sacramental perspective does as well. We frequently think of ourselves as the only ones in the world who can respond, but that is simply another witness to our anthropocentrism. Medieval Catholic analogical thought knew better: each being, in its particularity and difference from all others, gives the creator glory as it fulfills its own being. Such response need not be fully intentional or even minimally so for a relationship to exist. A body such as the entire universe, with creatures like ourselves who are highly responsive as well as others of minimal response, could provide personal counterparts to which God relates as mother.

Before we close out this brief sketch of God as the mother of the universe, we need to note that as creator this God is also judge. For as the giver of life, God judges those who thwart the nurture and fulfillment of her beloved creation. And here we see a clue to the nature of sin. God as mother is creator and judge in a way quite different from the way in which God as artist is envisioned to be creator and judge, for in the picture of God as artist, God is angry because his good, pleasing creation is spoiled by what upsets its balance and harmony, or because what he molded rebels against the intended design, whereas in the picture of God as mother, God is angry because what comes from her being and belongs to her lacks the food and other necessities to grow and flourish. The mother-God as creator is necessarily judge, at the very basic level of condemning as the primary (though not the only) sin the inequitable distribution of basic necessities for the continuation of life in its many forms. In this view, sin is not "against God," the pride and rebellion of an inferior against a superior, but "against the body," the refusal to be part of an ecological whole whose continued existence and success depend upon a recognition of the interdependence and interrelatedness of all species. The mother-God as creator, then, is also involved in "economics", the management of the household of the universe, to insure the just distribution of goods.

Conclusion

In closing I would remind the reader of the *kind* of project this paper is: the kind of theology being advanced here is metaphorical or heuristic theology which experiments with metaphors and models. Its claims are small. As remythologization, such theology acknowledges that it is, as it were, painting a picture. The picture may be full and rich, but it *is* a picture, an elaboration of a few key metaphors and models. Nonetheless, admitting that this kind of enterprise is mainly elaboration, we claim that some imaginative pictures are better than others, both for human habitation and as expressions of the gospel of Christian faith at a particular time. So we try out different models and metaphors in an attempt to talk about what we do not know how to talk about: the relationship between God and the world, from a Christian perspective, for our time. We flesh out these metaphors and models sufficiently to see their implications and the case that can be made for them. Hence, although this theology "says much," it "claims little." It is a postmodern, highly skeptical, heuristic enterprise, which suggests that in order to be faithful to the God of its tradition — the God on the side of life and its fulfillment — we must try out new pictures that will bring the reality of God's love into the imaginations of the men and women of today. That task must be attempted and attempted again. My contribution is a modest experiment with a few metaphors; other experiments with other metaphors are appropriate and needed.

NOTES

[1] This paper is based on material from my book, *Models of God: Theology for an Ecological, Nuclear Age* (Philadelphia: Fortress Press, 1987). In that work I experiment with the models of God as mother/creator, lover/redeemer, friend/sustainer of the world understood as God's body. The present essay is written in two tracks: the central argument, which appears as the text, and the scholarly discussion, especially as regards issues pertinent to this conference, which appears in the notes.

[2] David Tracy and Nicholas Lash, ed., *Cosmology and Theology* (Edinburgh and New York: T. & T. Clark and Seabury Press, 1983). Tracy and Lash contrast the collaborative model with two others, described as confrontational and concordist, neither of which is approppropriate for our time. In a similar fashion, Ernan McMullin asks for "consonance" between scientific and theological views "How Should Cosmology Relate to Theology?" in *The Sciences and Theology in the Twentieth Century*, ed. A. R. Peacocke (Notre Dame: University of Notre Dame Press, 1981) 52. Likewise, A. R. Peacocke seeks a "congruence" between science and theology in which they are mutually enriching while autonomous and distinctive. It is in this spirit that the present essay is written. However, those of us concerned to find such relationships between distinct fields should heed the cautious word of Cambridge physicist Sir Brian Pippard when he says that each field thrives by virtue of its own methods and not by aping those of others. "Insta-bilitity and Chaos: Physical Models of Everyday Life," *Interdisciplinary Science Reviews*, 7 (1982) 95-96.

[3] Present-day concern among theologians with anthropocentrism or homocentrism is wide-spread. James M. Gustafson, in the first volume of *Ethics from a Theocentric Perspective*, (Chicago: University of Chicago Press, 1981) 82, states the concern succinctly with his pithy remark that while human beings are the *measurers* of all things, they are not the *measure* of all things. Our anthropocentrism can, he believes, be overcome only by a profound acknowledgement of the sovereignty of God, a consent to divine governance which sets limits to human life and in which we "relate to all things in a manner appropriate to their relations to God" (p. 113). Only then will human beings, he says, "confront their awesome possibilities and their inexorable limitations" (pp. 16-17). Stephen Toulmin echoes these sentiments in an elegant statement on the cosmos understood on the model of our "home." See *The Return to Cosmology: Postmodern Science and the Theology of Nature* (Berkeley: University of California Press, 1982) 272. Sigurd Daecke finds anthropocentrism to be deeply embedded in Protestant theologies of creation reaching back to Luther ("I believe that God has created *me*") and Calvin (nature is the stage for salvation history) and finding a twentieth-century home in the humanistic individualism of Bultmann as well as the Christocentrism of Barth ("the reality of creation is known in Jesus Christ"). Even more recent church statements, including ecumenical ones from the WCC 1979 "Faith, Science and the Future" conference do not have a unitary view of human beings, nature, and God, speaking only in terms of "relation" and "connection" (see his essay "Profane and Sacramental Views of Nature" in *The Sciences and Theology . . .*, ed. Peacocke). In a somewhat different vein, Tracy and Lash, while agreeing that the anthropic principle is untenable in science, find a certain kind of anthropocentrism appropriate in theology: 1) human beings are both products of and interpreters of the evolutionary process; 2) human beings are responsible for much of our world's

ills: "if we are the 'center' of anything, we are the center of 'sin,' of the self-assertive disruption and unraveling of the process of things, at least on our small planet" (*Cosmology and Theology,* 280).

[4] Rosemary Ruether, *Sexism and God-Talk: Toward a Feminist Theology* (Boston: Beacon Press, 1983) 89.

[5] Tracy and Lash define "cosmology" in a variety of ways. "The term can refer to theological accounts of the world as God's creation; or to philosophical reflection on the categories of space and time; or to observational and theoretical study of the structure and evolution of the physical universe; or, finally, to 'world views:' unified imaginative perceptions of how the world seems and where we stand in it" (*Cosmology and Theology,* p. vii). Peacocke finds a similarity of intention in religious and scientific cosmologies: "Both attempt to take into account as much of the 'data' of the observed universe as possible and both use criteria of simplicity, comprehensiveness, elegance, and plausibility.... Both direct themselves to the 'way things are' not only by developing cosmogonies, accounts of the origin of the universe, but also in relation to nearer-at-hand experience of biological and inorganic nature." See *Creation and the World of Science* (Oxford: Clarendon Press, 1979) 31. The *intention* of my modest effort with the model of God as mother of the universe falls within these parameters.

[6] Many philosophers of science claim that science is also an imaginative activity. Max Black insists that the exercise of the imagination provides a common ground between science and the humanities. See *Models and Metaphors* (Ithaca, NY: Cornell University Press, 1962) 243. Mary Hesse suggests that "art" or "play" characterizes some aspects of scientific problem-solving. See "Cosmology as Myth," in *Cosmology and Theology,* ed. Tracy and Lash, 50. See also my *Metaphorical Theology: Models of God in Religious Language* (Philadelphia: Fortress Press, 1982) Ch. 3, for a treatment of the role of the imagination in science and theology.

[7] Paul Tillich and A. R. Peacocke have suggested female imagery for God the creator in order to underscore the immanence of God in the world. Hidden away in the third volume of his *Systematic Theology,* Tillich says that the symbolic dimension of the "ground of being" points to the mother-quality of giving birth, carrying, and embracing ([Chicago: University of Chicago Press, 1963] 293-94). He goes on to say that the uneasy feeling that many Protestants have about the first statement about God — that God is the power of being in all being — arises from the fact that their consciousness is shaped by the demanding father image for whom righteousness and not the gift of life is primary. What the father-God gives is redemption from sins; what the mother-God gives is life itself. Peacocke claims that in a panentheistic understanding of the relationship between God and the world, God is not "in" the world, but the world is "in" God. He goes on to say that most understandings of God as creator are dominated by stress on the externality of God's creative acts, with God "regarded as creating something external to himself, just as the male fertilizes the womb from outside. But mammalian females, at least, create within themselves and the growing embryo resides within the female body and this is a proper corrective to the masculine picture — it is an analogy of God creating the world within herself, we would have to say.... God creates a world that is, in principle and in origin, other than him/herself but creates it, the world, within him/herself" (*Creation and the World of Science,* 142).

[8] Dennis Nineham writes that it is "at the level of the *imagination* that contemporary Christianity is most weak." He goes on to say that people "find it hard to believe in God because they do not have available to them any lively imaginative picture of the way God and the world as they know it are related.

What they need most is a story, a picture, a myth, that will capture their imagination, while meshing in with the rest of their sensibility in the way that messianic terms linked with the sensibility of first-century Jews, or Nicene symbolism with the sensibility of philosophically-minded fourth-century Greeks" (John Hick, ed., *The Myth of God Incarnate* [Philadelphia: Westminster Press, 1977] 42).

[9] What is a stake here is not a sentimental love of nature or a leveling of all distinctions between human beings and other forms of life but the realization, as Teilhard de Chardin says, that his and everyone else's "poor trifling existence" is "one with the immensity of all that is and all that is still in the process of becoming" (*Writings in Time of War,* trans. René Hague [London: William Collins Sons, 1968] 25). We are not separate, static, substantial individuals relating in external ways — and in ways of our choice — to other individuals, mainly human ones, and in minor ways to other forms of life. On the contrary, the evolutionary, ecological perspective insists that we are, in the most profound ways, "not our own": we belong, from the cells of our bodies to the finest creations of our minds, to the intricate, constantly changing cosmos. Wallace Stevens says poetically what contemporary evolutionary science says technically: "Nothing is itself taken alone. Things are because of interrelations or interconnections" (*Opus Posthumous,* ed. S. F. Morris [New York: Alfred A. Knopf, 1957] 163).

[10] Arthur R. Peacocke, *Creation and the World of Science,* 16-17.

[11] An outstanding example of theology as hermeneutics is the work of David Tracy, especially his *The Analogical Imagination: Christian Theology and the Culture of Pluralism* (New York: Crossroad; London: SCM Press, 1981). A fine illustration of theology as construction is the work of Gordon D. Kaufman, especially his *The Theological Imagination: Constructing the Concept of God* (Philadelphia: Westminster Press, 1981).

[12] How that power is understood involves specifying the material norm of Christian faith. It involves risking an interpretation of what, most basically, Christian faith is about. My interpretation is similar to that of the so-called liberation theologies. Each of these theologies, from the standpoint of race, gender, class, or another basic human distinction, claims that the Christian gospel is opposed to oppression of some by others, opposed to hierarchies and dualisms, opposed to the domination of the weak by the powerful. This reading is understood to be commensurate with the paradigmatic story of the life, message, and death of Jesus of Nazareth, who in his parables, his table fellowship, and his death offered a surprising invitation to *all,* especially to the outcast and the oppressed. It is a destablizing, inclusive, nonhierarchical vision of Christian faith, the claim that the gospel of Christianity is a new creation for *all* of creation — a life of freedom and fulfilment for all. As Nicholas Lash has said in a variety of contexts, the story as told must be "a different version of the same story, not a different story" See *Theology on the Road to Emmaus* (London: SCM Press, 1986) 30, 44.

[13] Robert P. Scharlemann uses this phrase to describe the kind of theology that constructs theological models and he sees it as an alternative to other kinds of theology. "It is free theology in the sense that it can make use of any of these materials — confessional, metaphysical, biblical, religious, and secular — without being bound to them" ("Theological Models and Their Construction," *Journal of Religion* **53** [1973] 82-83).

[14] The relationship between image and concept which I support is articulated by Paul Ricoeur, whose well-known phrase, "The symbol gives rise to thought," is balanced by an equal emphasis on thought's need to return to its rich base in symbol. See especially "Biblical Hermeneutics," *Semeia* **4** (1975); and Study 8 in

The Rule of Metaphor: Multi-disciplinary Studies of the Creation of Meaning in Language, trans. Robert Czerny (Toronto: University of Toronto Press, 1977).

[15] Dennis Nineham, *The Myth of God Incarnate*, p. 201-02.

[16] Debate concerning personal models for God is widespread in theological circles. For a discussion of this debate see *Models of God*, pp. 78-87. However, it appears that a "congruence" between science and the Judaeo-Christian tradition may emerge at this point, if one says "yes" to Peacocke's question: "Does not the continuity of the universe, with its gradual elaboration of its potentialities, from its dispersal c. 10 thousand million years ago as an expanding mass of particles to the emergence of persons on the surface of the planet Earth (perhaps elsewhere as well) imply that any categories of 'explanation' and 'meaning' must at least *include* the personal?" He goes on to say an affirmative answer implies that "the source and meaning of all-that-is .... is least misleadingly described in supra-personal terms" (*Creation and the World of Science*, p. 75).

[17] There are probably as many definitions of metaphor as there are metaphoricians. I am grateful to Janet Martin Soskice for her straight-forward, uncomplicated definition of metaphor: "Metaphor is a figure of speech in which one entity or state of affairs is spoken of in terms which are seen as being appropriate to another" (*Metaphor and Religious Language* [Oxford: Clarendon Press, 1985] 96).

[18] My position here is very close to that of Ricoeur, as found in his *The Rule of Metaphor* and elsewhere.

[19] The conversation between science and theology on the matter of metaphors and models is a long and interesting one, with our conference as one of its results. I am especially indebted to the work of Ian Barbour, Mary Hesse, Frederick Ferré, E. H. Hutten, Rom Harré, Max Black and N. R. Hanson, among others, for their interpretations of this conversation. For my modest contribution to it, see *Metaphorical Theology*, Chs. 3 and 4.

[20] I find Ian Barbour's definition of theoretical models in science would serve as well in theology: ".... theoretical models are novel mental constructions. They originate in a combination of analogy to the familiar and creative imagination in creating the new. They are open-ended, extensible, and suggestive of new hypotheses.... Such models are taken seriously but not literally. They are neither pictures of reality nor useful fictions; they are partial and inadequate ways of imagining what is not observable" (*Myths, Models and Paradigms: A Comparative Study in Science and Religion* [New York: Harper and Row, 1974] 47-48).

[21] Jacques Derrida, "White Mythology: Metaphor in the Text of Philosophy," *New Literary History* **6** (1974) 42.

[22] "Demythologizing and the Problem of Validity," in *New Essays in Philosophical Theology*, ed. Antony Flew and Alasdair MacIntyre (London: SCM Press, 1955) 237.

[23] Michael Arbib and Mary Hesse, *The Construction of Reality*, (Cambridge: Cambridge University Press, 1986) 5.

[24] This perspective acknowledges with Nelson Goodman that, as Ernest Gombrich insists, "there is no innocent eye. The eye comes always ancient to its work.... Nothing is seen nakedly or naked" (*Languages of Art: An Approach to a Theory of Symbols* [Indianapolis: Bobbs-Merrill, 1968] 7-8). This means, of course, that we are always dealing in interpretations of reality (the reality of God or anything else); hence, there are no "descriptions" but only "readings." Some readings, however, are more privileged than others and this judgment will be made by the relevant community. New readings are offered in place of conventional or accepted ones, not with the view that they necessarily correspond more adequately to the reality in question *in toto*, but that they are a discovery/creation of some

aspect of that reality overlooked in other readings, or one especially pertinent to the times, etc.

[25] The heavily pragmatic view of truth suggested here is similar to that of some liberation theologians and rests on an understanding of "praxis" not simply as action vs. theory, but as a kind of reflection, one guided by practical experience. Praxis is positively, "the realization that humans cannot rely on any ahistorical, universal truths to guide life" (Rebecca Chopp, *The Praxis of Suffering: An Interpretation of Liberation and Political Theologies* [Maryknoll, NY: Orbis Books, 1986] 36). It assumes that human life is fundametally practical; hence, knowledge is not most basically the correspondence of some understanding of reality with "reality-as-it-is," but it is a continual process of analysis, explanation, conversation, and application with both theoretical and practical aspects. This understanding is not new: Aristotle's view of life in the *polis* as understood and constructed is similar: such knowledge is grounded in concrete history within the norms, values, and hopes of the community. Likewise, Augustine's *Confessions* is not a theoretical treatise on the nature of God, but a history, his own concrete, experiential history, of God acting in his life. On the present scene we see a clear turn toward pragmatism in the work of Richard Rorty, Michel Foucault, Richard Bernstein and others. While I would not identify my position with the extremes of pragmatism, it is, nonetheless, a healthy reminder that religious truth, whatever may be the case with other kinds of truth, involves issues of value, of consequences, of the quality of lived existence.

[26] Elizabeth Clark and Herbert Richardson, eds., *Women and Religion* (New York: Harper and Row, 1977) 164-65.

[27] For a fuller treatment of this point see *Metaphorical Theology*, Ch. 5.

[28] Lash claims in *Theology on the Road to Emmaus* that affirming Jesus as "the Son of God" is tantamount to declaring parenthood as a divine attribute. Jesus as Son — and we also as sons and daughters — is not only "produced" by God but "indestructibly, absolutely cherished." "Loving production" and being "cherished with a love that transcends destruction" is, Lash claims, the essence of true parenthood (pp. 165-66). Thus, the model of parenthood, when applied to the source of all-that-is, is a signal of hope, of life on the other side of death. How does one square this hope with the scientific prognosis of the decline and "end" of the universe, either through heat death or a reversal of its beginning in the Big Bang? Does the parental model offer any insights to the discussion, both scientific and religious, of hope for the future?

[29] The phrase is from Josef Pieper's book *About Love*, trans. Richard and Clara Winston (Chicago: Franciscan Herald Press, 1974) 22.

[30] The parental model in its siding with life has two aspects. It is concerned with all species, not just human beings, and it is concerned with the nurture and fulfillment of life, not just with birth. On the first point: whereas we as biological or adoptive parents are interested in only one species — our own — and with particular individuals within that species, God as the mother of the universe is interested in all forms of life. One indication of human pride is our colossal ego in imagining that of the millions of forms of life in the universe, we are the only ones that matter.

[31] While the focus of these reflections is on creation, the perspective is that of redemption; that is, the *direction* of creation is given by trust in the creator, a trust generated for Christians by looking to that paradigmatic figure, Jesus of Nazareth.

[32] Anders Nygren, with his much-discussed book *Agape and Eros*, trans. Philip S. Watson (Philadelphia: Westminster Press, 1953), initiated the twentieth-century conversations on the issue, taking the extreme view that the two kinds of love are totally unrelated and incommensurable, with eros as the corruption of

agape — the self-interest that creeps into distinterested love. Gene Outka summarizes the four points in Nygren's position most influential to Protestants: agape is spontaneous and unmotivated; it is indifferent to value; it is creative of value making the worthless human being worthy; and it is the initiator of fellowship with God (*Agape: An Ethical Analysis* [New Haven: Yale University Press, 1972]). In this picture, God gives all and we take all; moreover, human beings cannot love God but can only serve as conduits of divine (agapic) love to the neighbors whom we, like God, love in spite of their unlovableness. One of the main critics of Nygren's position is M.C. D'Arcy, a Roman Catholic, according to whom agape and eros exist in balance in human beings (and hence the ideal love relationship is friendship). Were we not capable of giving as well as receiving, says D'Arcy, the human agent would be eliminated and God would be simply loving the divine self through us. See his *The Mind and Heart of Love: Lion and Unicorn — a Study in Eros and Agape* (New York: Henry Holt and Co., 1947).

[33] There are other ways of being parental besides being a biological parent, a point that needs to be stressed at the outset, for much of the power of the model in terms of its influence on human behavior rests on its extension beyond its physical and immediate base. One can, of course, be an adoptive parent as well as a biological one, but even more important for our purposes is that all human beings have parental inclinations. These tendencies are so basic, wide-spread, and various that it is difficult to catalogue all the ways they are expressed. Some of the ways that come most readily to mind, such as teaching, medicine, gardening, and social work, are only the tip of the iceberg, for in almost any cultural, political, economic, or social activity, there are aspects of the work that could be called parental.

[34] Another reason is the Christ-centeredness of the tradition, which overlooks the first birth because it wants to stress the second birth. In promoting Christ's mission of redemption, the tradition has failed to appreciate fully the gift of creation.

[35] The Judaeo-Christian tradition has carried imagery of gestation, giving birth, and lactation as a leitmotif that emerges only now and then over the centuries. For Hebraic uses of the "breasts" and "womb" of God as metaphors for divine compassion and care, see Phyllis Trible, *God and the Rhetoric of Sexuality* (Philadelphia: Fortress Press, 1978) Ch. 2. Another well-known case of such imagery is among the mystics. See, for instance, Caroline Bynum, *Jesus as Mother: Studies in the Spirituality of the High Middle Ages* (Berkeley and Los Angeles: University of California Press, 1982).

[36] The model of God as mother, giving birth to the universe, suggests several points of congruence with contemporary scientific understandings of the composition and beginning of life, most notably that there is a common source of all life — we share a common genetic code and biochemical metabolic mechanisms with all other forms of life — and that the universe, as a matrix that eventuated in life, had a beginning (so claims the Big Bang theory). Thus, language of birth, growth, and death of the *whole* of the universe appears to be appropriate. Such language is common in statements summarizing the scientific perspective on the nature and beginning of the universe, such as this one from the World Council of Churches' consultation in 1975: "The universe as a whole and everything within it is now seen to have a history. Everything is born, develops, and ultimately must die" ("The Christian Faith and the Changing Face of Science and Technology" [*Anticipation*, May 1976, No. 22]).

[37] Norman Pittenger makes the accompanying theological point: "Thus we wish to speak of God as the everlasting creative agency who works anywhere and everywhere, yet without denying the reality of creaturely freedom — hence we point toward God as Parent" (*The Divine Triunity* [Philadelphia: United Church Press, 1977] 2).

[38] God as nurturer or sustainer of the universe is congruent with the notion of *creatio continua* — the cosmos as in the process of producing new emergent forms of matter. The model of God as mother of the universe suggests that creation is not, as in the craftsman model, a once-for-all completed artistic whole, but is a constantly changing, unbelievably complex and rich matrix that both produces new forms of life and supplies support for present forms.

[39] The problem of "evil" in its many forms, including both natural evil and human sin, emerges here, for it is obvious that not all species, let alone all individuals in any species, *do* flourish, and this for a variety of reasons. A gospel of inclusive fulfilment for *all* of creation, must face what the physical and biological sciences must also face: both the Second Law of Thermodynamics as well as current evolutionary theory underscore what Robert John Russell calls "a world of dissipation, decay and destruction." See "Entropy and Evil," *Zygon* [December 1948] 449-68. The way that many other theologians (Moltmann, Peacocke, Barbour, process thinkers, etc.) as well as philosophers of science speak to this issue is through the concept of the suffering God who participates in the pain of the universe as it gropes to survive and produce new forms. Here, Gethsemane, the cross, and the resurrection are important foci for underscoring the depths of God's love, who, in creating an unimaginatively complex matrix of matter eventuating finally in persons able to *choose* to go against God's intentions, nonetheless grieves for and suffers with this beloved creation, both in the pain its natural course brings all its creatures and in the evil that its human creatures inflict upon it. I find this discussion rich and powerful; nonetheless, I would raise a caveat concerning what it tends to underplay — human sin and responsibility. By locating the discussion of evil in the context of the entire cosmic complex, one may overlook the particularly powerful role that human beings increasingly play in bringing evil to their own species and to other species as well. Divine suffering for the cosmos must not obscure human responsibility for a tiny corner of it — our earth.

[40] As with much religious language, the prime analogate is God, not human beings. Here we see the circular character of divine predication: if love is predicated analogously of God, then God is seen as the source and definer of love (God is "love itself"); God is the prime analogate, for love refers properly to God. However, since we do not know how to make such a proper or literal reference, we turn to our human loves to give content to divine love and in so doing we predicate love of God metaphorically. Metaphorical predication, however, demands recognizing when a model falters, and the parental model falters at the point of inclusiveness.

[41] See Julian N. Hartt's analysis of this consensus in his essay "Creation and Providence," in *Christian Theology: An Introduction to Its Traditions and Tasks,* ed. Peter C. Hodgson and Robert H. King (Philadelphia: Fortress Press, 1985) 144ff.

[42] Ian Barbour notes that *ex nihilo* creation was first propounded in the intertestamental period and elaborated by Irenaeus and Augustine in order to counter the idea that matter was the source of evil. This worthy motive, however, does not deflect the criticism that the doctrine of *ex nihilo* creation supports the separation of God and the world. As Barbour notes, "An additional motive in the *ex nihilo* doctrine was the assertion of the total sovereignty and freedom of God" ("Teilhard's Process Metaphysics," in *Process Theology: Basic Writings,* ed. Ewart H. Cousins [New York: Newman Press, 1971] 339).

[43] This image, radical as it may seem for imagining the relationship between God and the world, is a very old one with roots in Stoicism and elliptically in the Hebrew Scriptures. The notion has tantalized many, including Tertullian and Irenaeus, and though it received little assistance from either Platonism or

Aristotelianism because of their denigration of matter and body (and hence did not enter the mainstream of either Augustinian or Thomistic theology), it surfaced powerfully in Hegel as well as in a variety of twentieth-century theologies, most notably, process theology. The metaphor, especially in its form as an analogy — self: body:: God: world — is widespread, particularly among process theologians, as a way of overcoming the externality of God's knowledge of and activity in the world. Grace Jantzen's position, e.g., is that, given the contemporary holistic understanding of personhood, an embodied personal God is more credible than a disembodied one and is commensurate with traditional attributes of God. See her book, *God's World, God's Body* (Philadelphia: Westminster, 1984). My view is very close to Jantzen's, both in her criticism of the craftsman model and of *ex nihilo*: "God formed it [the world] quite literally 'out of himself' — that is, it is his self-formation — rather than out of nothing" (p. 135). See my *Models of God,* Chap. 3. See also Peacocke's use of the model in *Creation and the World of Science,* 133ff.

[44] While I have criticized the craftsman model as supporting dualism and distance between God and the world, another artistic model, Peacocke's metaphor of God as "Improvisor of unsurpassed ingenuity," is a nice complement to the maternal/parental model. The model of God as composer, orchestra leader, or bell-ringer is particularly suggestive when dealing with the interplay of chance and necessity, randomness and determinism, that permits all the potentialities of the universe to develop — the process of continuing creation resulting in new forms of life. See *Intimations of Reality: Critical Realism in Science and Religion* [Notre Dame: University of Notre Dame Press, 1984] 73; see also *Creation and the World of Science,* 105ff.) Thus, the parent who gives birth to the universe delights in her creation and by playing with all its potentialities, helps to bring it to fulfillment.

[45] *The Four Loves* (New York: Harcourt, Brace, and Co., 1960) 176.

[46] Clifton Wolters, ed., Julian of Norwich, *Revelations of Divine Love* (Harmondsworth, Middlesex: Penguin, 1966) 166-67.

ON CREATING THE COSMOS

TED PETERS, Pacific Lutheran Theological Seminary, Berkeley

Introduction

We are living in a time ripe with opportunity to seek significant rapprochement between science and theology. The unlocking of nature's secrets by the physical sciences seems to be opening up new doors for common exploration. British scientist Paul Davies says that "science has actually advanced to the point where what were formerly religious questions can be seriously tackled." [1] On the religious front, too, we see a healthy enthusiasm. The Second Vatican Council acknowledged the need for academic freedom and declared the "legitimate autonomy of human culture and especially the sciences." [2] Pope John Paul II has gone considerably further. To the Pontifical Academy of Sciences meeting at Castel Gandolfo on September 21, 1982, the Holy Father announced that "there no longer exists the ancient opposition between true science and authentic faith." He went on to say to the scientific community, "the Church is your ally." [3] In short, there now exists an atmosphere of readiness on the part of many in both laboratory and church to explore avenues toward rapprochement.

It is in this atmosphere, conducive to fruitful conversation, that we undertake the explorations of this paper. Our thesis will be that the Christian doctrine of creation out of nothing (*creatio ex nihilo*) is sufficiently intelligible to warrant continued probings for complementary notions in the natural sciences. We will open by identifying our methodological stance as one of *hypothetical consonance* between theology and the sciences, a stance which corrects the excesses of the dominant two-language theory. We will then proceed to cosmology proper by tracing the theological origins of the idea of creation out of nothing. We will argue that the Christian idea of the creation of the whole world derives from the basic experience of divine redemption within history, especially the resurrection of Jesus on Easter. What is at stake in cosmology for the Christian theologian, then, is an understanding of the cosmos which is consistent with our understanding of a redeeming God as revealed in the event of Jesus Christ. This will lead to an examination of the logic of *creatio ex nihilo* and the possible consonance of this religious idea with the second law of thermodynamics and Big Bang cosmogony in physics. In particular, we will focus on the question of the relationship between the concept of *ex nihilo* and the temporal beginning of the cosmos.

As we proceed, we will assume two things about the Christian doctrine of *creatio ex nihilo*. First, in its abstract form it stresses the ontological dependence of all things upon God. Second, one concrete form for expressing this dependence is the cosmological assertion that, although God is eternal, the created universe began at a point of temporal initiation, i.e., the world has not always existed. In this paper we intend to get to the idea of dependence through the idea of beginning. It is, of course, possible for a theologian to speak metaphysically about the utter dependence of the creation on its creator without reference to a temporal beginning. However, it is the very idea of a temporal beginning which in our generation draws us toward possible consonance with scientific cosmology. The scientist cannot, within the canons of the discipline of physics, say anything about the utter dependence of the cosmos upon God. But the scientist can intelligibly discuss the possibility of a temporal initiation to all things, and this in turn raises the question of creation out of nothing in such a way that the theologian might be called upon.

We will then review arguments raised by some contemporary theologians which are contrary to *creatio ex nihilo* and in favor of the notion of continuing creation (*creatio continua*). We will criticize these arguments on two grounds: first, these are false alternatives and they do not exclude one another; and, second, the theological idea of creation out of nothing — especially in the form of a temporal beginning — is just as consonant with contemporary science as is continuing creation. We will conclude that a healthy contemporary theology should advocate both *creatio ex nihilo* as well as *creatio continua* and seek possible consonance with science on both counts.

Hypothetical Consonance

Just what kind of accord may be established between lab stool and pew is still too far beyond the horizon to see. Yet we need to start somewhere. What I suggest is that we begin by seeking *hypothetical consonance*, that is by listening for the sounds of consonance, for those moments when we sense a harmony between disciplines. We begin by listening for some preliminary resonating sounds. Then we proceed with the hypothesis that further accord can be discerned. We spell out the possibilities with the assumption that both scientists and theologians are seeking to understand one and the same reality; therefore, we should hope for, even expect, some sort of concord to arise from serious conversation.

The method of hypothetical consonance can be distinguished from the two-language theory — what Ian Barbour calls elsewhere in this volume the "independence" relationship — which seems to have been the operative assumption of most serious scholars for much of this century. This is the assumption that the language of science and the language of faith exist in independent domains of knowledge and that there is no overlap. One version of the two language theory is the commonly accepted separation of fact from value. Albert Einstein held this view. On the occasion of addressing a conference at Princeton University, Einstein said that "science can only ascertain what *is*, but not what *should be* ... Religion, on the other hand, deals only with evaluations of human thought and action; it cannot justifiably speak of facts and relationships between facts." Note the use of the word "only" here. Each mode of knowing can speak "only" in its own domain. There is strict segregation.[4]

Perhaps the strongest advocate of the two-language theory among today's theologians is Langdon Gilkey. It is not only the difference between fact and value which distinguishes the two modes of discourse, according to Gilkey; there is also the difference between proximate (or secondary) causation and ultimate (or primary) causation. There is no translation between them.

All modern religious discourse, according to Gilkey, is limited to speaking about limit experiences, to the dimension of ultimacy in human experience. Religious or mythical language speaks only about "ultimate or existential issues," he says. This means that it speaks only to us as persons. It does not speak about the world. Theology "... possesses no legitimate ground to interfere with either scientific inquiry or scientific conclusions, whether in the fields of natural or of historical inquiry."[5] Religious truths do not contain information. They are best classified as myths or symbols which make no authoritative assertion about concrete matters of fact. Gilkey's position represents the paradigm example of neoorthodox dualism which has confined matters of faith to the transcendent-personal axis and consigned all other matters dealing with the world we live in to the province of secular science.

What about the language of science according to the Gilkey scheme? Scientific language is informative. It seeks to inform us regarding facts which are measurable, objective, and publicly shareable. Science seeks to explain the facts of experiences in terms of laws which are automatic and blind. These laws can appeal only to natural or human causes and powers, forces which exist within the confines of the finite world. Science cannot appeal to supernatural forces nor even to purposes or intentions or meanings. It can support its conclusions only through testing of repeatible experiments, not through speculation about one-time historical events. In short, "the language of science is quantitative, mathematical, precise... it is limited to describing the impersonal system of relations between the things or entities around us."[6] If Pope John Paul II is correct that there is no opposition between science and faith, then Gilkey would say this is because the two cannot talk to one another.

Now the point of establishing the two-language theory is to make it possible for a religious person to speak both languages without cognitive dissonance. By confining scientific language to the sphere of the finite and observable world, it is disqualified from making judgments regarding the existence or non-existence of God. Inherently, science is neither theistic nor atheistic. It is neutral. It is objective. "It is because science is limited to a certain level of explanation that scientific and religious theories can exist side by side without excluding one another, that one person can hold both to the scientific accounts of origins and to a religious account, to the creation of all things by God."[7]

But I believe that we must now ask for more than simple avoidance of cognitive dissonance. I believe we sould seek for cognitive consonance.[8] What I am advocating here comes close to the version of the two language theory we find in the work of Ian Barbour. Barbour recognizes the two languages but he will not accept a strict segregation. He wishes to explore the ways in which the two languages are complementary. This means, first, that we search for "significant parallels" in the methods of science and theology. Second, we look for ways to construct "an integrated world-view." Third, we defend the importance of a "theology of nature." Fourth, we permit the scientific understanding of nature to help us reexamine our ideas of God's relation to the world.[9] What Barbour means here by "complementary languages" is akin to what I mean by "consonance". We should look for those areas of correspondence and then spell out the possibilities which would permit what science says to illumine theological understanding and *vice versa*.

With this methodological commitment in mind, we will turn our ears now in the direction of resonating sounds regarding the creation of the universe. We will ask if there might exist an edifying consonance between scientific and religious concerns regarding the origin of the cosmos, especially the idea of creation out of nothing.

Creation Out of Nothing

Some say that Christians should give up the idea of creation out of nothing (*creatio ex nihilo*), especially when it is formulated in terms of an original beginning of time and space. Because the concepts of ongoing change and evolutionary development have so imbued our modern scientific culture, the argument is that *creatio ex nihilo* is now an anachronism. It is out of date. It is no longer intelligible to a mind which has been influenced by the scientific worldview. I disagree. I submit that there is surprising and salutary consonance between this theological concept and contemporary astrophysics, especially thermodynamics and the Big Bang cosmogony, and that we should not compromise on this theological commitment.

Where does the Christian idea of creation out of nothing come from? It does not come initially from speculation regarding the origin of the cosmos. What provokes the idea is, in fact, the experience of divine redemption. It is the intra-cosmic experience of God's redeeming activity which leads eventually to the idea of God's act of cosmic creation.

In the Old Testament, for example, Hebrew consciousness begins with the Exodus, with the creation of Israel, not with the creation of the world. "The Lord brought us out of Egypt with a mighty hand and an outstretched arm," we find in the credo statement of Deuteronomy 26:5-9. Remembering the Exodus comes first in Hebrew consciousness; thinking about the ordering of the cosmos comes later. But it does come. The book of Genesis does get written. Genesis gets written because what we speculate about the creation must be consistent with what we have experienced with redemption. Psalm 136 opens by offering doxologies to the creator who "spread out the earth upon the waters.... who made the great lights.... the sun to rule over the day.... the moon and stars to rule over the night." Then the Psalm follows immediately by telling the Exodus story, how God "brought Israel out .... with a strong hand and an outstretched arm .... and gave their land as a heritage." No one in Israel experienced the actual creation of the cosmos at the beginning. Rather, the biblical writers described creation on the basis of their experience with redemption.[10]

The key point of continuity between redemption and creation is the idea that the future can be different from the past, i.e., the key is eschatology. More abstractly put, God does new things. The prophets constantly reiterate the theme of newness: there will be a new Exodus, a new covenant, a new Moses. Gerhard von Rad uses the term "eschatology" to describe the structure of the prophetic message. It is a message which draws us toward a "break which goes so deep that the new state beyond it cannot be understood as the continuation of what went before."[11] What this means is that reality is not dependent merely upon its past. God can cut it free from the principles established at the point of origin. All ties to a mythically conceived cosmos where the paradigms are fixed *in illo tempore* are cut. The God of our future salvation — the God beyond the present state of reality — is not dependent upon what already exists. Looking backward toward the beginning, then, God must not have been dependent upon any past before there was a beginning. The origin of the cosmos was not limited to making order out of a pre-existing chaos. The origin was itself the advent of something new. This is the point made by II Maccabees 7:28, which emphasizes that God did not create heaven and earth out of anything that already existed.[12] This is reiterated by St. Paul in the New Testament who describes God as calling "into existence the things that do not exist" (Rom. 4:17b).

Turning to the New Testament, we can further reconstruct the movement from redemption to creation. Here the Gospel is the experienced power of new life in the Easter resurrection that provides the foundation for our faith and trust in God to fulfill his promise to establish a new creation in the future.

The world as we know it is replete with death, with the precedent that dead people remain dead. But now something new has happened. God has raised Jesus to eternity, never to die again, and God promises us a share in this resurrection when the consummate Kingdom of God comes into its fullness. Now we can ask: What does it take to raise the dead? What does it take to consummate history into a new and everlasting kingdom? It takes mastery over the created order. It takes a loving Father who cares, but who is also a creator whose power is undisputed and unrivaled.

The Gospel begins with the story of Jesus told with its significance. Its significance is that in this historical person, Jesus Christ, the eternal God who is the creator of all things has acted in the course of time to bring salvation to all the things he has created. Salvation consists here in the forgiveness of sins and the promise of a final redemption from evil to be attained through the eschatological resurrection of the dead. The logic here is: the God who saves must also be the God who creates. Nothing less will do. Langdon Gilkey expresses it well:

> ... it is because of the knowledge of the love of God gained in Jesus Christ that the meaning and purpose of creation are known, and it is because of the power of God as Creator that redemption through Jesus Christ can be effected and our faith in Him made valid.... Thus the promise of the Gospel that nothing can separate us from the love of God depends upon the belief that all powers in nature and history are, as we are, creatures of God and so subject to his will. Only a creator of all can be the guardian of [our] destiny.[13]

Here we have the seeds of what will flower into the idea of *creatio ex nihilo* and its corollaries: asymmetrical time (a one-way arrow), the historical character of nature and God's activity in the world, and the promise of an eschatological new creation. What fertilized the seed and caused it to sprout was the challenge of an alternative viewpoint, namely, the belief that the material of the universe had always existed. This challenge came from two competitors to the Christian view in the early centuries of the church: dualism and pantheism.

The heart of dualism is the belief that God or the gods create the cosmos by ordering pre-existing matter — the word "cosmos" means order. For Plato, it was the demiurge which fashioned the stuff of the world into an ordered habitat. This is dualistic because it posits two or more equally fundamental or eternal principles, the world stuff as well as the divine being.

The heart of pantheism (or monism) is that everything is fundamentally identical with the divine. But, by identifying God and the world, pantheism collapses all the plurality and multiplicity of the cosmos into a singular unity, and this singularity finally denies the independent reality of the world and its history.

In apologetic reaction to dualism and pantheism the early Christian thinkers proffered the concept of *creatio ex nihilo*. Against the dualists, the apologists held that God is the sole source of all finite existence, of matter as well as form. There is no pre-existing matter co-eternal with and separate from the divine. If the God of salvation is truly the Lord of all, then he must also be the source of all. Theophilus of Antioch in the middle of the second century, for example, praised Plato for acknowledging that God is uncreated. But then he criticized Plato for averring that matter is coeval with God, because that would make matter equal to God. "But the power of God is manifested in this, that out of things that are not He makes whatever He pleases."[14]

Against the pantheists, in parallel fashion, the Christians held that the world is not divine. It is a creation, brought into existence by God but something separate from and over against God. The world is not equa-eternal with God, because it has an absolute beginning and is distinct from God. Irenaeus put it this way:

> But the things established are distinct from Him who has established them, and what have been made from Him who has made them. For He is Himself uncreated, both without beginning and end, and lacking nothing. He is Himself sufficient for this very thing, existence; but the things which have

> been made by Him have received a beginning.... He indeed who made all things can alone, together with His Word, properly be termed God and Lord; but the things which have been made cannot have this term applied to them, neither should they justly assume that appellation which belongs to the Creator.[15]

This led the apologists to distinguish between generation and creation. "Generation", coming from the root meaning to give birth, suggests that the begetter produces out of its essence an offspring which shares that same essence. But, in contrast, terms such as "creating" or "making" mean that the creator produces something which is other, i.e., a creature of dissimilar nature. The patristic apologists applied the term "generation" to the *perichoresis* within the divine life of the Trinity but not to creative activity without. Hence, John of Damascus could state emphatically that the creation is not derived from the essence of God, but it is rather brought into existence out of nothing.[16]

The upshot of all this is that, for the Christian, creator and created are not the same thing. And, more importantly, what is created is fully dependent upon its creator. The cosmos is not ontologically independent. One way to make this point is to draw a contrast between eternity and time: God is eternal, whereas the cosmos is temporal. The world, which is not God, has not existed for all eternity alongside of God. Thus, says Theophilus and Irenaeus, there needs to be a initial point of origin, a point at which something first appears, i.e., an absolute beginning. Following in this train, Augustine can write doxologically:

> ...in the Beginning, which is of you, in your Wisdom, which is born of your substance, you created something, and that something out of nothing. You made heaven and earth, not out of yourself, for then they would have been equal to your Only-begotten, and through this equal also to you.... There is nothing beyond you from which you might make them, O God, one Trinity and triunal Unity. Therefore, you created heaven and earth out of nothing....[17]

Thus, the creation is just that, a creation, which had a definite 'sunrise' and could, if God were so to will, also have a final 'sunset'. For Augustine, the creation of all things from nothing includes the phenomenon of time. Time is not eternal. Time comes into existence when material in motion comes into existence. Neither time nor space are containers into which we dump the course of events; rather, they themselves belong to the finitude of the created order. Time starts when space starts. The result is that *creatio ex nihilo* — looked at from inside the creation, our only perspective! — has come to refer to a singular beginning of time and space, as well as to the matter and form out of which all the things of the world are made.

In saying this it is essential to look back and note the path we have taken: we began with the experience of a God who redeems, who creates a free people out of slavery and who raises the dead to life. On the basis of these intracosmisc events, we have drawn inferences regarding God's relation to the cosmos as a whole. The motive of the Christian theologian is not in the first instance to produce a general theory of the origin of the universe. Rather, when the question of the origin of the universe is raised, the answer offered must be consistent with what we know to have been revealed by God in the event of raising Jesus from the dead on Easter. We need to keep in mind just what stake the theologian has in the discussion of cosmology.

The Bare Logic of Creatio Ex Nihilo

Suppose for a moment we disregard the historical stake Christian theology has in the doctrine of creation out of nothing and ask about the bare logic of the concept. What do we find?

The fundamental axiom is that the creature is entirely dependent upon the creator in the act of creating. The creative act begins with nothing, yet something created is the result. But more than the created product is the result; so also is the relationship of creator to what is created. The asymmetrical relation whereby the creator becomes the creator and the created becomes dependent upon the creator is established in the event of creation. Prior to the act of creating, God is not yet a creator. He becomes a creator God only by creating a creature. The act of creating is the hinge on which swings the mutually defining terms of creator and creature. This may lead eventually to the notion that, in a certain sense, the creation has a determining affect upon the creator. Just *how* we understand God to be the creator will depend upon the actual course of events which the history of the creation takes. The fundamental axiom —that the creature is dependent for its existence upon the creator — does not necessarily preclude a temporal reciprocity whereby the creator may also be affected by the history of creation.

Next, the movement from nothing to something is puzzling.[18] To be nothing (no-thing) is to be indeterminate. To be something (some-thing) is to be determinate. To be determinate is to exist in spacetime. The act of creation signals a shift from the indeterminancy of nothing to the spacetime determinancy of the things which constitute the universe. This leads to the question: is the event of creation itself a temporal event? At first, it would seem that it must be temporal, because for one thing to have a determinate effect on another thing they both must share a single spacetime continuum. But if space and time are themselves the result of the creative act, then the creative act itself cannot be subject to the same spacetime determinancy. So, perhaps it is better to speak of the creative act itself as eternal rather than temporal. By "eternal" here we do not mean simple everlastingness but rather supratemporality.[19] As eternal, God's act of creation is tangential to time and related to time, yet it is not subject to determinancy by time save in the sense already mentioned, that is, in the reflexive sense that the eternal creator is so defined as a result of the existence of temporal creation. In short, the event of creation marks the transition from eternity to time.

If we explore the notion of eternity a bit further, we note how the concept of eternity need not necessarily imply that creation from nothing must occur in a single instant, in a single moment or all in a flash. To say so would presuppose that eternity is subject to measurement by a temporal continuum, which is just what we tried to avoid by introducing the concept of eternity in the first place. This has three implications. First, the concept of eternity stretches us to the limits of our language. We cannot literally speak of the point of origin as the first "moment" or the act of creating as an "event". We cannot make sense out of talking about what God was doing "before" the event of creation, as Augustine has already observed.[20] Such terms are already time-dependent. There is no way to speak univocally about the point of origin at which eternity had a determinate impact on temporality.

A second implication of this is that we might not have to confine *creatio ex nihilo* to the onset of the whole of the cosmos at the temporal beginning. A a higher level of abstraction, what the apologists wanted to stress was that the world is utterly dependent upon God and, conversely, God is utterly independent of the world. In principle, one could say the world is infinite in time as long as it can be shown that the world is dependent for its being on the activity of God. To depict *creatio ex nihilo* as an act of creation at a singular temporal moment is one vivid way of making this otherwise more abstract point.

Thirdly, the concept of *ex nihilo* may be relevant for understanding newness within the ongoing course of intracosmic events. As we will see later, Fred Hoyle can use the idea of *ex nihilo* to describe what happens within the flow of natural events. Thus, the idea could in principle have some value for interpreting ongoing newness as well.

Perhaps now, considering what we have just said about the limits of language, we should ask about the nature of eternity. It seems that we ought not to define "eternity" as everlastingness. Everlastingness simply means more time, an infinite temporal succession. But if by "eternity" we wish to refer to the transition from indeterminate nothingness to determinate spacetime events, then it cannot in any simple way be subject to the temporal continuum. Eternity — along with God's power to create — must be able to survive the termination or elimination of spacetime.

There is another way to look at this logic. Let us ask: need one assume there was an agent prior to creation? Need one assume that there was a divine being before the creator-creature relationship was established? Could we work simply with the notion of a primordial nothingness as the ground of both creator and creature? This is the suggestion offered by philosopher of religion Robert C. Neville in an attempt to build a bridge between Christianity and Buddhism.[21] Neville believes one could see the agent of creation as the result, not the antecedent, of the event of creating. The creator's character derives from the character of the world created. But, according to the logic of *creatio ex nihilo*, we cannot actually know the ground of being. What we can know is the creation relation, which only conditionally applies to the creator-ground relation. Thus, the ontological ground is never an object of knowledge. Nothingness is not an object to be known. What Neville seems to be doing here is identifying what he calls the "transcendent ground" with nothingness. The problem is that in doing so Neville makes this nothing into a something. I believe this is misleading. It would be closer to the logic of *creatio ex nihilo*, in my judgment, to identify the "transcendent ground" with the event (difficult as it may be to use the word "event" here) of creating. It is nonsense to identify the ground with nothing. The *nihilo* in *ex nihilo* functions as a complementary idea to the fundamental axiom that all created things are utterly dependent upon God their creator. Neville, in effect, has the *nihilo* creating God. Now, we have already granted that there is a sense in which the history of determinate creation may have an effect on our understanding as to *how* God is the creator; but this does not in any way imply that God is the creature of nothingness.

The theologian's stake in this is to seek an understanding of the cosmos which has consonance with the Christian experience of divine redemption. It is this which sent the patristic theologians in the direction of *creatio ex nihilo* and to its accompanying notions of a point of origin, temporal history, and consummate eschatology. Let us now turn to contemporary conversations in the natural sciences, where we shall find that these ancient Christian notions are by no means rendered unintelligible by the emerging and reigning scientific cosmology.

Consonance with Thermodynamics and Big Bang Cosmology

The last three decades of scientific research have witnessed increasing support for a cosmology that includes a specific point of origin, the contingency of natural events, an overall irreversible direction of temporal movements, and the forecast of an eventual dissipation or heat death for the cosmos.[22] In particular, the application of the second law of thermodynamics measured in terms of entropy to the macrocosmos leads to the notion of temporal finitude. If the universe in its entirety is moving irreversibly from order to disorder, from hot to cold, from high energy to dissipative equilibrium, then we may draw two significant inferences. First, the universe will eventually die. Even though in far-from-equilibrium sectors or microcosms within the larger whole we will find creative activity and the emergence of new structures, the overall advance of the cosmos is in the direction of eventual dissipation and heat death. Second, the universe must have had a point of origin. It has not always existed. It could not have existed with an infinite past, otherwise it would have suffered thermal death a long time ago. Such scientific speculations open up to intelligibility questions regarding an original creation and a final eschatology.

So also does the theory of an expanding universe, the standard Big Bang model. When we retrace the trail of the expansion backward in time, we eventually find ourselves able to speculate about a point of origin, about the beginning *of* time (not a beginning *in* time). We can surmise that the expansion we witness today is the result of an explosion which occurred yesterday, a bang which began it all. Astrophysicists believe they have advanced our knowledge to a time as small as 10^{-35} or perhaps even 10^{-43} seconds after the very onset of the creative movement.[23] Furthermore, the complementary research in both astronomy and physics has led to the strong hypothesis that at the beginning the universe was completely singular. The idea of an initial singularity characterized by infinite density and temperature is produced by extapolating backwards from the currently observed expansion of the cosmos.[24] The bang, or initial singularity, is the event at which space and time were created. Now this marks the end of the line for scientific research, because astrophysicists cannot within the framework of their discipline talk about the singularity, let alone what was going on before it.

We may not be talking about the very beginning, however. We are not yet talking about the "origin" of the original singularity. There are initial conditions which have an ontological (though perhaps not a temporal) priority. The Big Bang model will not permit us to do what Augustine forbade, namely, to ask intelligibly about what was happening before the beginning. Scientifically speaking, we can go as far back as the initial singularity, not to the nothingness which may or may not have preceded it. As C.J. Isham makes clear in his essay elsewhere in this volume, as we move closer and closer to the posited singularity, time becomes more and more unusual.[25] Conventional time concepts become less and less useful, because as we approach the more quantum mechanical state of the universe it becomes difficult to think of anything evolving in time. What is significant for us here is this: what both Augustine and quantum mechanics can say about the very early universe is that time is determined by the motion of things; and this gives matter an ontological status prior to that of time. Although it is intelligible to speak of the beginning of the universe, it is difficult to speak of the origin of the singularity.

We have reached a limit to scientific method. Although we can point to a beginning, it is difficult to say much about it. If scientific explanations are grounded in the principle of sufficient reason, then to speak of an absolute beginning for which there is no explanation is to exceed the boundaries of the method. Thus, a phrase such as "the beginning of the cosmos" must be considered a form of expression which points to the limit of the standard Big Bang theory. Nevertheless, though we can acknowledge the limits of scientific discourse here, we have entered a conversation in which questions of ultimate origin have become intelligible. The principle of hypothetical consonance does not require that science and theology produce a single coherent worldview at the outset; it requires only that we find sufficient commonality so as to pursue respective questions in an intelligible dialogue. This we have on the question of temporal origin.

By speaking of *creatio ex nihilo* at this point the theologian can achieve some consonance without appealing to a crass God-of-the-gaps method. It is not the acknowledged limit to scientific conceptuality which is the point of departure here. Rather, it is the material content of the standard Big Bang theory. What we can say is this: the universe as we know it has not always existed in the past. It has come to be. Discussions of *creatio ex nihilo* make sense. Here the *nihilo* can refer to two things. It can refer first to the absolute non-existence out of which the divine power may have wrought the initial singularity. This is a specific way in which we might be able to speak of the world's total dependence upon God its creator. Or, secondly, it can refer to nothingness (no-thingness) in the sense of the not-yet-determinedness of things, i.e., it can refer to newness, to the contingent character of the path followed by the bang and subsequent cosmic expansion.

The expansion continues. According to the Big Bang theory, our universe started out very hot and has been in an overall one-way process of cooling off ever since. The temperature of radiant heat declines in proportion to the expanding region of space: double the radius and cut the temperature in half. When the temperature decreases past a certain threshold a so-called "freezing out" takes place. Each freezing out involves the appearance of new forms of matter and energy. At the very hot beginning we did not have such things as molecules, atoms, or even nuclei. These appeared at specific points in the thermal history of the universe. The things (and laws of nature that govern the things) of our universe were produced rapidly but unpredictably: When a volume of water freezes and expands, we know for certain that it will crack. Where it will crack cannot be predicted. In the dissipative macrosystem that is our universe the course of events has been unpredictable.

And, we should note, there is even more unpredictability in far-from-equilibrium subsystems within the universe where energy is concentrated so that creative things happen. Our sun and the stars, for example, are centers sponsoring continuing creativity. On the earth, living organisms draw energy from the sun and produce new and higher forms of order. As living beings, we survive by exchanging energy and material with our environment. We might say there is a flow of energy through our bodies which results in a concentration — if not creation — of order. This growth in order is paid for by the dissipation of energy in the wider environment. The negative entropy necessary to support life locally is but an aspect of the net entropy increase cosmically. The results are temporal events of ongoing creativity. To put it as does Ilya Prigogine, chaos within the cosmos is capable of producing new forms of order.[26] Time brings change, and change brings newness.

What this means is that what exists now is largely contingent, i.e., it is not simply the working out of eternal principles already present at the point of origin. It means, in short, that nature herself has a history. We can on this basis anticipate that things might occur in the future which may be different from those occurring in the past. The events of nature's history are constitutive of what nature is. Not only can we apply the word "creation" to the point of origin, the primal singularity at the beginning of all things, but it applies as well to the ongoing activity of finite natural events. We may speak intelligibly of *both* a beginning creation and a continuing creation.

The Scientific Debate: Creation out of Nothing vs. Continuing Creation

We have already discussed how the Christian doctrine of *creatio ex nihilo* developed through a process of explicating implications inherent in the ancient Hebrew experience of God's saving acts in history. In our own epoch, characterized by modern science and an emerging postmodern culture, we are also engaged in interpretive explication. Therefore we must ask: does *creatio ex nihilo* help make the Gospel intelligible today? It is my own position that it does. However, not everyone agrees. We must acknowledge that some contemporary thinkers believe the doctrine is outdated due to the change in worldview. Because we moderns allegedly have a more dynamic understanding of reality than did the ancients, many are recommending that *creatio ex nihilo* be replaced by one or another version of *creatio continua*. I do not believe we need to choose between them. I believe these two concepts are complementary and that we need not substitute one for the other. This complementarity is true for both Christian theology and natural science.

We make the observation here that the debate between creation from nothing and continuous creation is not limited to theologians. It occurs among scientists as well. For several decades astronomer Fred Hoyle, for example, argued for a theory of continuons creation under the banner of the "steady state theory." He thereby opposed any notion of an absolute beginning. Rather than think that all the matter in the universe appeared at a given point of origin, his position was that matter is always coming into being uniformly throughout infinite time and infinite space. Hydrogen atoms are appearing *de novo* at a constant rate throughout space, condensing, combining, and giving birth to new stars.

Hoyle argued against the Big Bang by saying that the theory of a unidirectional expanding universe rests on a time-singularity beyond which the history of the universe can not be traced; but Hoyle's opponents countered by showing how his spontaneous creation of hydrogen atoms violates the laws of local conservation of mass and energy and, further, that the phenomenon of continuing creation is as yet unobserved. For most scientists the debate was decisively won in 1965 with the discovery of the cosmic background radiation by Robert W. Wilson and Arno A. Penzias. Their discovery confirmed earlier predictions that such a universal microwave radiation would be a relic of an early stage in cosmic expansion. Hoyle has sought since to revise his approach by constructing other cosmologies in competition with the Big Bang model, but most scientists cede the final victory to some variant of the Big Bang view.

Why has Fred Hoyle been so adamant, especially when the preponderance of scientific evidence favors the Big Bang cosmology? It appears that Hoyle has religious as well as scientific reasons. He opposes the Christian religion. Like so many other scientific humanists of the modern world, he defines "religion" as escapism: "... religion is but a desperate attempt to find an escape from the truly dreadful situation in which we find ourselves."[27] What he does not like about the Big Bang theory, curiously enough, is that it looks to him like it might support Jewish and Christian theology. He opposes the idea of a point of origin. He opposes *creatio ex nihilo*. Over against the theologians he likes to quote the Greek Democritus, who said "nothing is created out of nothing" *(ex nihilo nihil fit)*. He seems to assume that Big Bang and *creatio ex nihilo* belong together, and to this he objects.

It appears clear that Hoyle wants to avoid giving even the slightest quarter to religious forces. What is significant for us here is that Hoyle assumes there exists a consonance between Big Bang cosmology and Christian theology. He recognizes an inherent connection, and this is what he does not like about it. Thus, as Ernan McMullin points out, the debate among scientists seems to press against the borders of their own disciplines and, further, it seems there is some tacit agreement that the notion of a point of origin with a subsequent history of nature has the greater religious relevance.[28]

The Theological Debate: Creation out of Nothing vs. Continuing Creation

Even though Hoyle has assumed the relevance of a singular beginning for Christian theology, not all Christian theologians see it this way. Process theologians of the Whiteheadian school, for example, reject what they call the "classical theism" of the apologists and, among other things, the idea of a beginning. Schubert Ogden, for example, advocates a Hartshornian version of panentheism according to which God is internally related to the world. God participates in the world's ongoing creative advance, though God did not bring the world into existence at a beginning in finite time. Ogden believes that, within this framemark, he can uphold the notion of the world's dependence upon God and, thereby, not violate the intention of the *creatio ex nihilo* doctrine.[29] John Cobb and David Griffin, however, go further than Ogden. "Process theology rejects the notion of *creatio ex nihilo*," they write.[30] By this they intend to reject not only a temporal beginning but also the notion of the utter dependence of the world upon God. Rather than the position of Theophilus and Irenaeus, they say they prefer Plato's notion of making order out of chaos. According to process theology, the term "creation" refers to the ongoing movement of the cosmos and not to something which initiated that movement in the beginning.

Because he deals with the scientific issues directly, the earlier work of Ian Barbour provides us with a better example of a theological position which downplays creation from nothing in favor of continuing creation. In the 1960s Barbour held that there are no strictly theological grounds for favoring either Big Bang or steady state theories. Both theories are capable of either a naturalistic or a theistic interpretation. Both theories push explanation back to an unexplained situation which is necessarily treated as a given — the primeval singularity which exploded in the case of the Big Bang or the constant creation of matter in the case of Hoyle's steady state. Neither theory asks about the pretemporal or eternal ground or framework for the natural events which occur within the stream of time. So, Barbour concluded:

> . . . we will suggest that the Christian need not favor either theory, for the doctrine of creation is not really about temporal beginnings but about the basic relationship between the world and God. The religious content of the idea of creation is compatible with either theory, and the debate between them can be settled only on scientific grounds, when further data are available."[31]

Now we might pause to ask: could this be an example of two-language segregation, according to which science is science and religion is religion, and each is consigned to its independent domain?[32] Barbour's position (at least until recently) has been that theologians have no particular investment in the winner of the debate between absolute beginning and continuous creation. Yet, should we not ask: why not both?

Barbour has said he does not want both. He wants only *creatio continua*. Why? He says *creatio continua*, not *creatio ex nihilo*, is the biblical view. He quotes Old Testament scholar Edmund Jacob, who wrote that the meager "distinction between the creation and the conservation of the world make it possible for us to speak of *creatio continua*."[33] But, on the basis of this, to make us choose between creation from nothing and continuing creation is, I believe, unwarranted. That the formulation *creatio ex nihilo* is itself post-biblical we have already granted. Yet, this should not lead us to deny that it has biblical roots. *Ex nihilo* is the result of evangelical explication, according to which the implications inherent in the compact experience of salvation witnessed to in scripture were drawn out by the apologists of the early church. Even if there are only a few references to *ex nihilo* in the Bible itself evangelical explication ought to count for something. To say that *ex nihilo* is not a biblical concept is exaggerated.

What Barbour actually advocates is a synthesis of creation and providence in the concept of continuing creation. This does not mean that he abandons the Christian commitment to the notion that the world is dependent upon God. What we have to give up, he says, is the idea of "*creatio ex nihilo* as an initial act of absolute origination, but God's priority in status can be maintained apart from priority in time."[34] What Barbour has done here is virtually equate *ex nihilo* with initial beginning, discard the idea of initial beginning, and thereby discard *ex nihilo*.

Arthur Peacocke comes close to the Barbour position here; but, whereas Barbour nearly eliminates *ex nihilo*, Peacocke keeps it. Peacocke believes that the essence of the doctrine of *creatio ex nihilo* is this: the creation owes its existence to God. Once this is affirmed, however, it makes no difference as to whether the cosmos began or not. He says that scientifically

> . . . we may, or may not, be able to infer that there was a point (the hot big bang) in space-time when the universe, as we can observe it, began ... But, whatever we eventually do infer, the central characteristic core of the doctrine of creation itself would not be affected, since that concerns the relationship of all the created order, including time itself, to their Creator — their Sustainer and Preserver.[35]

Note that Peacocke does not dismiss *creatio ex nihilo* per se. He keeps it. But he removes from its stipulated definition any commitment to a point of origin. He then goes on to commit himself to a doctrine of *creatio continua* following an evolutionary model, according to which nature consists of a process producing new emergent forms of matter.

Both Barbour and Peacocke reject the relevance of an initial origin. Both affirm the dependence of the creation upon God its creator. Both advocate *creatio continua*. Yet there is a slight difference. Whereas Barbour nearly gives up on *ex nihilo*, Peacocke affirms it.

Why are we so quick to give up the idea of an initial origin? Or, to put it more precisely, why does a temporal beginning seem to be so expendible when explicating our theological concept of creation out of nothing? To reduce *creatio ex nihilo* to a vague commitment about the dependence of the world upon God — though accurate — does not help very much. It simply moves the matter to a higher level of abstraction. We still need to ask: just what does it mean for the world to owe its existence to God? One sensible answer is this: had God not acted to bring the spacetime world into existence, there would be only nothing.

Furthermore, it makes sense to talk about the temporal point of origin. The assertion that the cosmos is utterly dependent upon God is familiar to theologians, but such an assertion lies outside the domain of scientific discourse. The idea of an initial origin, however, does lie within the scientific domain. The point I am making here is this: for theologians to raise again the prospects of *creatio ex nihilo* understood in terms of a beginning to time and space is to be consonant with discussions already taking place within scientific cosmology. We have an opportunity here to bridge the gap between disciplines.

Nevertheless, this opportunity seems to be ignored. Most theologians in our own period are inclined to invest their energies in *creatio continua*, while either rejecting or at least sidetracking *creatio ex nihilo*. Theologians seem to assume that the idea of continuing creation has the greater scientific credibility. But, it is not clear yet just what continuing creation could mean for a theologian. Could it mean what Fred Hoyle means by it? Hardly. We will now explore the meaning of the phrase "continuing creation," and we will do so by first asking about the relationship between creation and change.

Creation and Change

Christian thinking has not always distinguished between creation from nothing and continuing creation in quite the same way we do today. The prevalent distinction has been that between creation and change. For Thomas Aquinas it was important to make the distinction between absolute creation and changing things which have already been created. In fact, the term "creation" refers solely to what appears *ab initio*, to God's bringing things into being from nothing. "Creation is not change," he writes, because "change means that the same something should be different now from what it was previously." [36] God's role as creator, then, was that of the first cause. If we were to translate Thomas directly into the present context of the Big Bang, we might say that God caused the singularity to explode, but only after creating the singularity itself, of course.

Thomas believes in a point of origin because it is biblical. For this reason he rejects two competing positions, those of Aristotle and

Bonaventure. On the one hand, Aristotle held that the cosmos is eternal and argued for it on philosophical grounds. While granting to Aristotle the credibility of his philosophical arguments, Thomas affirms a point of origin and a finite time to the world on scriptural grounds. One could, in principle, hold to *creatio ex nihilo* while affirming either an eternal cosmos or a temporally finite cosmos and remain philosophically coherent. Nevertheless, special revelation decides the issue for Thomas. On the other hand, Bonaventure favored the idea of an initial origin and argued for it on philosophical grounds. Thomas agrees with Bonaventure's conclusion but disagrees with his method. For Thomas, the metaphysical arguments alone cannot settle the issue as to whether the world is eternal or temporally finite. He seems to assume that the biblical position is consonant with what he knows philosophically, but it is the biblical commitment itself which is decisive. The result is a doctrine of *creatio ex nihilo* with the specific meaning: the cosmos has a point of initial origin.[37]

For Thomas, God transcends the cosmos. As the uncaused cause, the cosmos is originally dependent upon God; yet God is not just one factor among others within the world system. The world process is itself a dynamic process in that it involves change, but in itself it does not create new things out of nothing. No created thing can create something absolutely. Only God can, and God did it already back at the beginning.

Langdon Gilkey criticizes Thomas for using the idea of cause in making the case for God. Gilkey believes the causal analogy for describing God's relation to the world is misleading for two reasons. First, it separates God from the world. Causality implies external relations. If God is the first cause and the world is his dependent effect, then God and world are set over against one another and God's immanence is denied. Second, Gilkey says Thomas compromises the transcendence of God by drawing him into the world system. God has become one more factor in the endless chain of cause and effect. Once we have placed God in the causal chain, there is no escape from the inevitable question: what caused God? Thus, the analogy drawn from the spacetime experience of cause and effect, when applied to the eternal divine, is a mistake.[38]

On the one hand, if God for Thomas transcends the world, then Gilkey faults Thomas for loss of immanence. On the other, if God for Thomas is a factor in the intracosmic process, then he is faulted for loss of transcendence. Why does Gilkey press this point? The answer is that Gilkey's own agenda is to avoid mixing science and religion. Gilkey says it is the task of science to answer the "how?" questions, such as "how did the cosmos begin?" It is the task of theology to answer the "why?" questions, such as "why did God create?" Gilkey's complaint against Thomas is that he sought to answer the "how?" question by saying that God had "caused" the world to come into being.

If we were to follow the path led by Barbour and Gilkey, we might end up making no definitive theological commitments whatsoever regarding whether the cosmos ever had an initial origin, or, if it did, just how God was involved in this origin. We would have to carry on our theological discussion in a field of discourse that would be fenced off from scientific speculations on origin and change in nature. Yet, as we shall see, few theologians in our time — including Barbour and Gilkey — in practice hold to keeping the fence very high. To illustrate, we will examine the widely accepted theological postulate that God's relationship to the world is best described in terms of *creatio continua*.

What Does Creatio Continua Mean?

To Fred Hoyle *creatio continua* means the constant process of bringing *de novo* into existence things which hitherto had not existed. Thomas did not use the term *creatio continua*. Had he accepted Hoyle's definition he might have argued that it still does not mean changing things which already exist. Hence Hoyle and Thomas would disagree as to when this continuous creation, as creation, occurs. Hoyle would say that there never was a beginning, that the cosmos is now and always has been in a steady state of creative activity. Although there are new beginnings every day, there never was an absolute beginning to all these absolute beginnings. Thomas, in contrast, would say that creation happened once at the beginning of all things, and that today's intra-cosmic events are watched over by God's conserving care (*conservatio*). For Hoyle there is no creator and creation is contemporary. For Thomas there is a creator and creation is past. If we were to avoid the strictures of Barbour and Gilkey and mix science and religion, then we would observe that the Thomistic view has greater consonance with Big Bang theory than it does with Hoyle's steady state theory.

Why then are theologians such as Barbour sympathetic with *creatio continua*? Oddly enough, one reason for advocating continuing creation has to do with re-mixing science and religion. Theologians today commonly assume that modern understandings of nature reveal a basically dynamic rather than a static worldview. Because it is assumed that the ancients who formulated *creatio ex nihilo* had lived in a static cosmos, and that we moderns now live in a dynamic cosmos, it follows that we need a modern understanding of creation that is more dynamic. *Creatio continua* seems at first glance to fit the bill. Barbour supports continuing creation by arguing that:

> ...today the world as known to science is dynamic and incomplete. Ours is an unfinished universe which is still in the process of appearing. Surely the coming-to-be of life from matter can represent divine creativity as suitably as any postulated primeval production of matter 'out of nothing'. Creation occurs throughout time.[39]

Is Barbour consistent? Here he asserts that our modern scientifically produced picture of a dynamic world is in fact relevant to the theological doctrine of creation. He is assuming that some sort of dynamism in theology should parallel the dynamism found in science. Having committed himself now to following the scientific lead, one would expect him to affirm a temporal beginning over against continuing creation. After all, that is where the preponderance of scientific evidence lies. But instead he reaffirms continuing creation and not *ex nihilo*.

What does Barbour mean by continuing creation. From the passage cited above, we can see that this is not *creatio de novo* as proffered by Hoyle. It is, following the model of biological evolution, the process of bringing life out of already existing matter. It is what Thomas would call "change." Barbour wants the doctrine of creation to refer to God's continuing activity *within* the world, not the creation of the world *per se*. What this amounts to, it appears to me, is a merging of creation with providence. Barbour is not alone in doing this. Gilkey also uses the term "continuing creation" to combine creation and preservation. "Creation is seen now to take place throughout the unfolding temporal process... thus, creation and providential rule seem to melt into one another.... The symbol of God's creation of the world points not to an event at the beginning...."[40] What theologians used to call preservation or providence has been renamed "creation".

Have we arrived at anything more important than a change in vocabulary, a change which tends to hide the issues? Whereas Thomas used the term "creatio," to refer to the ultimate temporal beginning of things and to distinguish this from ongoing change, theologians such as Barbour, Gilkey, and Peacocke use "creation" to refer to the process of change within already existing creation.[41] The apparent motive for the switch is to merge creation with preservation or providence, but the result risks a total elimination of any theological commitment to a temporal beginning. In fact, such a beginning cannot even be discussed theologically, because we have lost the word for it. For temporal beginnings we must listen to the scientists.

Conclusion

Perhaps one of the ironic values of seeking consonance between religious and scientific discourse will be the impetus for Christian thinkers to return to the classic commitment to *creatio ex nihilo* while, at the same time, gaining a deeper appreciation for *creatio continua*. It simply makes sense these days to speak of $t=0$, to conceive of a point at which the entire cosmos makes its appearance along with the spacetime continuum within which it is observed and understood. If we identify the concept of creation out of nothing with the point of temporal beginning or perhaps even the source of the singularity, we have sufficient consonance with which to proceed further in the discussion. Contemporary scientists do not support either a dualist or pantheist alternative, nor do they favor the idea that the stuff of the universe as we know it has an infinite past. On this particular issue, the scientific community of today is not the adversary to Christian theology that the pagan philosophies of ancient Greece and Rome were. Christian theologians can approach the matter with the positive anticipation that further inquiry may lead to constructive results.

The idea of continuing creation may obtain a more profound meaning through Prigogine's usage of the second law of thermodynamics as it combines the irreversibility of time with the creation of order out of far-from-equilibrium chaos. Cosmic entropy is complemented by local creativity. What happens locally is that genuinely new things appear. The structures of reality are not reducible to, nor fully pre-determined by, the existence of past material,. Thus, what Thomas Aquinas understood as mere change in already existing things is qualified: though the cosmic conservation of energy remains intact, there really do arise events in which new structures occur. We might call these new things "transformations" of reality, but the degree of unpredictable newness certainly exceeds what the medieval mind of Thomas conceived.

The primary reason for defending the concept of *creatio ex nihilo* in concert with *creatio continua* is that the primordial experience of God doing something new leads us in this direction. The Hebrew prophets promised that God would do something new in Israel. The New Testament promises us that God will yet do something new for the cosmos on the model of what God has already done for Jesus on Easter, namely, establish a new creation. What these things imply is that, when looking backward to the beginning of all things, we speculate that God's initial act of creation was not dependent upon anything which preceded it. To speak of creation out of nothing is a way of emphasizing this point. Similarly, creative activity, whether divine or natural, has by no means ceased. It continues. Creation is not simply a thing but rather a whole course of natural and historical events in which new things happen every day, a course of events which is bound by its finite future. The end of the cosmos will be something new too. The question which remains is whether the anticipated heat death constitutes a sort of cosmic Good Friday, and whether it makes sense to hope that beyond it lies an Easter for the universe.

NOTES

[1] Paul Davies, *God and the New Physics* (New York: Simon and Schuster, 1983) p. ix.

[2] *Gaudium et Spes*, 59.

[3] "Science Must Contribute to True Progress of Mankind," *L'Osservatore Romani* (October 4, 1982) 3.

[4] Einstein himself had to qualify this strict segregation, because he believed the ecstatic experience of reason has a religious quality to it. In the address cited above, the scientist went on: "whoever has undergone the intense experience of successful advances made in this domain [reason], is moved by profound reverence for the rationality made manifest in existence. By way of the understanding he achieves a far-reaching emancipation from the shackles of personal hopes and desires, and thereby attains that humble attitude of mind towards the grandeur of reason incarnate in existence, which, in its profoundest depths, is inaccessible to man. This attitude, however, appears to me to be religious, in the highest sense of the word."

[5] Langdon Gilkey, *Religion and the Scientific Future* (New York: Harper & Row, 1970) 18. A contemporary Roman Catholic example of the two-language theory would be Hans Küng who, after describing the Big Bang cosmogony, writes: "the language of the Bible is not a scientific language of facts, but a metaphorical language of images.... The two planes of language and thought must always be clearly separated...." *Does God Exist*? (Garden City, New York: Doubleday, 1980) 639.

[6] Langdon Gilkey, *Creationism on Trial: Evolution and God at Little Rock* (San Francisco: Harper & Row, 1985) 113.

[7] *Ibid.*, 117. Whereas theologian Gilkey advocates that the individual be bilingual and speak both the language of theology and science, some scientists require that we choose. Physicist Paul Davies cited above, for example, works with the traditional stereotype of the open-minded scientist, whom we should emulate, and the outdated narrow-minded theologian, whom we should reject. *God and the New Physics*, 6f.

[8] The term 'consonace' comes from an essay by Ernan McMullin, "How Should Cosmology Relate to Theology?" in *The Sciences and Theology in the Twentieth Century*, edited by A.R. Peacocke (Notre Dame: University of Notre Dame Press, 1980). McMullin may be saying two things. On the one hand, he advocates a two language position, arguing against any God-of-the-gaps reasoning from scientific gaps. He contends further that the theologian has no stake in any particular scientific theory. On the other hand, he is concerned about compatibility, consonance, and coherence between scientific and theological statements. Theologians should aim for some sort of "coherence of world-view" to which both science and theology contribute. This means, among other things, that theology may have to be reformulated. "When an apparent conflict arises between a strongly supported scientific theory and some item of Christian doctrine, the Christian ought to look very carefully to the credentials of the doctrine. It may well be that when he does so, the scientific understanding will enable the doctrine to be reformulated in a more adequate way." *Evolution and Creation*, ed. by Ernan McMullin (Notre Dame: University of Notre Dame Press, 1985) 2.

[9] Ian G. Barbour, *Issues in Science and Religion* (New York: Harper & Row, 1966) 4f, 51. In a later work Barbour seeks to overcome the science-theology

language split by looking for commensurability and by applying the notion of "critical realism," drawn from science, to theology. *Myths, Models, and Paradigms: A Comparative Study in Science and Religion* (San Francisco: Harper & Row, 1974).

[10] This is the point made by Gerhard von Rad: Israel only discovered the nature of creation when it connected it with its experience of saving history. *Old Testament Theology*, 2 Volumes (New York: Harper & Row, 1957-65) I,136.

[11] Ibid., II:115.

[12] It might be noted that the context of this, the earliest reference to *creatio ex nihilo*, shows that its purpose is not initially cosmological; rather, it stresses God's power as vastly superior to that of an oppressive king. The cosmological significance of the power of Israel's God grows.

[13] Langdon Gilkey, *Maker of Heaven and Earth* (New York: Doubleday, 1959, 1966) 269, 279.

[14] Theophilus, *Autolycus*, II:4.

[15] Irenaeus, *Against Heresies*, III:X,3; cf., II:10,4; cf. also Gilkey, *Maker of Heaven and Earth*, 44-66.

[16] John of Damascus, *Exposition of the Orthodox Faith*, I:vii.

[17] Augustine, *Confessions*, Book 12, Chap. 7.

[18] What is nothing? It is common among contemporary theologians to follow the Platonic school and distinguish the nondialectical from the dialectical concepts of nothing. *Ouk on* is the nothing which has no relation at all to being; it is the absolute negation of being. *Me on*, in contrast, is the nothing which has a dialectical relation to being. This second or dialectical nothing does not yet have being but has the potential for becoming something. It constitutes the temporal quality which erodes present being and makes possible future reality. It is not-yet-being. Paul Tillich believes that the Christian doctrine of *creatio ex nihilo* constitutes a rejection of *meontic* matter. The *nihil* out of which God creates is *ouk on*, the undialectical negation of being. *Systematic Theology*, 3 Volumes (Chicago: University of Chicago Press, 1951-63) I:188. Yet, when we consider the ongoing work of God as creator, the not-yet quality of *me on* seems to be a necessary construct. Jürgen Moltmann agrees. *God in Creation* (San Francisco: Harper & Row, 1985) 75ff. I believe we need to work with both notions of nothing.

[19] Paul Tillich defines eternity as the "power of embracing all periods of time." In doing so, he comes close to advocating the Greek penchant for abstracting eternity *from* time. For Tillich, "presence is not swallowed by past and future; yet the eternal keeps the temporal within itself. Eternity is the transcendent unity of the dissected moments of existential time." *Systematic Theology*, I:274. That idea that temporality is kept within eternity is important if we are to understand God as involved in the world of spacetime. This is further developed in the trinitarian theology of Robert Jenson who defines "eternity" as God's "faithfulness *through* time." *The Triune Identity* (Philadelphia: Fortress, 1982) 68.

[20] Augustine, *Confessions*, Book 11, Chap. 12. In Q. 46 of his *Summa Theologica* Thomas Aquinas acknowledges that for Aristotle the world is eternal and that good arguments can be raised against the idea that the world has a beginning in time. Nevertheless, Christians should hold that the world is created — i.e., it has not always existed — as an article of faith based upon revelation.

[21] Robert Cummings Neville, "Creation and Nothingness in Buddhism and Christianity," paper delivered to the inter-religious conference, "Buddhism and Christianity: Toward the Human Future," Graduate Theological Union, August 10-15, 1987.

[22] Due to space and thematic constraints, we will limit the present discussion to the standard Big Bang cosmology, leaving aside for the time being modifications

such as the eternal oscillating model, the mother universe, the many worlds thesis, etc. Each deserves its own attention but is beyond the scope of the present essay.

[23] Not everyone means exactly the same thing by the Big Bang, nor do all intend the same moment. Here we will assume the following chronology: (1) $t=0$ at the singularity; (2) Planck time up to 10^{-43} seconds; (3) start of the Big Bang; (4) end of Big Bang period and formation of galaxies at 300,000 years; and (5) present era of expansion.

[24] See James S. Trefil, *The Moment of Creation* (New York: Charles Scribner's Sons, 1983) 156f.

[25] C.J. Isham follows the work of J.B. Hartle and S.W. Hawking who hold that spacetime is finite yet without boundary or edge. Here is an attempt to eliminate all contingency with respect to the initial conditions and, hence to eliminate the notion of a moment of creation. Isham proceeds to argue that there is no actual point of creation but rather that four-dimensional space simply is. Theologically this means there is no initial creative event which is different from others. Instead, we should understand God's creative work as his sustaining work. Wim B. Drees is critical here, suggesting that Isham's view is consonant only with a Spinozistic view of the relation between God and the world. "Beyond the Limitations of Big Bang Theory; Cosmology and Theological Reflection," *CTNS Bulletin* 8:2 (1988). But there is an alternative. According to the Oxford cosmologist and colleague of S.W. Hawking, R. Penrose, the initial conditions must have been special in order to obtain the particular cosmic history we have had. The Penrose theory retains the sense of contingency, posits an arrow of time, and even suggests the work of a Creator.

[26] Ilya Prigogine and Isabelle Stengers, *Order Out Of Chaos* (New York: Bantam, 1984) 12.

[27] Fred Hoyle, *The Nature of the Universe* (New York: Mentor, 1950) 125.

[28] McMullin, "How Should Cosmology Relate to Theology?" 32ff.

[29] Schubert M. Ogden, *The Reality of God* (New York: Harper & Row, 1966) 62f, 213.

[30] John B. Cobb, Jr., and David Ray Griffin, *Process Theology* (Philadelphia: Westminster, 1976) 65. Sallie McFague also rejects the idea of *creatio ex nihilo*, but she does so because it is dualistic, i.e., she rejects the implication that the world is totally different from God and that there exists a hierarchy between God and the creation. *Models of God* (Philadelphia: Fortress, 1987) 109.

[31] Barbour, *Issues in Science and Religion*, 368. Since this book was written, decisive evidence in favor of Big Bang has come in. Barbour is much more willing now to favor this theory, but his motive is clearly scientific.

[32] Even though we cited Barbour above as best representing the position of hypothetical consonance advocated as the method for this paper, at this point one wonders if Barbour himself sinks back into the two language theory.

[33] Barbour, *Issues in Science and Religion*, 384.

[34] *Ibid.*, 458. What Barbour specifically rejects is a beginning to time. Nevertheless, he wants to keep the "intent of *ex nihilo* by saying that novelty is as such not traceable to its antecedents." *Ibid.* This was in 1966. More recently Barbour has clarified his position, showing considerably more theological interest in the Big Bang theory and reaffirming not only *creatio ex nihilo* in the sense of ontological dependence but even opening the door to the possibility of an initial beginning. See: Ian Barbour, "Creation and Cosmology" in *Cosmos as Creation*, edited by Ted Peters (Nashville: Abingdon Press, 1988).

[35] Arthur R. Peacocke, *Creation and the World of Science* (Oxford: Clarendon Press, 1979) 79.

[36] Thomas Aquinas, *Summa Theologica*, I, q. 45, art. 2.

[37] David Kelsey recognized that for Thomas the "theological issue at stake... was faithfulness to biblical revelation." Then Kelsey himself tries to deal with the issue operating with the major premise that we need to be faithful to scripture. He then adds a minor premise: the doctrine of creation is "never merely a restatement of cosmologies found in the Bible." This premise has a corollary: the Christianness of the doctrine is "partly warranted by extrabiblical truth." For Kelsey, evidently, being biblical also means being extra-biblical. With these premises in place, Kelsey can now argue for the "abandonment of the theological claim about creation as a singular event." He wants to keep *creatio ex nihilo* while jettisoning a beginning to the world. He starts where Thomas does, interprets the same scripture, yet ends up with the opposite conclusion. "The Doctrine of Creation from Nothing," in McMullin, *Evolution and Creation*, 184-190.

[38] Gilkey, *Maker of Heaven and Earth*, 70. Paul Davies expends considerable effort attacking Thomas Aquinas' causal proof for the existence of God arguing, as does Gilkey, that the logic of the first cause only pushes the question back a further step: what caused God? Or, to put it another way, why should God be an exception to logic? *God and the New Physics*, 33-40. The answer is that God is transcendent to the world of cause and effect; and, though involved in the world, God is not determined by the world of cause and effect.

[39] Barbour, *Issues in Science and Religion*, 385.

[40] Langdon Gilkey, *Message and Existence* (New York: Seabury, Crossroad, 1980) 90; cf., *Maker of Heaven and Earth*, 312.

[41] The idea of contuining creation for Barbour and Peacocke seems to be drawn from consonance with biological evolution, whereas the idea of a point of origin overlaps with astrophysics. Hence, these are not mutually exclusive by any means. We might observe further that the notion of emergence as Peacocke and Barbour employ it probably represents a more thoroughgoing understanding of change than was conceived by Thomas and, hence, properly deserves the title "continuing creation". A still more radical picture is drawn by Jürgen Moltmann. He affirms that "creation" applies to the beginning of things and that time's structure is asymmetrical. Yet, the process of creation still goes on. God, he says, "...stands in the Becoming of the still open, uncompleted process of creation.... Creation at the beginning is the creation of conditions for the potentialities of creation's history.... It is open for time and for its own alteration in time." *The Future of Creation* (Philadelphia: Fortress, 1979) 119f. Creation will be complete only in the eschaton.

Interesting point!

HOW TO DRAW CONCLUSIONS FROM A FINE-TUNED COSMOS

JOHN LESLIE, Department of Philosophy, University of Guelph

Introduction

The argument from design is an argument for God's reality based on the fact that our universe looks very much as if it had been designed. The argument for multiple worlds starts from the same fact. But it concludes instead that there exist many small-u universes inside the large-U Universe which is the whole of reality. These "universes," "mini-universes," "worlds," can be of immense size. There may be immensely many of them. And their properties are thought of as very different. Sooner or later one or more of them will have life-permitting properties — and obviously it will be in a life-permitting universe that living beings such as we will find themselves. Our universe may indeed look as if it were designed. In reality, though, it is merely the sort of thing which should be expected sooner or later. Given sufficiently many years with a typewriter even a monkey would produce a sonnet.[1]

While the multiple worlds hypothesis seems to me impressively strong, the God hypothesis is a viable alternative to it. If God exists, then of the various ways in which he may act on the universe there are only two with which my arguments will deal. First, God makes the universe obey a particular set of basic laws of nature, also "sustaining" it in existence if this is necessary. Secondly, God creates its initial state in such and such a fashion. He starts it off with exactly this or that number of particles in exactly this or that arrangement; or at least he does this just so long as it has not been done already through his specifying its laws. (It might be that the laws themselves dictated the number and arrangement of the particles.)

If God is real then his reality seems to me most likely to be as described by the Neoplatonist theological tradition. He is then not an almighty person but an abstract creative force which is "personal" through being concerned with creating persons and acting as a benevolent person would. To defend this theme would take us into complexities far removed from the main arguments of this paper. Any readers intrigued by it will need to turn elsewhere.[2] It might instead be that God was a divine person creating everything else, a person owing his existence, perhaps, to the ethical need for it. This alternative would allow for a divine creative freedom which was, to some extent, exercised arbitrarily. Still, it is not unreasonable to believe that a divine person would have wanted any scheme of things which he created to be life-containing.

The Fine-Tuned Universe

Our universe can seem spectacularly "fine-tuned" for life. Tiny changes in it would rule out not only human life but, so it is argued, absolutely all life. Today's cosmologists make many claims like these.[3]

Large regions coming out of a Big Bang could be expected to be causally uncoordinated, since light would not have had time to link them. When they first made contact, tremendous turbulence would occur, yielding a cosmos of black holes, or temperatures which stopped galaxies from forming for billions of years, after which everything would be much too spread out for them to form. Placing a pin to choose our orderly world from among the physically possible ones, God would seem to have been called upon to aim with immense accuracy. This is the *smoothness* problem. The cosmos threatened to recollapse within a fraction of a second or else to expand so fast that galaxy formation would be impossible. To avoid these twin disasters its rate of expansion at early instants needs to be fine-tuned to perhaps one part in 10^{55}, making space extremely flat. This is the *flatness* problem.

The smoothness and flatness problems might be avoided by inflation. Exponentially fast expansion at very early times would mean that everything visible to us had grown from a single region whose original parts were all causally linked and which, therefore could yield smoothness. A highly expanded space could be very flat like the surface of an extremely inflated balloon. However, inflation itself might seem to have required fine tuning in order to occur at all and to yield irregularities neither too small nor too great for galaxies to form. Thus, besides having had to select a Grand Unified Theory very carefully, God would have had to have fine-tuned the universe with an accuracy better than one part in 10^{50}.

Had the nuclear weak force been appreciably stronger, the Big Bang would have burned all hydrogen to helium. There could then be neither water nor long-lived stable stars. Making the nuclear weak force appreciably weaker would also have destroyed the hydrogen; the neutrons formed at early times would not have decayed into protons. Again, this force needed to be chosen appropriately if neutrinos were to interact with stellar matter both weakly enough to escape from a supernova's collapsing core and strongly enough to blast its outer layers into space so as to provide material for making planets.

For carbon to be created in quantity inside stars the nuclear strong force must be within ninety-nine per cent of its present value. Increasing its strength by maybe two per cent would block the formation of protons (so that there could be no atoms) or else bind them into diprotons so that stars would burn about a billion times faster than our sun. Decreasing it by roughly five per cent would unbind the deuteron, making stellar burning impossible. (Increasing Planck's constant by over fifteen per cent would be another way of preventing the deuteron's existence. So would making the proton very slightly lighter or the neutron very slightly heavier, since then it would not be energetically advantageous for pairs of protons to become deuterons.)

With electromagnetism very slightly stronger, stellar luminosity would fall sharply. The main sequence (on which stars spend most of their lives) would consist entirely of red stars, probably too cold to encourage life's evolution and unable to explode as the supernovae one needs for creating elements heavier than iron. Were it very slightly weaker all main sequence stars would be very hot and short-lived blue stars. Changes in strength by only one part in 10^{40} could spell disaster. A slight strengthening could transform all quarks (essential to atoms) into leptons or make protons repel one another strongly enough to prevent the existence of atoms even as light as helium. A strengthening by one per cent could have doubled the years needed for intelligent life to evolve, by making chemical changes more difficult. A doubled strength would have meant that 10^{62} years were needed. By increasing the electromagnetic fine structure constant to above 1/85 (from its present 1/137) we would have too many proton decays for there to be long-lived stars, let alone living beings who would not be destroyed by their own radioactivity. The need for electromagnetism to be fine-tuned if stars are not to be all red or all blue can be rephrased as a need for the fine-tuning of gravitation (because it is the ratio between the electromagnetic and gravitational forces which is crucial). Gravitation also needs fine-tuning for stars and planets to form and for stars to burn stably over billions of years. Gravity is roughly 10^{39} times weaker than electromagnetism. Had it been only 10^{33} times weaker, stars would be a billion times less massive and would burn a million times faster.

Various particle masses had to take appropriate values for life of any plausible kind to stand a chance of evolution: (i) If the neutron-proton mass difference — about one part in a thousand — were not almost exactly twice the electron's mass, then all neutrons would have decayed into protons or else all protons would have changed irreversibly into neutrons. Either way, there would not be the several hundred stable nucleides on which chemistry and biology are based; (ii) Superheavy particles were active very soon after the Big Bang. Fairly modest changes in their masses could have led to disastrous alterations in the ratio of matter particles to photons, giving a universe of black holes, or else of matter too dilute to form galaxies. Further, the superheavies had to be very massive to prevent rapid decay of the proton; (iii) The intricacy of chemistry (and the existence of solids) depends on the electron's being much less massive than the proton; (iv) The masses of a host of scalar particles could have determined whether the cosmological constant would ever be the right size for inflation to occur and whether it would later be small enough to allow space to be flat (failing which space would be expanding or contracting very violently). Today the cosmological constant is zero to one part in 10^{120}; (v) The "screening", "anti-screening" and "vanishing" which give forces so much of their oddity (the nuclear strong force, for instance, is repulsive at extremely short ranges while at slightly greater ones it is first attractive and then it disappears entirely) are crucially dependent on particle masses. The present masses make possible intricate checks and balances which underlie the comparatively stable behaviour of galaxies, stars, planets and living organisms.

While some such claims may well be wrong, others seem about as well established as those about the reality of quarks, black holes, neutron stars, or the Big Bang itself. Remember, too, that clues heaped upon clues can constitute weighty evidence despite doubts about each element in the pile.

Explanations of the Fine-Tuning

As indicated earlier, one way of accounting for the seeming evidence of fine-tuning would be to suppose that there exist vastly many "worlds" or "universes" with varied properties, ours being one of the rare ones in which life evolves. (There is no need to replace "vastly many" by "infinitely many," though people often write as if this were essential.)

Ways of getting multiple universes include these: (a) the universe oscillates: bang, squeeze, bang, and so on. Each oscillation could count as a new "world" or "universe" because of having new properties, or because the oscillations are separated by knotholes of intense compression in which information about previous cycles is lost (or in which time breaks down so that talk of "previous" cycles is nonsense); (b) many-worlds quantum theory is usually understood as giving us a large-U Universe which branches into more and more "worlds" (small-u universes) which interact hardly at all; (c) worlds, small-u universes, could occur as quantum fluctuations, as suggested by E.P.Tryon. They could be fluctuations in an ever-expanding space or in a "space-time foam" existing "before" or "outside" (or at any rate neither after nor inside!) the regions in which space and time are well-structured; (d) if space is "open" (instead of "closed" like the surface of a sphere), then in the most straightforward models it is infinitely large and contains infinitely much material. Large regions much beyond the horizon, set by how far light can have travelled towards us since the Big Bang, could well be counted as "other universes," especially if their properties were very different; (e) even a "closed" cosmos could be of any size, and the nowadays very popular inflationary cosmos would be in fact gigantic. A. H. Guth and P. J. Steinhardt suggest that even our domain, characterized by a particular way in which early symmetries chanced to break (see our later discussion), stretches 10^{25} times farther than we can see, and that it is only one of very many equally huge domains.[4]

Even granted ideal conditions, life might evolve only with great difficulty. (Maybe its first beginnings depend on tremendous luck with molecular combinations in some primeval soup.) If so, then multiple universes could help produce it by sheer force of numbers. (Toss a hundred coins sufficiently often and some day the lot will land heads together.) But a multiplicity of universes could be all the more helpful if the universe varied widely, so making it more likely that conditions would be ideal somewhere. Now, modern unified theories do suggest that wide variations could be expected.

Why? Well, at early times there may have been only a single force and a single type of particle. As the universe cooled, this unity would have been destroyed by symmetry-breaking phase transitions. It would have become energetically advantageous for a scalar field (or more probably fields) to take a non-zero value (or values). The choice of any such value may have been a random affair. A commonly used image is of a ball rolling off a hilltop, the position in which it ends up being settled by chance. Or perhaps the values vary from one gigantic region to another, not randomly but deterministically, as envisaged by, e.g. P.C.W. Davies and S.D. Unwin.[5] Now, interacting with a field can make particles take on mass, and particle masses, besides being of perhaps great direct importance to the possibility of the evolution of life, also underlie the differences between the strengths of Nature's four main forces. Hence any theory giving us multiple universes might also fairly readily provide different combinations of force strengths and masses. If many scalar fields were involved in the symmetry-breaking and if each such field affected different particles in different ways, the range of variations would be enormous.

This way of looking at things is favoured by, for instance, A.D. Linde, who speaks of inflation's cosmos as "a lunch at which all possible dishes are available."[6] We cannot hope to see regions in which the force strengths and masses are different from those found locally. Inflation has pushed them far beyond the reach of our telescopes.

According to some theories absolutely all physical possibilities would be realized somewhere, some time; but this is in no way indispensable to making life's presence unmysterious. A monkey could produce a page of poetry unmysteriously without having to type all possible pages.

Let us ask, however, whether life really does stand in special need of explanation, and, if so, whether a multiplicity of "worlds" or "universes" with varied properties could provide a satisfying explanation. An initial point to note is that neither a multiple worlds explanation nor an explanation by reference to God would supply a substitute for a long, causal account of life's evolution. What these explanations could instead provide would be a causal, or some other, insight into how it came to be inevitable, likely, or very possible, that there would be, somewhere, a situation characterized by force strengths and masses such as made life's evolution inevitable, likely or very possible. Next, I find it helpful to tell a succession of stories.

First comes the Fishing Story. You know a lake contained a fish 23.2576 inches long, for you have just caught the fish in question. Does that fact about the lake specially need explanation? Of course not, you tend to think. Every fish must have some length! Yet you next discover that your fishing apparatus could accept only fish of this length, plus or minus one part in a million. Competing theories spring to mind: (a) that there are millions of fish of different lengths in the lake, the apparatus having in the end found one fitting its requirements; and (b) that there is just the one fish, created by a deity who wanted you to catch it. Either explanation will serve; and so, for that matter, will the explanation that the deity created so many fish of different lengths that there would be sure to be one which you could catch. (God and multiple worlds are far from being flatly incompatible.) In contrast, that the one and only fish in the lake just happened to be of exactly the right length is a suggestion to be rejected at once.

The tale has countless variations: for instance, the Poker Game Story (a nice response to those who say that the "improbability" of our universe is no more impressive than that of any hand of cards, every possible hand being equally improbable). You seem to see mere rubbish in your opponent's poker hand of an eight, six, five, four and three. It is natural to assume that chance gave it to him. But you then recall that poker has many versions; that you had agreed on one in which his "little tiger" ("eight high, three low, no pair") defeats your seemingly much stronger hand; that a million dollars are at stake; and that card players have been known to cheat. At once your suspicions are aroused.

Again, an old arch collapses exactly when you pass through. You congratulate yourself on a narrow escape from purely accidental death, until you notice your rival in love tiptoeing from the scene. Again, a tale by Ernest Bramah about an ingenious merchant: "Mok Cho had been seen to keep his thumb over a small hole in a robe of embroidered silk"; now, "although the tolerant-minded pointed out that in exhibiting a piece of cloth even a magician's thumbs must be somewhere. . . ."

The main moral must by now be plain. Our universe's elements do not bear labels announcing whether they are in special need of explanation. A chief (or the only?) reason for feeling that something stands in such need, i.e., for reluctance to dismiss it as just how things happen to be, is that one actually glimpses some tidy way in which it might be explained. In the case of catching the 23.2576-inch fish, a fish of the only length which can be observed, the first of the two tidy explanations which suggested themselves could be called a "fish ensemble" explanation. It runs parallel to the "world ensemble" (or multiple universes) explanation of how it came to be at all likely that anyone would be able to observe a cosmos.

There are subsidiary morals too. Thus, notice how you cannot account for catching your fish by considering many *merely possible* fish, remarking that only one of just about exactly 23.2576 inches could be caught, and then declaring that this would sufficiently explain the affair even if yours had been the only fish in the lake. What you instead need is either a benevolent, fish-creating deity or else a lake with many *actual* fish of varying lengths. The fish, really existing fish, of lengths which cannot be caught, help to render unmysterious the catching of the fish which can be.

Is this a dizzying paradox? Not at all. Firing an arrow at random into a forest you hit Mr. Bloggs: persuasive evidence, surely, that the forest is full of people, despite how the other people gave Bloggs no greater chance of being hit. You need a well-populated forest to have much chance of there being somebody precisely where the arrow lands. You need fish of many different lengths to have much chance that one will be of precisely the right length. (When the fish is captured then the details of how it came to be captured and of how it came to be of just the right length will form a long causal story which will perhaps be entirely unaffected by the other fish in the lake. The complex details of how Bloggs came to stand precisely where he stood may be unaffected by the others in the forest. But I have already drawn attention to this kind of point. I said, remember, that a multiple worlds explanation would not be a substitute for a long, causal account of life's evolution. Instead it would offer insight into why it was inevitable, likely or very possible that life would evolve somewhere.)

Would you protest that, if fish appeared one after another, with randomized lengths, then there would be nothing particularly unlikely in the right length's being had by the very first fish of all? You would be trading on an ambiguity. Yes, the very first is no more unlikely to be "just right" than is the second or the millionth. In that sense its just-rightness "isn't particularly unlikely." But it can still be particularly unlikely where this means that its unlikelihood is very small. Hence no just-right fish is likely to exist unless there are many fish.

But are there not infinitely many infinitesimally different fish lengths which the apparatus could accept? — just as many, in fact, as the fish lengths which it would reject? Well, there being infinitely many points inside a bull's-eye is no ground for optimism that an arrow will hit this tiny target. (One often meets with a flat announcement that there could be nothing impressive in the supposed evidence of fine tuning unless among all possible sets of force strengths and particle masses only one could lead to intelligent life. I see no excuse for such an announcement. Surely the fine tuning could become impressive as soon as the life-encouraging possibilities constituted, say, only one thousandth of the total range of possibilities. To deny this is almost as bad as announcing that the evidence could be impressive only if every single aspect of our universe were fine tuned, or only if the fine tuning made life's evolution one hundred per cent certain.)

Yet, you exclaim, are we not in fact virtually compelled to accept the God hypothesis? The alternative is to assume, so to speak, that the lake contains many fish and that we had been waiting until a catchable fish — a universe we could observe — came along. Yet surely we were not disembodied spirits lying in wait until there came to be a universe containing bodies for us. So are we not forced to believe in a divine hand which made our universe one in which life was likely to evolve?

Not so, I think. Let us agree that in God's absence our births could only be a matter of tremendous luck. Let us suppose, for instance, that if our universe's symmetries had broken very slightly differently life could never have evolved in it. So what? The multiple worlds hypothesis shows how it could be likely that some set of beings should have the immense luck of being born. While they could be extremely lucky, their luck would not be unbelievably amazing.

Here we could tell a story of a lottery. When the lottery tickets were being printed one of them was given a number which made it worth a million dollars. Most of the tickets were actually sold. No one winning the million dollars should feel compelled to seek some very special explanation for having won: some explanation of a kind inapplicable to just any other winner. Yes, the absence of such an explanation would mean that he or she had had immense good fortune; but it was very likely that somebody would have it. However, this particular lottery story fails to reflect an important extra element in the cosmological case: namely, that it is a case in which (so to speak) the winning of a lottery is a prerequisite of observing anything. Given this extra element one cannot argue in the following style: "Though it would not be unbelievably amazing that somebody had won a million dollars by mere chance, it could still be very amazing to me that the somebody should be me; not, presumably, unbelievably amazing, as one presumably ought to be reluctant to say that no matter who wins a lottery by mere chance, that person ought to be flatly unwilling to believe that it was chance that settled the affair; but still amazing enough to make me doubt that chance really did give me my victory; since what I should expect to be observing is a situation in which I am holding a non-winning ticket." One cannot argue in that style, because in the cosmological case an observational selection effect guarantees that a "non-winning ticket" — a lifeless universe — will never be seen by anyone.

To highlight this extra element we might tell a new version of the Fishing Story. A mad scientist allocates numbers to millions of human ova, fertilized and then frozen. She fishes for ten seconds with an apparatus able to catch only a 23.2576-inch fish. If unsuccessful she destroys ovum number one. She then fishes for another ten seconds on behalf of ovum number two; and so on. Any test-tube boy-baby born because "his" fishing led to success can (on reaching adulthood) be thankful that he survived this savage weeding. He has been extremely lucky. But not unbelievably lucky. He presumably ought not to feel compelled to reject the mad scientist's report of how he came to be born. For with respect to believability this report is, I suggest, very much like a report that the scientist had fished repeatedly on behalf of the same one ovum, for successive ten-second periods, until success crowned her efforts. (Notice, though, that the two cases differ markedly with respect to how lucky he is to have been born. In the case where there were many ova it was only through immense luck that his ovum gave rise to a conscious being.) If, in contrast, the mad scientist reported that she had set aside only a single ovum for the fishing experiment and fished for just one ten-second period, then he should reject this. It would not be enough for him to comment, "If that ovum had not had such immense luck then I shouldn't be here to ask whether to be surprised, so there's nothing for me to be surprised at."

A variant is the Firing Squad Story. When the fifty sharpshooters all miss me, "If they hadn't missed then I shouldn't be considering the affair" is not an adequate response. What the situation demands is, "I'm popular with those sharpshooters — unless immensely many firing squads are at work and I am one of the very rare survivors."

But — you protest — we have no firm reason to think that universes really could have had any of a wide range of features much as a fish could have had many lengths. Might not only the one kind of universe be possible? Or might not only universes like ours be at all likely?

We need not linger over the idea that only the one kind of universe is logically possible. Today, "the logically possible" means what could be described without self-contradiction. Now, it might conceivably be: (a) that only a single kind of universe was compatible with physical laws of the general sort which rule our universe; and perhaps also (b) that a universe would have to obey either exactly those laws or else very different ones, because any attempt to vary the laws just slightly would lead to contradictions. Yet the fact remains that there would be nothing self-contradictory in a universe's obeying very different laws, or even nothing worthy of the name law. There could be universes obeying magical principles, in which butter was produced by shouting "Rettub!" There could be universes so chaotic that there would be little point in stretching the word "laws" until it fitted them. How, though, should we react to the idea that there is something about Nature's actual laws and physical constants which makes them alone really possible or really likely?

While it looked to us as if God had very skilfully hit a bull's-eye, a tiny "window" of life-encouraging force strengths, particle masses, etc., might it not have been hard to avoid missing the window? When we represented the situation on graph paper could we not be using the wrong kinds of scale? Might not a truly appropriate graph show the so-called window as filling almost the entire field of real possibilities?

My response is that all this might conceivably be so, but that it ought not to trouble us very much. A wild example could illustrate the point. Suppose that the words **made by God** are found all over the world's granite. Their letters recur at regular intervals in this rock's crystals. Two explanations suggests themselves. Perhaps God put the words there, or perhaps very powerful visitors from Alpha Centauri are playing a practical joke. Both explanations could account for the facts fairly well, yet along comes a philosopher with the hypothesis that the only "really possible" natural laws are ones which make granite carry such words. And in that case, says he, there is no need to "fine tune" anything in order for there to be such words. Nothing else is genuinely possible! Explanation fully provided! The so-called bull's-eye or tiny window in fact fills the entire field! Yes, there are countless logically possible laws but the only really possible ones are the laws which yield electrons, pebbles, stars and **made by God**.

Surely this would be ingeniously idiotic. We must not turn our backs on tidy explanations, replacing them by a hand waved towards the obscure notion of "limits to what is really possible." Prior to our discovering that there are messages in granite or that any of a hundred small changes in force strengths, particle masses, etc., would seemingly have prevented life's evolution, prior to our discovering this, I agree, it could be attractive to theorize that only one kind of granitic crystal pattern or one set of strengths and masses "is really possible." But afterwards? Surely the attractiveness has vanished. Blind necessity must be presumed not to run around scattering messages or making a hundred different factors each look exactly as if chosen in order to produce life.

It might still be that force strengths, particle masses, etc., were dictated by the laws which applied to our cosmos, laws cohering elegantly in some totally unified theory. For these laws could be due not to blind necessity but to divine selection of a totally unified theory which provided automatically the results which seemed to us to need fine tuning. (Rather similarly, a very carefully chosen theory might conceivably yield granitic messages automatically. We could then perhaps say that "the real fine tuning" was carried out by God's very careful choice.) Or again, it might just be that immensely many such totally unified theories were valid, each in a different universe, and that we existed in one of the rare life-permitting universes.

Concluding that it was not any blind necessity which gave life-permitting values to a hundred factors, we should be relying on the first moral drawn from the Fishing Story: viz., that one reason for thinking that things need to be explained is that tidy explanations spring to mind. Yet we should need to rely also on a cheerful disregard of the possibility that it was *a priori* tremendously likely that blind necessities were operative, necessities firmly dictating various life-permitting factors, or else making them very probable (so that, in case after case, the seeming needs for fine tuning were mere artifacts of graphs wrongly scaled), or else perhaps setting up a situation in which these factors, though seemingly so multitudinously distinct, in truth formed a web such that every attempt to ruin life's prospects by changing one factor would only produce compensatory changes in others. Yet such a cheerful disregard can be reasonable even if we grant that some clear sense can here be attached to *a priori* tremendous likelihood; for, as always, tidiness of explanation should weigh heavily with us. (The Story of the Granite is an attempt to show how very reasonable this sort of disregard could sometimes be. Again, consider this case. Feeling two balls in an urn but knowing nothing about their colours, you draw a ball, replace it, draw again, replace, and so on for a hundred draws. Every single time a red ball is drawn. A tidy explanation suggests itself: that both the balls are red. Would you resist this on the grounds that maybe it was *a priori* tremendous likely that one of them was blue?)

But, you object, is it not silly to suppose that we could, even in thought, inspect all possible universes so as to find what proportion would be life-permitting? I answer that the Story of the Fly on the Wall shows that we need inspect only those possible universes which are much like ours in their basic laws though differing in force strengths, particle masses, expansion speeds, etc. A wall bears a fly (or a tiny group of flies) encircled by a large empty area. The fly (or one of the group) is hit by a bullet. We can at once fairly confidently say. "Many bullets are hitting the wall and/or a marksman fired this particular bullet." We need not bother about whether distant areas of the wall are thick with flies. All that is relevant is that there are no further flies locally.

When telling this story I have sometimes suggested that the alternative to the marksman hypothesis would be that many bullets were hitting the wall near the fly. This was a blunder. For suppose the wall carried many solitary flies, each surrounded by a large fly-free area. There would now be a good chance of some bullet's hitting some solitary fly, provided only that many bullets were hitting the wall. So from the fact that there is only one fly locally (only one life-permitting kind of possible universe inside the local group of possible universes, those much like ours in their basic laws), one has no right to conclude that in the absence of a marksman (God) there are probably many bullets locally (many actually existing universes much like ours in theirs basic laws). It need only be supposed that there are many bullets hitting the wall at varying places: many actually existing universes with differing characteristics. Although the basic laws of these other universes could plausibly be thought to be much like those of our universe, they might instead be very different.

It is often said that only one universe is open to our inspection and that judgements of probability cannot be made on the basis of a single trial. The Telepathized Painting Story is a suitable response to this. Having done his best to paint a countryside, Jones tries to transmit the horrid results to Smith by mere power of thought. Behold, Smith reproduces every messy tree and flower and cloud! Whereon a philosopher reacts as follows: "Can't conclude anything from that! Must have more than one trial!" Faced by such a reaction we ought to protest that Smith's painting is complex. Though only a single painting it is many thousand blobs of paint. Much could be learned from it. Now, experiencing many thousand billion parts of our universe, might we not rather similarly gain some right to draw conclusions about the whole? After learning about ordinary messages we could be justifiably reluctant to dismiss as mere chance, or even as neither probable nor improbable because we have not experienced other universes, any **made by God** messages which we found written on it. And after some experience of physics and biology we could fairly confidently perform thought-experiments showings how dim life's prospects would have been had various force strengths and particle masses been slightly different; now, this could encourage us to believe in God or in multiple worlds. (Yet philosophers have argued solemnly that a Creator would find it logically impossible to leave any signs of his creative action because, poor fellow, he would from the very definition of "universe" be limited to showing us just a single universe. Hence, one presumes, even writing **made by God** all over it would have no tendency to prove anything! And if that were so, then, of course, the mere fact of its containing life could give us no reason to believe anything dramatic.)

Let us now turn to a story apparently damaging to my case: the oft-told tale of the Great Rivers Passing Through the Principal Cities of Europe. What superb evidence of the Creator's action! A variant points to the Mississippi. See how wonderfully it threads its way under every bridge! Another concerns pond life. The rotifers of Little Puddle marvel at the deity who has provided filthy water and mud. Had their ancestors evolved in arsenic-filled waters then they would be marvelling at the Creator's benevolence in supplying arsenic. An atrocious case of thinking backwards! What blindness to Darwinian theory! What parochial concern with the prerequisites of rotiferhood!

I reply: (a) Even those defending the unfortunately named anthropic principle often take pains to deny that their concern is only with the human race. As was made plain enough by B. Carter, who baptized the principle, what is involved is a possible observational selection effect stemming from the nature not of humanity but of *observerhood.* The principle reminds us that, if there were many actual (small-u) universes most of which had properties hostile to the evolution of intelligent life then, obviously, we intelligent products of evolution would be observing only one of the rare universes in which intelligent life could indeed evolve. People who tell sarcastic tales about rotifers are missing the point; (b) Or perhaps they do not miss the point but have minds dominated by the curious belief that life can evolve just about anywhere — for instance, in frozen hydrogen or on neutron star surfaces or in the interiors of ordinary stars or in interstellar gas clouds. But they then invite the responses, first, that there are quite powerful reasons for thinking that frozen hydrogen, neutron stars, etc., would be inhospitable environments; second, that, if life were as easily achieved as they fancy, then Fermi's "where are they?" puzzle (of why we have no evidence of extraterrestrial life) becomes very hard to solve; and third, that in any case there would be no frozen hydrogen, neutron stars, ordinary stars or gas clouds if the Big Bang had been followed by recollapse within ten seconds or if various other unfortunate happenings had occurred — happenings seemingly avoidable only by very accurate fine tuning.

"If rotifers could talk..." may, however, be replaced by "if carbon could talk...." The sceptic might say that the prerequitites of intelligent life are just whatever are the prerequisites of carbon, of water, of long-lived stable stars and maybe of a handful of further things. Now, how would matters look to a philosophical club consisting of carbon atoms, water molecules, long-lived stable stars and so forth? Instead of an anthropic principle wouldn't there be a carbonic principle? Instead of worshipping a Creator benevolent towards humans, wouldn't club members pray to one who loved a particular liquid?

In reply it can be helpful to insist that intelligent life seems to depend on a very long list of things. When the philosophical club came to its grand conclusion that carbon, water, long-lived stars, etc., are what are truly important here, or are at any rate just as important as the intelligent life which so obsesses humans, then surely the length of the list — and the fact that the things which were on it were on it because of being prerequisites of intelligent life — would show the wrongness of this.

Still, suppose for argument's sake that nothing but carbon was required for producing living intelligence. The prerequisites of carbon and of living intelligence thus being identical, might it not be arbitrary to concentrate on the latter? Why not forget about the difficulty of generating intelligence? Why not talk instead of how hard it is to produce carbon? Now, the existence of carbon might indeed act as a selection function picking out our kind of universe from the field of all possible universes. Many scientific theories might fail through being incompatible with the observed fact of there being carbon. Yet — says the sceptic — this is all very ordinary science. Compare this with how the theory that rock becomes fluid at a pressure of two tons per square inch is refuted by the existence of Mt. Everest. There is nothing in this to justify talk of a benevolent Creator, multiple worlds, or "the Mt. Everest principle"!

This seems to me very wrong. It overlooks the point of the Fishing Story, the Poker Game Story, the Collapsing Arch Story and the Story of the Silk Merchant's Thumb. It forgets that carbon particles do not talk, observe nothing, and could not plausibly be loved for their own sakes by a benevolent deity. What is so special about observerhood or about being such as a benevolent deity could well love? It is that these suggest tidy explanations. Every thumb must be somewhere, but the placement of the silk merchant's is special because it suggests an explanation — a love of making money — for its being there rather than elsewhere. Likewise, the reason why a 23.2576-inch fish is special is that nothing else can be observed with the help of your fishing apparatus and that this, when combined with belief in many fish of varied lengths, very tidily explains why the fish is being observed. True, the catching of the fish also gives reason for believing in a benevolent, fish-creating deity. But such double suggestiveness need not dismay us. Smith's empty treasure chest, on an island whose only inhabitants are Smith, Brown and Jones, can fairly powerfully suggest that Brown is a thief despite also suggesting just as powerfully that theft has been committed by Jones.

Concluding Remarks

Contemporary religious thinkers often approach design arguments with a grim determination that their churches shall not again be made to look foolish. Recalling what happened when churchmen opposed first Galileo and then Darwin, they insist that religion must be based not on science but on faith. Philosophy, they announce, has demonstrated that design arguments can have absolutely no force.

I hope to have shown that philosophy has demonstrated no such thing. Our universe, which these religious thinkers believe to be created by God, does look (though this may come as a surprise to them) very much as if it were created by God. Consideration of many parables ("stories") shows that this conclusion draws strong support from the type of reasoning that serves us well in ordinary life, and we should continue to trust such reasoning even here. The question of whether our universe is God-created is indeed no ordinary question, but that cannot in itself provide any strong excuse for abandoning ordinary ways of thinking. Theology is not a call to reject common sense.

Still, one must bear in mind two main points: (1) world ensemble plus observational selection effect could provide a powerful means of accounting for any fine tuning which we felt tempted to ascribe to divine selection. Now, this does not at all mean that belief in God can gain no support from fine tuning. (Remember the case of the Empty Treasure Chest.) Still, fine tuning would not point towards God in an unambiguous way. Of my various parables, not one gives any support to the God hypothesis which it does not also give to the world ensemble hypothesis; (2) A world very obviously God-made would tend to be one not of freedom but of puppetry. This can give grounds for thinking that God would not make his creative role entirely plain.

It would be quite another matter, though, for God to avoid every possible indication of his existence even when this meant selecting physical laws which were *prima facie* far less satisfactory than others he would otherwise have chosen. A God of that degree of deviousness comes uncomfortably close to the kind of deity who creates the universe in 4004 B.C. complete with fossils in the rocks.

Finally, here is a puzzle in probability theory. The fifty bullets of the firing squad all miss. You jump to the conclusion, (x), that you are popular with the sharpshooters. But you then reflect, (y), that there may be vastly many squads at work, making it virtually certain that somewhere there would be someone asking, "How did they all manage to miss me?". Question: Is (y) as attractive as (x)? Here one's intuitions can tug in conflicting directions. Other things being equal, should the popularity hypothesis (which corresponds to the God hypothesis) be preferred because it gave you yourself a greater probability of being alive to ask, "Why did they all miss?" Yes, I feel tugged to say, after retelling the Lottery Story. Someone was bound to win and only that someone would be asking, "How did I come to win?" Yet ought not the winner to suspect strongly that his girlfriend who works at Lottery Company headquarters has secretly ensured his win? For that would explain why he in particular was able to say, "The winner is standing here, in my shoes."

On the other hand I consider the apparent evidence that we are the only intelligent beings in our galaxy. Query: do intelligent beings inhabit many other galaxies among the perhaps many hundred billion in the visible universe? Ought I to argue, (x_1), that if each galaxy stood a fair chance (say 30%) of containing intelligent beings then, just as it could be quite to be expected that anyone with an appropriately placed girlfriend should win a lottery or that someone popular with God or with the sharpshooters should be missed by a firing squad, so also it was quite to be expected that intelligent life would evolve in this galaxy; and that this scenario ought to be preferred to one in which the chances of its evolving here had been only one in several hundred billion? Or ought I to be content with the idea, (y_1), that there was a reasonable probability (let us again say 30%) of intelligent life's evolving at least once in the history of the universe, in some galaxy or other? Ought I to argue that provided there was such a reasonable probability it would be absurd to puzzle over why it evolved here, since wherever intelligent life evolved would be "here" to the intelligent beings who lived there?

In fact it is in the second way, way (y_1), that I feel inclined to argue. But this means that my intuitions now tug otherwise than they did in the Girlfriend case. And this suggests that the God hypothesis has no advantage over multiple worlds.

NOTES

[1] I consider various aspects of the arguments for multiple worlds in many papers, beginning with "Anthropic Principle, World Ensemble, Design," *American Philosophical Quarterly* **19** (April 1982) 141-151. (Some misprints are corrected in the October number).

[2] The theme is developed in my *Value and Existence* (Oxford: Basil Blackwell, 1979) and in several articles: most recently in "Mackie on Neoplatonism's 'Replacement for God'," *Religious Studies* **22** (1987) 325-342 (a reply to J. L. Mackie's chapter discussing my work in his *The Miracle of Theism*).

[3] Reviewed, with full references, in my "The Prerequisites of Life in Our Universe," in *Newton and the New Direction in Science*, eds. G. V. Coyne, M. Heller, and J. Życiński (Vatican City State: Vatican Observatory, 1988) 229-254.

[4] "The Inflationary Universe," *Scientific American* **250** (May 1984) 116-128.

[5] "Why is the Cosmological Constant so Small?" *Proceedings of the Royal Society of London*: A **377** (1981) 147-149.

[6] Page 245 of "The New Inflationary Universe Scenario," In *The Very Early Universe*, eds. G. W. Gibbons, S. W. Hawking, S. T. C. Siklos (Cambridge: Cambridge University Press, 1983) 205-249.

THE OMEGA POINT THEORY: A MODEL OF AN EVOLVING GOD

FRANK J. TIPLER, Max-Planck-Institut für Physik und Astrophysik, Garching b. Munchen, and Department of Mathematics and Department of Physics, Tulane University.

1. *Introduction*

Science is now considered by many to have refuted the fundamental tenets of the Christian religion. Many twentieth century theologians (and many scientists) have attempted to avoid conflict between science and religion by claiming that science and religion deal with wholly different forms of knowledge: the realm of science is the natural world, while the realm of religion is human morality and religious experience. But this division must ultimately fail. The starting point of morality is an understanding of humankind's place in nature, something that is obviously a scientific question. Our scientific understanding of our relation to the natural world must necessarily affect religion. Many important Roman Catholic authorities have recognized this by actually taking a stand on cosmological questions. They have claimed that Catholic doctrine requires the physical universe to have begun a finite time ago.[1] Furthermore, it is obvious that religious experience is truly meaningful only if there *really is* a God out there who is the source of this experience; no Christian believes for a moment that the experience of the presence of God is merely the subject matter for a specialist in abnormal psychology. Throughout the whole of human history, religion has been inextricably entwined with the science of the day, and this will never change.

In this paper I shall discuss two recent developments in physics which have important implications for religion. The first is the realization that we humans are present in the Universe at an exceedingly early time in its history. Almost all of universal history, and possibly almost all of the history of life, lies in our future. If most of life is in the future, then it is exceedingly unlikely that *Homo sapiens* is the most advanced form of life that will ever evolve in the cosmos; rather, our species should expect to be replaced one day by another. Traditional religion must come to grips with the fleeting existence of our species in universal history. It is our relative insignificance in time, not space, which is the real challenge posed by modern cosmology for traditional religion.

I shall show that this view leads naturally to a physical theory for an evolving God, which I term the Omega Point Theory. I shall outline this theory in Section 3.

The second development is the possibility of a Theory of Everything (TOE). A TOE might imply that there is only one logically possible universe. This would refute both the Cosmological Argument and, more importantly, its premise that God had some freedom of choice in creating the universe. The traditional God would be made superfluous, but an evolving God might be made necessary. The possibility of a TOE and its implications for religion will be discussed in Section 4.

2. *The Idea of an Evolving God*

It is the purpose of this paper to provide an argument for the existence of a Supreme Being who is also a Person. My analysis will be carried out entirely within physics itself, and although I shall feel free to use terminology from religion — omnipotence, transcendence and immanence, omniscience, and omnipresence, for example — I shall regard these notions as physical concepts, and accordingly define them in physics. However, the God whose existence I shall claim arises naturally in modern cosmology is not the traditional unchanging Deity, nor the wholly other Being of modern 20th century theology, but rather an evolving God somewhat like the God of Schelling, Alexander, Whitehead, and Teilhard de Chardin. An evolving God is very much in the world, creates it, and is created by it. The created and the creator are the same entity seen from different temporal perspectives, and described in different modes. How this works will be made clear in Section 3, where I shall outline the Omega Point Theory. A fuller development of this theory can be found in chapter 10 of *The Anthropic Cosmological Principle*[2] (hereafter referred to as ACP), which I co-authored with John D. Barrow. The theological implications of the Omega Point Theory were strongly de-emphasized in the book, however. There, the theory was presented as a purely physical theory — as in fact it is. But in this paper I shall adopt a theological point of view and present the Omega Point Theory as a model of an evolving God. Here I use the word "model" as physicists use it: a simplified picture expressed in mathematical symbols whose essential features are believed to correspond to reality. The "standard model" and the "Friedmann model" are two examples of this use of the word in cosmology. I am sure that my model of an evolving God is incorrect in its details, but I am also sure that *any* fully consistent concept of an evolving God who is a Person must resemble my model in its essential features. Indeed, I will go further: after I define "person", "soul", and "mind" in Section 3 in terms of modern computer theory, it will be clear that it is in the basic nature of "persons" to evolve — to change in time — so that the adjective "evolving" in "evolving God" is redundant.

My model of an evolving God is, of course, dissimilar in many respects to the traditional concept of the Supreme Personal Being. Since my model assumes that at the most basic ontological level there is nothing but physics and the "stuff" studied by physics, my model can conversely be regarded as a challenge from physics to the traditional idea of Deity; it is a claim that not only does the traditional God not exist, He is superfluous.

Although the average lay person may be more convinced by the argument from design, professional theologians are theists because they feel there is much to be said for some version of the cosmological/ontological argument: there is some entity — a Supreme Being — which *necessarily* exists in the sense that Its nonexistence would be a logical contradiction. The existence of such a Being is believed to be the answer to the questions of "Why is there something rather than nothing?" and "Why *this* universe rather than some other universe?" I shall consider what modern physics has to say about these two questions — and the cosmological argument — in Section 4. The reader is referred to Sections 2.9, 4.7, and 6.14 of ACP for more details. In essence, the answers which many physicists are giving to these two questions are: (1) that the physical universe in its own right necessarily exists, and further, (2) there is only one logically possible universe, for only one solution exists to the equations of physics and there is only one consistent set of equations. Here is what I regard as the greatest challenge to traditional theism: the possibility that the physical universe might necessarily exist in its own right. If true, this would mean that God is at best superfluous unless He is in the world. However, this possibility would not be a challenge but instead a nice completion to my model of an evolving God, in which the Deity and the entire physical universe are two aspects of the same thing, just as a certain collection of atoms acting under blind physical laws from one point of view is also a human being from another point of view. Both modes of description of a human being are equally valid, but epistemologically, neither can be completely reduced to the other, although ontologically a human being *is* at the most basic level a collection of atoms and nothing else. That is, I believe in ontological reductionism but epistemological anti-reductionism in the sense defined by Ayala.[3] I shall assume the truth of this position in what follows. (See ACP, Section 3.2, for a defense of this position). My model of an evolving God is most decidedly *not* a variant of pantheism. God and the physical universe are *not* two words for exactly the same thing.

The postulate from which I shall deduce in Section 3 an evolving God is fundamentally a moral one: value is something connected with life, and thus, if value is to remain in the universe, life must persist indefinitely; the laws of physics must permit forever the continued existence of life. Thus my argument for an evolving God has a certain family resemblance to Kant's moral argument. Furthermore, this continued existence of any sort of life will imply, as I shall argue in Section 5, not merely a continued existence of a low form of life, but also progressive evolution without limit in spacetime: the limit of both cosmological and biological evolution is a point beyond space and time, the Omega Point. We thus recover a progressive evolution in the large, something which has been forever banished from evolutionary biology. Teleology, although removed from terrestrial biology, reappears when biology is combined with cosmology. I shall develop these ideas, putting the Omega Point Theory in its historical perspective, in Section 5.

3. *The Omega Point Theory*

The crucial fact upon which the Omega Point Theory is based is that we are observing the universe at a very early time in its history. The universe is 10 to 20 billion years old, and our Earth is 4.5 billion years old. But as large as these numbers are relative to human lifetimes, they are insignificant in comparison to the length of time the universe will continue to exist: even if the universe is closed, bounds on the rate of expansion and the matter density imply at least 100 billion years until the final singularity, and if the universe is open or flat, then it will continue to exist forever. Now life has existed on our planet for at least 3.5 billion years — microfossils of what appear to be quite advanced forms of bacteria have been found which are that old, so life itself must be even older. Probably life of some form can continue to exist on the Earth for as least as long as the Sun remains on the main sequence, some 5 billion years. Thus we would expect life to continue to exist for longer than it already has existed. This lower bound on life expectancy is much longer than the mere 100,000 years modern man (*Homo sapiens*) has existed.[4] It is also much longer than a typical mammalian species survives, which is about a million years. So if our species survives as long as does the average mammalian species, it can expect to continue to exist for only one five-thousandth of the future of life on this planet. Furthermore, the future history of life on this planet is itself only a tiny fraction of the future history of the universe. These numbers put the human race in its proper perspective in the history of the cosmos.

It is important to emphasize that the above lower bounds on the length of time the universe will continue to exist are *very* solid. That the universe will continue to exist for at least 5 billion more years must be regarded at least as certain as the fact that it has already existed for at least 5 billion years. There is simply no way our knowledge of physics could be so wrong as to falsify this prediction of longevity. Thus, any religious appraisal of the nature and destiny of humankind must take into account this longevity. Almost all Christian theologians adopt a much shorter temporal perspective. This is as great an error — and as great a misunderstanding of humankind's place in nature — as believing that the universe was created a few thousand years ago.

Let us consider the implications of this longevity by assuming that life will continue to exist as long as the physical universe does. Note that this is basically a moral postulate. More precisely, the existence of life is the prior requirement for there to be any morality at all: lifeless and dead matter is neither good nor bad. Furthermore, a universe in which life and intelligence evolved, but in which life (and hence intelligence) and all its works disappeared forever would in my judgement be ultimately meaningless. One can of course adopt other definitions of "ultimate meaning" (see Section 3.7 of ACP for example; traditional Christian theism is one example), but I think we can agree that *if* ultimate meaning is to reside somehow in the physical universe itself, then a necessary condition is for life of some sort to continue to exist. Thus, indefinite survival is a necessary condition for a naturalistic ethics to be possible. If life must die out, then a naturalistic competitor to Christian ethics is not possible. Furthermore, whatever one's views as to the source of ultimate meaning, it is extremely important to investigate whether it is physically possible for life to exist as long as the universe does, for the answer is central to understanding humankind's place in nature.

In order to investigate whether life can continue to exist forever, I shall need to define "life" in physics language. I claim that life is a form of information processing (the converse is not true), and that the human mind — and the human soul — is a very complex computer program. This is *not* to say that life is *nothing* but information processing. This naive reductionist view I would strongly reject. All I am claiming is that at the most basic level of physics, life is simply information processing. But there are higher levels of epistemological description. Human beings love others, they have emotional needs and deep feelings. These very real aspects of human life cannot be reduced to simple theorems of information theory and physics. (In principle, these aspects are equivalent to extremely complex theorems, but such theorems would be humanly incomprehensible and effectively undiscoverable. This is ontological reductionism combined with epistemological irreductionism.) The crucial point is that the higher levels must be consistent with the physics level; any discussion of human feelings must be consistent with the general limitations on human minds deduced from physical information theory which is applied assuming minds are computer programs. This is in the end no different from the requirement that a moral philosophy or a work in literary criticism must not contradict the brute physical fact of people having to eat in order to live. I find it fascinating — and one of the most important ideas I hope to convey in this paper — that far-reaching and unexpected conclusions about human destiny can be drawn from the physics level alone.

A complete justification for my claim that the mind is a computer program would fill a book. A central argument is the Turing Test. I thus refer the reader to several books on the Turing Test.[5] See also Sections 3.2, 3.5, 7.2, and 10.6 of ACP. Instead of reviewing the Turing Test, let me give here a religious justification for this claim: I shall justify the computer/information processing model of life and mind simply by pointing out the astonishing similarities between the mind-as-computer-program idea and the traditional Christian concept of the "soul". Both are fundamentally "immaterial": a program is a sequence of integers, and an integer — 2, say — exists "abstractly" as the class of all couples. The symbol "2" written here is a *representation* of the number 2, and not the number 2 itself. In fact, Aquinas and Aristotle defined the *soul* to be "the form of activity of the body." In Aristotelian language, the *formal* cause of an action is the abstract cause, as opposed to the material and efficient causes. For a computer, the program is the formal cause, while the material cause is the properties of the matter of which the computer is made, and the efficient cause is the opening and closing of electric circuits. For Aquinas, a human soul needed a body to think and feel, just as a computer program needs a physical computer to run.

Aquinas thought the soul had two faculties: the agent intellect (*intellectus agens*) and the receptive intellect (*intellectus possibilis*), the latter being the ability to acquire concepts, and the former being the ability to retain and use the acquired concepts. Similar distinctions are made in computer theory: general rules concerning the processing of information coded in the central processor are analogous to the agent intellect; the programs coded in RAM or on tape are the analogues of the receptive intellect. (In a Turing machine, the analogues are the general rules of symbol manipulation coded in the device which prints or erases symbols on the tape vs. the tape instructions, respectively.) Furthermore, the word "information" comes from the Aristotle-Aquinas' notion of "form": we are "informed" if new forms are added to the receptive intellect. Even semantically, the information theory of the soul is the same as the Aristotle-Aquinas' theory.

The point I am trying to make is that in a sense the mind-as-a-program idea is just old wine in a new bottle; it poses no challenge to the traditional view of the physical nature of man. But thinking of the human mind as a computer program, and more generally, regarding all thought as a species of information processing, is a conceptual advance of enormous significance, for it allows us to turn many philosophical problems about the scope and limits of human thought (or the thought of any possible intelligent being, for that matter) into formal problems of mathematical computer theory. For example, new light is thrown on the old issues of reductionism vs. irreductionism and determinism vs. indeterminism by thinking what these mean to a computer (see Section 3.2 of ACP for more discussion). More importantly, in the language of information processing, it becomes possible to say precisely what it means for life to continue forever. I shall say that "life" can continue forever if: (1) information processing can continue indefinitely along at least one world line γ all the way to the future "boundary" of the universe — that is, until the end of time; (2) the amount of information processed between now and this future boundary is infinite in the region of spacetime with which the world line γ can communicate; (3) the amount of information stored at any given time T within this region can go to infinity as T approaches its future limit (this future limit of T is finite in a closed universe, but infinite in an open one).

The above is a rough outline of the more technical definition given in Section 10.7 of ACP. But let me ignore details here. What is important is the physical (and ethical!) reason for imposing each of the above three conditions. The reason for condition 1 is obvious; it simply states there must be at least one history in which life (= information processing) never ends.

Condition 2 tells us two things. First, that information processed is "counted" only if it is possible, at least in principle, to communicate the results of the computation to the history γ. This is important in cosmology, because in most model universes event horizons abound. In the Friedmann universe, every comoving observer at some point loses the ability to send light signals to every other comoving observer, no matter how close. Life obviously would be impossible if one side of one's brain became forever unable to communicate with the other side. Life is organization, and organization can only be maintained by constant communication between the different parts of the organization. The second thing condition 2 tells us is that the amount of information processed between now and the end of time is potentially infinite. I claim that it is meaningful to say that life exists *forever* only if the number of thoughts generated between now and the end of time is actually infinite. But we know that each "thought" corresponds to a minimum of one bit being processed. In effect, this part of condition 2 is a claim that time duration is most properly measured by the thinking rate, rather than by proper time as measured by atomic clocks. The length of time it takes an intelligent being to process one bit to think one thought — is a direct measure of "subjective" time, and hence is the most important measure of time from the perspective of life. A person who has thought 10 times as much, or experienced 10 times as much (there is no basic physical difference between these options), as the average person has in a fundamental sense lived 10 times as long as the average person, even if the chronological age is shorter than the average.

The distinction between proper and subjective time crucial to condition 2 is strikingly similar to a distinction between two forms of duration in Thomistic philosophy. Recall that Aquinas distinguished three types of duration. The first was *tempus*, which is time measured by change in relations (positions, for example) between physical bodies on Earth. *Tempus* is analogous to proper time; change in both human minds and atomic clocks is proportional to proper time, and, for Aquinas also, *tempus* controlled change in corporeal minds. But in Thomistic philosophy, duration for incorporeal sentient beings - angels - is controlled not by matter, but rather by change in the mental states of these beings themselves. This second type of duration, called *aevum* by Aquinas, is clearly analogous to what I have termed "subjective time." *Tempus* becomes *aevum* as sentience escapes the bonds of matter. Analogously, condition 2 requires that thinking rates are controlled less and less by proper time as T approaches its future limit. *Tempus* gradually becomes *aevum* in the future. The third type of Thomistic duration is *aeternitas*: duration as experienced by God alone. *Aeternitas* can be thought of as "experiencing" all past, present, and future *tempus* and *aevum* events in the universe all at once.

Condition 3 is imposed because, although condition 2 is necessary for life to exist forever, it is not sufficient. If a computer with a finite amount of information storage — such a computer is called a *finite state machine* — were to operate forever, it would start to repeat itself over and over. The psychological cosmos would be that of Nietzsche's Eternal Return. Every thought and every sequence of thoughts, every action and every sequence of actions, would be repeated not once but an infinite number of times. It is generally agreed (by everyone but Nietzsche) that such a universe would be morally repugnant or meaningless. Augustine argued strongly in Book Twelve of *The City of God* that Christianity explicitly repudiates such a world view: "Christ died once for our sins, and rising again, dies no more."[6] The Christian cosmos is progressive. Only if condition 3 holds in addition to condition 2 can a psychological eternal return be avoided. Also, it seems reasonable to say that "subjectively", a finite state machine exists for only a finite time, even though it may exist for an infinite amount of proper time and process an infinite amount of data. A being (or a sequence of generations) that can be truly said to exist forever ought to be physically able, at least in principle, to have new experiences and to think new thoughts.

This raises a fundamental problem for the view of eternal life held by many Christians. There is no question but that an individual human being is a finite state machine. His brain is limited in the number of memories it can store. We are unaware of this because a rough calculation shows we would have to live at least a thousand years before the limit of capacity would be reached at the maximum memory storage rate recorded in psychological experiments. However, a thousand years is but an infinitesimal fraction of eternity (defined as infinite subjective time). It is possible to have only a finite number of new thoughts and new experiences after being raised from the dead at the Last Judgement. At normal subjective time rates, only a thousand years worth of new experiences are possible if the old memories are retained. It is logically impossible for "eternal" life to be eternal in an experiential sense, unless we imagine the fundamental finiteness of humanity is abolished upon resurrection. This is no solution, for a being which has and uses a potentially infinite memory would be utterly non-human. Our humanity is defined in part by our basic limitations. A finite memory is one of these.

Implicit in the above argument is the idea that living, feeling, thinking, etc., necessarily involve a change from one state to another. This is a definite consequence of the mind-as-a-program concept. But I claim it is a reasonable consequence. Consider a standard science fiction scenario, that of placing a person in suspended animation. No mental or any other changes occur to the person while she is frozen solid. Consistent with this lack of change, I will suppose that in fact the person when revived remembers nothing of the period while in suspended animation. Question: was that person "alive" while in suspended animation? Certainly the program that codes personality was not running during that time. That person was quite literally in limbo while in suspended animation. I claim there was no self-awareness during that time, because self-awareness means analyzing a mental model you have of yourself, and analyzing means mental change. That person was dead by most current legal definitions during the suspended animation period, for these definitions are based on neurological or other bodily activity (i.e., change of some sort). Nevertheless, I would conjecture that most people would be reluctant to consider her dead, because she was by assumption reanimated. But what if she were *never* reanimated? Suppose for some reason we discover we can't reanimate her even in principle. Even if the program which coded her personality were never erased, his self-awareness, by assumption, would never return. Isn't this what we mean by death? Isn't this the actual state — the lack of self-awareness for all future time — that the legal definitions of death are attempting to capture? So a program that cannot change, that is forever static in principle, cannot be a person no matter how complex it is. Nor can it be "intelligent" in any meaningful sense, because the essense of intelligence [7] means the ability to learn from experience, and this again is a species of change, of information processing.

Let us now consider whether the laws of physics will permit life/information processing to continue forever. Von Neumann and others have shown that information processing (more precisely, the irreversible storage of information) is constrained by the first and second laws of thermodynamics. Thus the storage of a bit of information requires the expenditure of a definite minimum amount of available energy, this amount being inversely proportional to the temperature. (See Section 10.6 of ACP for the exact formula.) This means it is possible to process and store an infinite amount of energy between now and the final state of the universe only if the time integral of P/T is infinite, where P is the power used in the computation, and T is the temperature. Thus the laws of thermodynamics will permit an infinite amount of information storage in the future, provided there is sufficient available energy at all future times.

What is "sufficient" depends on the temperature. In the open and flat ever-expanding universes, the temperature drops to zero in the limit of infinite time, so less and less energy per bit processed is required with the passage of time. In fact, in the flat universes, only a *finite* total amount of energy suffices to process an infinite number of bits! This finite energy just has to be used sparingly over infinite future time. On the other hand, closed universes end in a final singularity of infinite density, and the temperature diverges to infinity as this final singularity is approached. This means that an ever increasing amount of energy is required per bit near the final singularity. The amount of energy required per bit actually diverges to infinity at the singularity. However, most closed universes undergo "shear" when they recollapse, which means they contract at different rates in different directions (in fact, they spend most of their time *expanding* in one direction while contracting in the other two!). This shearing gives rise to a radiation temperature difference in different directions, and this temperature difference can be shown to provide sufficient free energy for an infinite amount of information processing between now and the final singularity, even though there is only a *finite* amount of proper time between now and the end of time in a closed universe. Thus, although a closed universe exists for only a finite proper time, it nevertheless could exist for an infinite subjective time.

But although the laws of thermodynamics permit conditions 1 through 3 to be satisfied, this does not mean that the other laws of physics will so permit. It turns out that, although the energy is available in open and flat universes, the information processing must be carried out over larger and larger proper volumes. This fact ultimately makes impossible any communication between opposite sides of the "living" region, because the redshift implies that arbitrarily large amounts of energy must be used to signal (this difficulty was first pointed out by Freeman Dyson). This gives the *first testable prediction* of the Omega Point Theory: *the universe must be closed.*

However, as I stated earlier, there is a communication problem in most closed universes — event horizons typically appear, thereby preventing communication. However, there is a rare class of closed universes which doesn't have event horizons, which means by definition that every world line can always send light signals to every other world line. Now Penrose has found a way to precisely define what is meant by the "boundary" of spacetime, where time ends. In his definition of the "c-boundary", world lines are said to end in the same "point" on this boundary if they can remain in causal contact unto the end of time. If they eventually fall out of causal contact then they are said to terminate in different c-boundary points. Thus the c-boundary of these rare closed universes without event horizons consists of a single point. For reasons given in Section 10.6 of ACP, it turns out that information processing can continue only in closed universes which end in a single c-boundary point, and only if the information processing is ultimately carried out throughout the entire closed universe.

Thus we have the *second testable prediction* of the Omega Point Theory: *the future c-boundary of the universe consists of a single point* — call it the Omega Point. (Hence the name of the theory.) It is possible to obtain other predictions. For example, a more detailed analysis of how the energy is used to store information leads to the *third testable prediction* of the Omega Point Theory: *the density of particle states must diverge to infinity as the energy goes to infinity, but nevertheless this density of states must diverge no faster than the energy squared.*

But these predictions [8] just demonstrate that the Omega Point Theory is a scientific theory of the future of life in the universe, and it is not my purpose to discuss the science in detail here. Rather, I am concerned here with the theological implications of the Omega Point Theory. That the theory has such implications will be obvious if I restate a number of the above conclusions in more suggestive words. As I pointed out, in order for the information processing operations to be carried out arbitrarily near the Omega Point, life must have extended its operations so as to engulf the entire physical cosmos. We can say, quite obviously, that life near the Omega Point is omnipresent. As the Omega Point is approached, survival dictates that life collectively gain control of all matter and energy sources available near the final state, with this control becoming total at the Omega Point. We can say that life becomes omnipotent at the instant the Omega Point is reached. Since by hypothesis the information stored becomes infinite at the Omega Point, it is reasonable to say that the Omega Point is omniscient; it knows whatever it is possible to know about the physical universe (and hence itself).

The Omega Point has a fourth property. Mathematically, the c-boundary is a completion of spacetime: it is not actually in spacetime, but rather just "outside" it. If one looks more closely at the c-boundary definition, one sees that a c-boundary consisting of a single point is formally equivalent to the entire collection of spacetime points, and yet from another point of view, it is outside space and time altogether. It is natural to say that the Omega Point is "both transcendent to and yet immanent in" every point of spacetime. When life has completely engulfed the entire universe, it will incorporate more and more material into itself, and the distinction between living and non-living matter will lose its meaning.

There is another way to view this formal equivalence of all spacetime and the Omega Point. In effect, all the different instants of universal history are collapsed into the Omega Point; "duration" for the Omega Point can be regarded as equivalent to the collection of all experiences of all life that has, does, and will exist in the whole of universal history, together with all non-living instants. This "duration" is very close to the idea of *aeternitas* of Thomistic philosophy. We could say that *aeternitas* is equivalent to the union of all *aevum* and *tempus*. If we accept my earlier argument that life and personhood involve change by their very nature, then this identification appears to be the only way to have a Person who is omniscient, and hence whose knowledge cannot change: omniscience is a property of the necessarily unchanging, not-in-time, final state, a state nevertheless equivalent to the collection of all earlier, non-omniscient changing states.

Thus the indefinitely continued existence of life is not only physically possible; it also leads naturally to a model of an evolving God.

4. *Is There Only One Possible Physical Universe?*

The idea that there may be only one logically possible actually existing universe is an old idea. Hume (or perhaps I should say, Philo) briefly toyed with it in his *Dialogues on Natural Religion*. Einstein often said that he became a physicist in order to find out "if the dear Lord had any choice when he created the universe." But it is only in the last few years, with the advent of the superstring theories, that the possibility of universal uniqueness began to be seriously discussed.

Now any philosopher of science can tell you that this idea is complete nonsense. Any scientific theory, indeed any logical system, is based on axioms which are themselves unjustified. Thus further scientific advance is always possible, for the axioms of the present day science can always be found to be consequences of even more fundamental axioms, and so on *ad infinitum*. A philosopher will tell you that one can always find alternatives to the present day theories which will account for the observations we have, just as well as the theories which are generally accepted by scientists. In other words, the axioms used to describe current observations are far from unique, if for no other reason than that we know very well the observations are not absolutely precise. Unavoidable experimental errors allow alternative theories, since many theories will be consistent with the data. The philosopher might also point out that physicists have occasionally claimed in the past they had the ultimate theory, only to see their world view collapse like a house of cards. So why do we find many famous contemporary physicists proclaiming that a unique physical theory is not only possible, but just around the corner?

The basic reason is that it is easy to *say* one can always find an alternative theory. It is extraordinarily hard to actually go out and find one. The database of observations is now so enormous that it is exceedingly difficult to construct a mathematical theory which is even roughly in agreement with experiment and which is fully self-consistent and universal.

The self-consistency problem is the most suggestive. It manifests itself primarily in the problem of infinities in quantum field theory. Almost all quantum field theories one can write down are simply nonsensical, for they assert that most (or all) observable quantities are infinite. Only two very tiny classes of quantum field theories do not have this difficulty: finite quantum field theories and renormalizable quantum field theories. Even before superstring theories became a major area of study, Steven Weinberg stressed how exceedingly restrictive the requirement of renormalization really is. It is really the renormalizability of Yang-Mills quantum fields that caused particle theorists to concentrate attention on this class of theories almost exclusively when attempting to model matter. But there is a countable infinity of possible renormalizable Yang-Mills theories. Any compact Lie group defines one. The Lie group $SU(2) \times U(1)$ gives the Weinberg-Salam unified theory of the weak and electromagnetic interactions, and SU(3) correctly describes the color force which binds nuclei. But these Lie groups were picked out of the pack by experiment, not by logic. Still, this is considerable progress. We now have consistent theories for three of the four known forces. Unfortunately, general relativity, which is the standard theory of gravity, the fourth force, gives a non-renormalizable theory. Furthermore, even the renormalizable field theories have not completely eliminated the nonsensical infinities; they have really only succeeded in hiding them from view.

This is where superstrings come in. Green and Schwarz were able to show in 1985 that, in the context of the standard way of adding Yang-Mills fields to superstring theories, only *two* Lie groups, $E_8 \times E_8$ and SU(32), would give a consistent theory. And as a bonus, these theories were not merely renormalizable, they were actually *finite*! (to first order, anyway; there are pious hopes that the theories are finite to all orders). It also appears that gravity and the other three forces are present in the low energy limit of superstring theories. Now this is real progress! Full mathematical self-consistency has reduced the range of possible theories from the countable infinity of the possible Yang-Mills theories to a mere two candidates.[9] Self-consistency is also important in other ways in superstrings.

The trend is clear. The more forces and phenomena we try to include in a single theory, the less freedom we have to construct one. And, side by side with this shrinking range of possible consistent theories, there are fewer and fewer phenomena not included in the theory. There are actually physical arguments to show that we may have seen most of the fundamental phenomena, in contrast to the situation at the end of the nineteenth century. For example, all known elementary particles (fermions) can be grouped into what are called "families". If there were more than about 4 families, the synthesis of elements in the Big Bang would be different from what it is observed to be. And we have already observed 3 families.

Is it any wonder many physicists have come to believe that this process of fitting a larger and larger set of possible data points to a smaller and smaller number of self-consistent theories will converge on a single unique physical theory, a Theory of Everything (TOE)?

A number of people have claimed that the Gödel incompleteness theorem shows a TOE cannot be true necessarily and *a priori*.[10] I think this claim is incorrect. Gödel has indeed proven that any theory which is sufficiently complex to contain all of arithmetic cannot be proven consistent by arguments inside the theory itself. But this just means that a self-justifying TOE must be simpler than the full theory of arithmetic. There are in fact branches of mathematics which can be proven decidable and consistent by reference to the branch itself. For examples, Euclidean geometry was proven decidable by Tarski and hyperbolic geometry was proven decidable by Schwabhäuser.[11] Nagel and Newman have given a proof of consistency of an important part of logic, the sentential calculus, or logic of propositions, in their popular-level book *Gödel's Proof*.[12] Even arithmetic with addition only can be proven decidable. It is quite possible that the TOE could lie in one of the decidable branches of mathematics.[13]

The important role self-consistency has played in the search for the TOE is one reason for believing that the TOE, if found, will be the only one logically possible. But it is not the only reason. After all, the TOE is so hard to find because it has to account for so many things. Why couldn't the universe have been much simpler? There are two answers to this question, both involving the Anthropic Principle. I shall give only one answer here,[14] the answer which involves an analysis of what the word "existence" means.

A thing can be said to exist only if it or its effects can be detected in some way. But the word "detected" itself presupposes the existence of something to do the detecting. Now an analysis of just what detecting or measuring means in physics shows that a measurement is carried out only if some piece of information is recorded. This in turn implies that a universe must be complex enough to permit the recording of information before it can have observers of any sort. In the ACP, Barrow and I devote some 400 pages to showing just how enormously complex this apparently simple requirement that observers exist within it makes the universe. In a nutshell, the universe must be as complex as it actually is in order to have observers of our complexity. Since we humans are not really *that* complex, this suggest it must be almost as complex as it is in order to have observers of any sort.

This brings us to the age-old philosophical problem of whether a universe which has no observers in it — and which has no detectable effect on a universe which does contain observers — can possibly be said to exist. My own inclination would be to say no, because there is no way I can say that anything inside such a universe exists; it is not possible to give meaning to the word "existence" in such a context. So with this understanding of the word "existence", it is quite plausible that only one Universe is logically possible — i.e., capable of existence — and we're in it. It is interesting that from this view of what existence means, it is the observers, or rather the possibility of observers and their observations, that permit the universe to exist. In a sense, the creatures inside the universe create both the universe and themselves.

Even if only one universe is logically possible, this does not mean that this unique universe actually exists. It would seem that a further assumption is required: the assumption that something exists. A reasonable assumption, to be sure, but nevertheless an additional assumption. However, it is not clear to me this additional assumption is actually required. Barrow and I develop at some length in Section 3.5 of ACP the fascinating idea that a perfect computer simulation of a universe would be indistinguishable from the real universe it simulates.[15] Now a simulation is just a sequence of natural numbers, and all sequences of natural numbers have mathematical existence, even though they may never have achieved the privilege of an actual physical representation in our actually existing physical universe. But if one of these corresponds to a perfect simulation of our physical universe, then as far as the humans simulated in the program can tell, it is real. Our copies behave no differently than we ourselves. Thus the existence (in the mathematical sense of the word) of these sequences of numbers is ultimately indistinguishable from existence in a physical sense, and mathematical existence comes ultimately from the laws of logic themselves!

In other words, the universe may very well be, in John Wheeler's phrase, a self-excited circuit. It may necessarily exist in its own right. *If* it does so exist — and I emphasize the word "if", because there are many gaps in the above argument — then the God whose existence is asserted by the cosmological/ontological argument, the wholly other God of Barth, and more generally any God who does not need the universe as much as the universe needs him, is quite superfluous. And further, this sort of God is superfluous in answering the very question for which his existence is invoked: why is there something rather than nothing; why *this* universe rather than some other universe?

5. *The Implications of the Omega Point Theory*

My favorite definition of "religion" appeared in an article by Miller and Fowler published in the *CTNS Bulletin*: "'Religion' and 'theology' are taken to refer to the following: anything is religious which is concerned with the meaning of personal place; and theology is interpretative reflection on and explicit articulation of the meaning of personal place."[16] Perhaps I like this definition because it turns the paper you are now reading, the ACP, and even Darwin's *The Origin of the Species* into religious tracts! But is this definition really that different from Tillich's view[17] that religion, in the widest sense of the word,[18] is that which deals with questions of "ultimate concern"?

Certainly "personal place" was the central focus of the preceeding two sections: in Section 3 the existence of an evolving God was inferred from the naturalistic ethical postulate that it must be possible for life never to die out in the universe, while in Section 4, it was argued that perhaps this never-dying life was, is, and shall be collectively responsible for the necessary existence of the universe itself (including the life within it). If the argument of Section 4 is accepted, then the ethical postulate of Section 3 is unnecessary; both it and the evolving God can be inferred as properties of the necessarily existing universe — but this universe owes its existence to the collectivity of (past, present, future) living things, and the collectivity of living things *is* the evolving God! The created and the Creator are inextricably bound up in one another.

Humankind's place in the scheme of things is that of an intermediate link; we cannot expect our species, *Homo sapiens*, to live forever. We could not possibly survive the great cold and great heat that await life in the far future. The history of life on the Earth to date is a preview of what will be the total history of life in the universe: all individual living species that evolve on Earth eventually become extinct, but life itself goes back in an unbroken chain, more than 3.5 billion years long, to the early youth of our planet. As we humans are descended from simpler one-celled organisms, long since extinct, so beings more complex than *Homo sapiens* will descend from us. And beings still more complex will in turn descend from them, up to the Omega Point.

This picture of the chain of life is strikingly similar to the medieval and Enlightenment view of life, which the famous historian of ideas Arthur O. Lovejoy termed "The Great Chain of Being."[19] In this view, all living things were arranged in a vast *static* hierarchy, with inorganic materials at the bottom, followed by plants and animals, mankind in the center, the angels higher still, and with God at the top. The Omega Point Theory is essentially a temporalized version of The Great Chain of Being. Not surprising, because as I emphasized in Section 3, life is fundamentally a temporal phenomena; this same insight is what underlies Darwin's *Origin of the Species*. "Origin" is itself a temporal word.

This temporally progressive Chain of Being, with one species being ultimately replaced by another coding more information (this is what is meant by "more complex" or "more advanced") is a consequence of the assumption of "progress" which is built into conditions 1 through 3 of Section 3. Our own species has limits; there is a limit to the knowledge that can be coded in a human brain. So if knowledge is to continue to increase, indeed to increase without limit, it must one day be coded in other than human brains. Judging from the present rapid development of computers, I would guess that our successor species will be quite literally "information processing machines," machines with minds superior to ours. Perhaps the molecular biologist Manfred Eigen is correct in saying tha DNA reaches with *Homo Sapiens* the limit of the complexity it can code. If so, if life is to gain in complexity and knowledge is to increase, then the leading shoot of life must move from one substrate — DNA — to another. Certainly this move must occur at some point in the future, because DNA-based life cannot survive in the high temperature environment near the final singularity. The extinction of our species is required both by the laws of physics and the inherent logic of eternal progress. But this should not horrify us. All religions agree that what is ultimately important is the eternal continuation of intelligent personality (ultimately God's), not the particular racial form it happens to take. If the Omega Point Theory is true, life shall not perish from the Cosmos, but shall grow into the Omega Point.

Acknowledgements

I should like to thank Professor Jürgen Ehlers for the hospitality at the Institut für Astrophysik of the Max-Planck-Institut für Physik und Astrophysik, where this paper was written, and to the Max Planck Society for financial support. I am grateful to Professor Frank Birtel for replacing books and papers lost by various airline carriers. Finally, I am extremely grateful to the participants of the Vatican Conference, particularly Michael Heller, John Leslie, Ernan McMullin, Bob Russell, and Bill Stoeger, for their valuable criticisms of an earlier version of this paper.

NOTES

[1] In 1909 the Pontifical Biblical Commission listed the creation of the entire universe at the beginning of time as one of the "fundamental truths" of the Genesis creation story. Pope Pius XII claimed in a major address delivered in 1951 that the Big Bang theory supported Catholic doctrine. See I. G. Barbour, *Issues in Science and Religion* (New York: Harper Row, 1971) 373-375, for a discussion of this view of the Roman Catholic position. It should be emphasized, however, that this position, although held by many influential Catholics, cannot be considered Catholic dogma. Although this position on the beginning of the Universe is not Catholic dogma, there are scientific-historical statements, such as the Resurrection of Christ, which definitely are. See Anthony Kenny, *A Path from Rome* (Oxford: Oxford University Press, 1986). And the Resurrection is the scientific foundation of Christianity: as St. Paul himself emphasized, if Christ did *not* rise from the dead, belief in *any* Christian tenet is in vain. Christianity rests, as do the natural sciences, on a matter of fact; Christianity *requires* that at least one "gap" in the natural order — the Resurrection — occurred in the past. I personally do not believe in the Resurrection, for reasons succinctly stated by David Hume in his work *On Miracles*, and also because I am an ontological reductionist: there are *no* gaps in the natural order. Furthermore, I think eternal life for an individual human being would be a bad thing, for reasons stated in Section 3. As Hume said to one of his biographers, individual eternal life would just lead to an accumulation of garbage in the Cosmos: errors and crimes made by individuals would *never* be forgotten. But collective eternal life can lead to unlimited progress.

St. Augustine recognized that the Resurrection, *qua* scientific fact, had far-reaching implications for scientific cosmology. The second half of the twelfth book of *The City of God*, devoted to showing the *uniqueness* of the Resurrection, implied that one of the central assumptions of Greek science, namely the Eternal Return, could not possibly be true: "For Christ died once for our sins, and rising again, dies no more." Thus, although Augustine was willing to allow natural science to tell us that some *unimportant* (for redemption) Biblical passages must be re-interpreted metaphorically, the Resurrection was definitely *not* open to such re-interpretation; rather, for Augustine, the Resurrection was an uncloseable "gap" in the natural order, and any acceptable scientific theory must be consistent with it. See McMullin's essay in this volume.

[2] John D. Barrow and Frank J. Tipler, *The Anthropic Cosmological Principle* (Oxford: Oxford University Press, 1986); hereafter referred to as ACP.See also Frank J. Tipler, *Essays in General Relativity* (New York: Academic Press, 1980) 21-37.

[3] Francisco J. Ayala, "Introduction," in *Studies in the Philosophy of Biology*, by Francisco J. Ayala and Theodosius Dobzhansky (Berkeley: University of Chicago Press, 1974).

[4] Eric Delson, "One Source, Not Many," *Nature* **325** (1988) 206.

[5] D. R. Hofstadter and D. C. Dennet, *The Minds I* (New York: Basic Books, 1981), is the best and most complete defense of the Turing Test as a test for the presence of a mind. This book provoked an exchange between the philosopher John Searle and Dennett in the pages of *New York Review of Books* over the validity of the Turing Test. I recommend reading this exchange, although I think Dennett won hands down. Searle simply cannot understand the enormous effective computer power of the human brain (10^{10} to 10^{15} bits of memory and a

computation speed between 10 and 1,000 gigaflops; see Section 3.2 of ACP. For comparison, the Cray-XMP has a memory of about 10^{10} bits and a speed of 1 gigaflop. The Cray *crawls* in comparison to the human brain). Searle's "Chinese room" thought experiment could not possibly work because it would be absolutely impossible for a human inside to move paper fast enough for the room to pass the Turing Test (in Chinese).

[6] See Note 1.

[7] I. G. Barbour, 1971, *op. cit.*, regards "intelligence" as one of the two most essential properties of God, if God is to be thought of as a Person. (The other essential property is "purpose".)

[8] An explicit assumption made in this analysis is that the Second Law of Thermodynamics holds in the large at all times, and more generally, that time direction is always defined, even arbitrarily close to the final singularity. (A time direction arising from the spacetime metric is not absolutely required, but a time direction defined by the Second Law is necessary.) Recent work in quantum cosmology has challenged both assumptions. Hawking has pointed out that his boundary condition on the wave function of the universe requires the universe to be spatially closed, but it also requires the entropy to decrease after the time of maximal expansion. This would make the continued progression of life impossible; knowledge could increase only to a finite maximum at the time of maximal expansion. In Hawking's universe, the history of the contracting phase would be identical to the history of the expanding phase, only run in reverse. Thus life would never continue to the end of time, for the end of time is really the same as the beginning. If the Omega Point Theory is to hold, Hawking's boundary condition must be incorrect. This is a fourth prediction of the Omega Point Theory, but not a significant one, because very few believe Hawking's cosmological model. A universal reversal of entropy seems too improbable.

A far more fundamental challenge to the Omega Point Theory is the possibility that time direction may not be defined when the spacetime metric is quantized. Isham discusses this possibility in his paper in this volume. Furthermore, Penrose's c-boundary is a classical concept, and it is not clear that an analogue of the c-boundary exists in quantum cosmology. An analogue of the singularity exists — a place where the radius of the universe is zero is still there even in quantum cosmology — but if time direction is not defined, we cannot distinguish between the initial and final singularities. Nevertheless, I think the Omega Point Theory can survive this challenge. Quantum cosmology is built on the Many-World Interpretation of quantum mechanics, and all the Omega Point Theory really requires is that a time direction based on entropy be defined in one branch universe, where conditions 1, 2, and 3 can hold. A time direction arising from the metric is not essential, as I said above, nor is it necessary for time to be globally defined for the entire collection of branch universes.

[9] Unfortunately, this reduction to a mere two theories is spoiled by the non-uniqueness of the vacuum state in superstring theories. Different vacua give different physics, and as yet there is no good reason to pick one vacua over another.

[10] See, for example, Stanley L. Jaki, "Teaching of Transcendence in Physics," *American Journal of Physics* **55** (1987) 884-888.

[11] J. Donald Monk, *Mathematical Logic* (New York: Springer-Verlag, 1976) 234.

[12] Ernest Nagel and James R. Newman, *Gödel's Proof* (London: Routledge and Kegan, 1971).

[13] My method of avoiding the limitations for a TOE of the Gödel Incompleteness Theorem is similar to Nobel laureate economist Paul Samuelson's

proposal for avoiding the democracy-is-impossible implications of the Arrow Impossibility Theorem. According to this Theorem, no social welfare function — a procedure for deciding which alternatives (among economic goods, among political leaders, among religions, etc.) society as a whole should choose — exists which satisfies four assumptions. The first assumption is nondictatorship: the social welfare function cannot consist of picking a single person (the dictator) and letting this person decide what the whole society will choose. The second is independence of irrelevant alternatives: if the social welfare function implies alternative A is preferred to alternative B, then a change in individual preferences which does not change any one individual's preferences between A and B cannot change the social choice of A over B. The third is that society cannot switch from A to B if a single individual switches in the other direction from B to A. That is, if more individuals start to prefer A to B, then the choice of society as a whole cannot switch in the opposite direction. Finally, the social welfare function must be consistent (transitive): If A would be chosen over B, and B over C, then A must be chosen over C. See Paul A. Samuelson, "Arrow's Mathematical Politics," in *Human Values and Economic Policy*, by Sidney Hook (New York: New York University Press, 1967) 41-51. For the Arrow Impossibility Theorem see David Friedman, *Price Theory* (Cincinnati: South-Western Publishing, 1986) and Jerry S. Kelley, *Arrow Impossibility Theorems* (New York: Academic Press, 1978).

[14] The other answer involves the Participatory Anthropic Principle, which was invented by John A. Wheeler. It draws on the Copenhagen Interpretation of quantum mechanics, which holds that many of the properties subatomic particles exhibit are determined by the observer's choice of what to measure. Following the logic of this interpretation, Wheeler conjectures that *all* the properties of *all* the particles in the universe are determined by the collection of all the acts of observer-participancy in the past, present, and future. In particular, these acts collectively bring into existence all the observers themselves. Thus in this answer also, the creatures collectively are responsible for creating the entire universe and themselves. But in this answer, the creation is more direct; the word "creation" is used in a sense closer to its everyday usage. See the ACP index for references to the Participatory Anthropic Principle. See also John Wheeler, "Probability And Determinism," *IBM Journal of Research and Development* **32** (1988) 4-15. He points out that the Participatory Anthropic Principle presupposes the Omega Point Theory, for only the enormously more powerful observer-participators of the far future can interact on the scale necessary to bring our enormous universe into existence.

[15] See also Douglas R. Hofstadter and Daniel C. Dennett, *The Mind's I* (New York: Basic Books, 1981).

[16] James B. Miller and Dean R. Fowler, "What's Wrong With the Creation/Evolution Controversy?" CTNS Bulletin **4** (Autumn 1984) 1-13.

[17] Paul Tillich, *The Dynamics of Faith* (New York: Harper and Row, 1957).

[18] See I. Barbour, 1971, *op. cit.*, 219, for a detailed discussion.

[19] Arthur O. Lovejoy, *The Great Chain of Being* (Cambridge, Mass.: Harvard University Press, 1936).

THE QUANTUM WORLD

JOHN POLKINGHORNE, Trinity Hall, Cambridge

Introduction

The discovery of quantum theory produced the most profound modification of Newtonian physics that has occurred since the publication of the *Principia*. The clear and determinate character of physical processes, as Sir Isaac understood it, has dissolved at its constituent roots into the cloudy and fitful quantum world. That was a transformation much more radical than the invention of the field concept (which in some ways was just an interpolation into action-at-a-distance) or even the relativising of time (since the guiding principle of relativity is, in fact, the absolute invariance of proper-time). Einstein was the last of the ancients; his uncompromising resistance to the insights of modern quantum theory was a stubborn clinging to the old familiar ways of thought. By and large, theologians have found his company more congenial than that of Schrödinger and Heisenberg, and so they too have proved reluctant to come to terms with the peculiar novelties of what quantum theory has to say. In casting that stone I am aware of the sound of tinkling in the scientific glasshouse, since the physicists themselves have only recently begun to wrestle afresh with the problematic interpretation of quantum theory. Between this contemporary activity and the early (but not wholly satisfactory) struggles of the heroic 1920s, lies fifty years of patient and successful exploitation, in which physicists were content to draw consequences from the theory without troubling themselves to ask profound questions about its interpretation.

My purpose in this paper is not to attempt a systematic account of basic quantum physics. I have tried elsewhere to do that for the general reader.[1] Rather, I want to draw attention to a number of issues which arise from the character of the quantum world and which seem to me to be of some significance for the metaphysician and the theologian. I also wish to repudiate a number of claims which have been advanced as consequences of quantum theory, but which do not seem to me to follow from it.

Issues Which Arise from the Quantum World

First let me address the more positive part of the task. I think there are nine issues worth our attention. I discuss each in turn.

Quantum theory has lent its aid to the death of mere mechanism. The Newtonian picture of the solar system appeared so precisely mechanical that the regular rotation of an orrery seemed a fitting representation of its character. Obviously that clockwork universe could not survive the dissolution of the picturable and predictable into the cloudy and fitful, produced by the advent of quantum theory. But, in fact, that theory has only had a minor part to play in the demise of the mechanical. The seeds of its actual decay lay within Newtonian theory itself. That fact is important, since quantum theory in general only manifests its idiosyncratic character in processes of a smaller scale than normally concerns us. For example, most neurophysiologists seem to think that the synoptic activity of the brain does not occur at a level which makes it an intrinsically quantum phenomenon. (Therefore the curious hope that the Heisenberg uncertainty principle gives us a basis for free will proves to be misplaced.) It is necessary that quantum theory should recapture the impressive successes of Newtonian dynamics for these larger systems, since otherwise it would only have succeeded in explaining the microworld at the expense of our understanding of the macroworld. The correspondence principle (the requirement that quantum theory turns smoothly into Newtonian physics for "large" systems) is a well-understood consequence of quantum mechanics. The real *coup de grâce* for mechanism comes at that "large" level, with the realization that predictable systems, like the orrery and the simple pendulum, are only very exceptional cases, even in the Newtonian account of physical processes. At the beginning of this century, more or less contemporaneously with the first intimations of the quantum world, Poincaré's exploration of the instabilities of classical dynamical systems began to reveal that they possess such an exquisite degree of sensitivity to particular circumstances as makes them intrinsically unpredictable. The celebrated fact that there is no analytic solution to the gravitational three-body problem is due to this very property. Recent investigations have considerably extended our understanding of the openness of complex dynamical systems, linking these properties with the irreversibility of time and the genuine novelty of the future.[2]

Quantum theory provides a striking instance of the general fact that exploration of the physical world often yields surprises, so that, if we are to do justice to the way things are, we need a release from an undue tyranny of common sense. The counter-intuitive character of a world governed by Heisenberg's uncertainty principle (which says that if we know where an electron is we don't know what it is doing, and *vice versa*) needs no stressing. Even things which at first sight seem contradictory — for example, that entities should manifest properties of both waves and particles — can turn out actually to be the case. I would like to emphasize that the apparent paradox of wave/particle duality has, since the invention of quantum field theory by Paul Dirac in 1927, been perfectly understood. It is *not* the case that we use a wave model at one time and a particle model at another and that is all we can say about it, as theologians sometimes allege. It *is* the case that we have a theory that combines wave and particle models without taint of paradox and which is open to our rational inspection.

Even logic finds a modification in the quantum world. The distributive law of Aristotelian logic does not hold for subatomic particles, and a new quantum logic is required to mirror their idiosyncratic character.

All these strange aspects of the quantum world (uncertainty, wave/particle duality, quantum logic) arise from what Dirac correctly identified, as the basic feature which distinguishes quantum from Newtonian physics, the superposition principle.[3] This states that in the quantum world we can mix together possibilities which in the Newtonian world are for ever separate and distinct. For Newton a particle is either here or there. In that clear, determinate world there can be no ambiguity about its position. Quantum theory, however, allows states in which a particle is a mixture (a superposition) of "here" and "there". Such states are not to be interpreted as corresponding to a particle's being in the middle, spatially between "here" and "there"; rather, they are to be interpreted probabilistically, as states in which the particle will sometimes be found "here" and sometimes be found "there". Thus the superposition principle underlies the unpicturability and statistical character of the quantum world.

Quantum theory helps us to distinguish reality from naive objectivity. The unpicturable quantum world certainly does not enjoy the objective character of the world of everyday experience. Ironically, when Dr. Johnson kicked the stone in his "refutation" of Bishop Berkeley, from the quantum mechanical point of view he was in contact with something which was mostly empty space, and for the rest a weaving of wave mechanical patterns. The entities of the quantum world are curiously elusive. Does that mean in fact quantum theory is just a peculiar manner of speaking about events in the everyday world of laboratory apparatus? Are there not really electrons? Some have been tempted to espouse the positivistic answer. The subject's grandfather, Niels Bohr, succumbed; he once said, "There is no quantum world. There is only abstract quantum physical description." I am sure he was wrong to say that. The beautiful patterns of the structure of the physical world, revealed by elementary particle physics, demand to be taken more seriously than that.

It was because he wanted to maintain a realistic view that Albert Einstein fought so strongly against the mature version of quantum theory which developed in the late 1920s. His basic instinct was right, but his error was to suppose that picturable objectivity — a clear and determinate world — was the only form that physical reality could take. In fact, the first duty of a realist is to respect the nature of that with which one has to deal. Quantum entities do not have the properties of simultaneously possessing exact position and momentum, of being visualisable. Following Werner Heisenberg,[4] I want to say that they possess the potentiality of position and momentum, one of which can be actualized in the act of observation, but not both simultaneously. (This potentiality for a variety of possible outcomes corresponds precisely to the superposition of those classically immiscible outcomes, which quantum theory permits.) On this view,[5] it is ontology which controls epistemology. (I suppose that is a definition of realism.) The uncertainty principle arises from the nature of the entities with which we have to deal, not from our lack of dexterity in investigating them. Those who are familiar with analyses of attempts to get round the uncertainty principle in thought-experiment schemes of measurements — a battle which Einstein lost to Bohr — will recall that these efforts only fail if quantum theory is applied consistently to all participating systems.[6] That remark precisely exemplifies the control of epistemology by ontology. It is a similar realist stance which leads me to interpret the intrinsic umpredictability of complex Nextonian systems as indicating an ontological openness to the future in the nature of such systems.

Heisenberg seems to have thought that his picture of quantum entities as the carriers of potentiality in some way made them less real than the picturable objects of everyday. I think he was wrong, as I shall argue next.

It seems to me that the view of quantum theory I am espousing encourages the position that it is intelligibility which is the ultimate guarantor of reality. My endorsement of the reality of the quantum world, and my rejection of Bohr's positivism, arose from the conviction that the rationally beautiful and transparent patterns of that world must be taken with the utmost seriousness. They cannot be downgraded into mere ways of speaking. As a quantum physicist I find myself in sympathy with Bernard Lonergan when he says: "since we define being by its relation to intelligence, necessarily our ultimate is not being but intelligence."[7] Lonergan was talking about God conceived as the unrestricted act of understanding. The unpicturable and Unpicturable have something in common.

I should be misleading you if I were to suggest that our understanding of quantum theory is complete. One of its lessons is that science can live with unresolved questions. The mystery yet to be clarified is the exact nature of the act of measurement. Here it is that the potential becomes actual; the electron which has been a superposition of "here" and "there", on being addressed with the experimental question "where are you?" has to settle for one or the other. How does that cloudy, fitful quantum world produce clear answers registered by the everyday measuring apparatus of the laboratory? It is perhaps the most striking of all the quantum paradoxes that, after sixty years of enormously successful cultivation of the subject, there is no agreed answer to that fundamental question. Broadly speaking, four lines of attack have been pursued.

One denies the validity of the question in the form in which I have put it. It supposes that quantum uncertainty arises, not from the intrinsically indeterminate character of the quantum world, as I have been alleging, but from an ignorance on our part of its detailed workings. What I have claimed to be ontological is asserted to be epistemological after all. There are undiscerned (perhaps indiscernible) causes ("hidden variables") which actually determine everything that happens. The cloudiness is in the eye of the beholder, who cannot make out what is going on. In this view quantum measurement is no different from Newtonian measurement. Both are the revealing of what has always been the case. An ingenious and instructive theory of this type has been constructed by David Bohm. I will give reasons later for rejecting it.

The other three lines of attack accept the indeterminate character of quantum theory.[8]

(a) A quantum measurement involves a chain of correlated consequences from the quantum entity (say, an electron) to the registration of the result of measurement by some macroscopic instrument (say, a Geiger-counter click). The point along the chain where the result gets "fixed" is claimed to be at the macroscopic level. It is here that a system is engaged which is sufficiently large and sufficiently complex to have an irreversible character. The celebrated Copenhagen interpretation, the orthodoxy prescribed by Bohr and his friends, espoused a relatively crude version of this idea. Bohr divided the world up into quantum entities (indeterminate) and classical measuring apparatus (the determinators). An experiment was an indissoluble combination of the two, in which the impingement of the latter on the former produced a definite result. The reason this won't do as it stands is that it is essentially a dualist description (quantum world and measuring apparatus) of a universe which is, in fact, a unity (the measuring apparatus is itself made out of quantum constituents). I think this approach looks in the right direction, but a fully satisfying answer would need to dissolve the duality by means of a much more extensive, and subtle analysis than it has so far proved possible to give.

(b) The problem for (a) is to distinguish what is large and determinating from what is small and indeterminate. All experiments of which we have knowledge involve the ultimate intervention of a conscious observer who notes the result. Some have felt it is at this final stage that things get fixed and that it is consciousness that plays the determining role. The proposal has a certain specious attraction, linking as it does the mystery of quantum measurement and the mystery of conscious thought, but it also has very strange consequences. Are we to suppose that the computer print-out of the result of a quantum experiment, stored away unread, only acquires a definite imprint months later when someone opens the cupboard and reads it?

(c) Even more bizarre is the many-worlds interpretation. This proposes that at every act of measurement the universe splits into parallel, disconnected, universes, in each of which one of the possible results of measurement is realized. There is a universe where the electron is "here" and another universe where it is "there". The world, and we with it, is being cloned at a prodigious rate. The stupendous prodigality of this proposal has meant that it has had substantially more appeal to the gee-whiz writers of popular science than to sober physicists. However, it enjoys some currency among cosmologists as a way of applying quantum theory to the whole cosmos, a project which may not be feasible or necessary.

Quantum theory affords some degree of support for an antireductionist stance. This surprising consequence of subatomic physics (which, after all, is methodologically a very reductionist subject) arises in two ways.

One is the famous EPR experiment.[9] Using quantum theory Albert Einstein, Boris Podolsky, and Nathan Rosen (EPR) pointed out that, when two quantum entities have once interacted, they retain a certain power to influence each other simultaneously, however widely they subsequently separate. EPR thought that this counterintuitive "togetherness in separation" must show an incompleteness in the theory and its need for amendment. Recent experiments, however, and particularly the beautiful work of Alain Aspect and his collaborators in Paris, have revealed that just such an effect of "nonlocality" is to be found in nature. Thus, even at the level of fundamental constituents, the world does not fall apart but, instead, exhibits a degree of mutual cohesion.

Secondly, it is conventional to say that quantum theory shows another integrationist tendency by refusing to separate observer and observed. This would be reflected, for instance, in Bohr's insistence on linking quantum entity and measuring apparatus in the description of the measurement process. That particular way of expressing it is, I think, enforced by Bohr's unacceptable dualism, which can only be salvaged by such a requirement of indivisibility. To my mind, a better way of looking at the matter is to emphasize that the determining apparatus itself arises out of an indeterminate quantum substrate. This is the emergence of a level autonomy within physics itself, quite as striking and quite as conceptually irreducible as the emergence of life from inanimate matter, or self-consciousness from animal being.[10] Here is certainly a profoundly antireductionist insight produced by quantum physics.

Quantum theory provides a significant testbed for claims in the philosophy of science. I have already dealt with some issues that relate to this. Let me refer to one other which relates to the underdetermination of theory by experiment. Clearly a theory which claims to cover an infinity of cases cannot be uniquely determined by a finite number of tested instances. However, an important sieve for rationally acceptable theories is provided by the requirement that one should not need continually to make *ad hoc* adjustments to keep the theory going. (If one had tried to preserve Ptolemaic ideas post-*Principia*, every new set of observations would have required a fresh batch of epicycles, whilst the Newtonian gravitational theory successfully coped for two hundred years, in a perfectly natural and unforced way, with every increase in accuracy. Still the problem of the perihelion of Mercury eventually showed that even the Newtonian theory was only verisimilitudinous.) So the question is not, "Are there ambiguities at any stage?" but, "Are there truly perplexing ambiguities, incapable of being resolved by rational criteria?" If we are to answer that in a meaningful way we shall have to look to fundamental science. It is not surprising, nor significant, that there are, say, conflicting views about chemical valency. Here, people are trying to explain a situation whose basic physics is understood, but the elucidation of its consequences is too complicated to calculate precisely. In such circumstances you have to do what you can, and competing and conflicting models result. They are all patently partial; but fundamental physics ought to be free from the ambiguity of expediency.

I can think of only one example of a significant clash in contemporary fundamental physical theory: Bohm's determinate, hidden variable, quantum theory versus conventional quantum mechanics. As I have said elsewhere, that clash seems like a "duck/rabbit" with a vengeance,[11] since both of these theories have the same experimental consequences, though they are so very different in their character. Yet almost all physicists espouse conventional quantum theory and reject Bohm's ingenious ideas. Why?

There are two answers. One lies in that fundamental requirement of fruitfulness for future development which is an important discriminator of theories. The conventional theory has been able to incorporate the requirements of special relativity in a natural and successful way which has so far eluded Bohm's approach. But even if this were not so, I think the majority verdict would still rightly fall to the conventional theory. Bohm's theory has an air of contrivance about it which does not commend it to many of us. Not only is it hard to believe that even so clever a man as he would have thought of his equations without having those of quantum theory first before him, but also, and above all, the way the statistical character is inserted into the theory has an arbitrary air about it.[12]

Though there is nothing absolutely inevitable about conventional quantum theory, its selection as the understanding of the nature of the subatomic world seems to me to be rationally motivated in the way I have described. It is an example of the fruitfulness of those skillful acts of judgment whose essential role in scientific inquiry was persuasively emphasized by Michael Polanyi.[13]

Quantum theory is also a significant testbed for claims in the history of science. The one I wish to refer to is the degree of influence exerted on scientific development by the general atmosphere of contemporary thought. I shall have to be brief, so let me say that I see no reason to suppose that the cloudy fitfulness of the quantum world was in any way related to the rootlessness of the Weimar republic, from which many of the pioneer papers originated. It is entirely and adequately understood as arising, rather, from the peculiar behavior of light and the statistical character of atomic decays. Though a fascist, authoritarian ideology emerged, physics could not abandon those necessary insights and restore a rigid predictability.

Finally, and most disturbingly, quantum theory suggests that there may be limits to rational inquiry. It is the belief of conventional quantum theories that individual quantum events are radically uncaused; only their overall statistical pattern is prescribed. No explanation is to be offered of why, on this occasion, the electron is found "here" rather than "there". What are we to make of that? It is important to recognize the surprising nature of the claim being made by the physicists. It is in no way comparable to the familiar philosophical problem of the uncertainty of the future, the indeterminate nature of the result of tomorrow's sea battle. Rather, it is the claim that there is no retrospective explanation to be offered of a particular occurrence's having happened.

Does this radical lack of physical causality represent the existence, even at the humblest levels of the universe, of a certain freedom granted to the creature? (Shades of A. N. Whitehead!) Or is God the ultimate Hidden Variable, skillfully exercising his room to maneuver at the rickety constituent roots of the world, whilst cleverly respecting the statistical regularity which his faithfulness imposes? We all know that William

Pollard[14] suggested this as a means for God's action in the world, though it seems a rather hole-and-corner sort of providence to me. Or does even physics have its apophatic element? A more profound understanding of the nature of the measuring process, irreversibly turning potentiality into actuality, is needed before we can make much progress with answering these questions.

The Non-Consequences of Quantum Theory

Quantum theory has been prayed in aid of so many diverse points of view that it seems necessary to conclude this paper by listing a number of its non-consequences.

Quantum theory is strange and counterintuitive, but it does not license the attitude that anything goes. Flimsy analogies have often been invoked in attempts to give quantum backing to non-quantum phenomena. The togetherness-in-separation of the EPR experiment does not itself tell us anything about the possibility of telepathic communication. The idiosyncratic oddness of the quantum microworld is not a basis for believing in the paranormal in the macroworld.

Quantum theory is not of itself a sufficient basis for a universal metaphysics. In their different ways Whitehead's process philosophy and Bohm's holomovement and implicit order present grand, even baroque, metaphysical schemes claiming some anchorage in the quantum world. Whatever the merits of these detailed proposals (and I am skeptical about both), they rapidly go beyond anything that a sober assessment of contemporary physical theory could be held to sanction.

Quantum theory does not endorse the essential rightness of Eastern religious thought. Popular books, such as Fritjof Capra's *The Tao of Physics*[15] and Gary Zukav's *The Dancing Wu Li Masters*,[16] have suggested the contrary. They seek to assert that the dissolving-yet-connected quantum world corresponds to the expectations of Eastern philosophy and contrasts with the uncompromisingly structured expectations of Western thought. The arguments are half-truths, since they depend upon a lopsided account of the quantum world. Although that world has its elusive character, all does not dissolve away. There is a clarity of form that remains. This finds expression in, for example, those symmetry principles which play so important a part in contemporary fundamental physics.[17] It is instructive that Capra unwisely dismissed symmetry as an out-moded hangover from Greek thought. According to him, in Eastern thought symmetry:

> "...is thought to be a construct of the mind rather than a property of nature, and thus of no fundamental importance.... It would seem, then, that the search for fundamental symmetries in particle physics is part of our Hellenic heritage, which is, somehow, inconsistent with the general world view that begins to emerge from modern science."[18]

Those words, written in 1975, have proved a strikingly false anticipation of the path fundamental physics was to take in the subsequent years. I have argued that Western thought (particularly in its striving for a balance between God's transcendence and immanence, and between being and becoming) provides the basis for a natural theology more in accord with the pattern and structure of the physical world than that which Eastern thought provides.[19]

Quantum theory does not approve the idea of an observer-created world. To be sure, the quantum act of measurement involves a subtle, and incompletely understood, interaction between the means of observation and the system observed. The consequences are, however, strictly limited according to the potentiality available to be made actual. It is another example of the unjustified typing of quantum mechanical insight to proceed from this to extravagant claims such as John Wheeler's Participatory Anthropic Principle (PAP): Observers are necessary to bring the universe into being. The gap between saying that the act of measurement determines whether an electron is "here" or "there" and the stupendous claim of the PAP seems to me to be quite unbridgeable. I cannot see any legitimate grounds for being as friendly towards this fanciful assertion as John Barrow and Frank Tipler seem to be in their discussion of the matter.[20]

I believe that the issues that quantum theory raises for theology are best treated in ways that are modest in metaphysical intent, rather than grandiose.[21] We are presented with a picture of the physical world that is neither mechanical nor chaotic, but at once both open and orderly in its character. A simple everyday notion of objectivity is too limited an account even for physical reality. The latter displays an elusiveness which is nevertheless rationally structured, though perhaps not exhaustively so. In the twentieth century scientists have had to be exceptionally flexible in their response to the way things are, abandoning cherished conceptions of what is reasonable in the face of the way things actually seem to be. The contingent rationality of the world so explored is consonant with its being the free creation of a reasonable Creator. However strange and unexpected the discoveries of quantum physics have proved to be, it is still the case that the "unreasonable effectiveness of mathematics" (in Eugene Wigner's phrase) continues to operate as a guide to the pattern of the physical universe. Indeed, I have argued that it is this very intelligibility of the quantum world which is the guarantee of its idiosyncratic reality. Perhaps that is the most important conclusion, for it allies physics with theology in a common endeavor to understand the many-leveled structure of the universe that we inhabit.

NOTES

[1] J.C. Polkinghorne, *The Quantum World* (Harlow: Longman, 1984, also Penguin and Princeton University Press).

[2] I. Prigogine and I. Stengers, *Order Out of Chaos* (London: Heinemann, 1984).

[3] P.A.M. Dirac, *The Principles of Quantum Mechanics* (Oxford: Oxford University Press, 1958, fourth edition) ch. 1.

[4] W. Heisenberg, *Physics and Philosphy* (London: Allen and Unwin, 1958).

[5] For a fuller discussion see Polkinghorne 1984, ch. 8.

[6] N. Bohr, *Atomic Physics and Human Knowledge* (New York: Wiley, 1958) ch. 4.

[7] B. Lonergan, *Insight* (London: Darton, Longmann and Todd, 1957) 677.

[8] For more detail see Polkinghorne 1984, ch. 6.

[9] *Ibid.*, ch. 7.

[10] J.C. Polkinghorne, *One World* (London: S.P.C.K., also Princeton University Press, 1986) ch. 6.

[11] *Ibid.*, 10-11.

[12] Bohm's theory is so constructed that, if one starts with a probability distribution corresponding to a wavefunction, then it will propagate in time as that wave function requires, but one has to impose that initial condition.

[13] M. Polanyi, *Personal Knowledge* (London: Routledge and Kegan Paul, 1958).

[14] W.G. Pollard, *Chance and Providence* (London: Faber, 1958).

[15] F. Capra, *The Tao of Physics* (New York: Wildwood House, 1975).

[16] G. Zukav, *The Dancing Wu Li Masters* (London: Fontana, 1980).

[17] See for example, P. Davies, *The Search for a Grand Unified Theory of Nature* (New York: Simon and Schuster, Inc., 1984).

[18] F. Capra, 1975, 272.

[19] J. Polkinghorne, 1986, 82-3.

[20] J. Barrow and F. Tipler, *The Anthropic Cosmological Principle* (Oxford: Oxford University Press, 1986) ch. 7.

[21] A proposed complementary mind/matter metaphysics will be presented in my forthcoming publication: *Science and Creation* (London: S.P.C.K., to appear).

QUANTUM PHYSICS IN PHILOSOPHICAL AND THEOLOGICAL PERSPECTIVE

ROBERT JOHN RUSSELL, Center for Theology and the Natural Sciences, Graduate Theological Union, Berkeley.

1. *Introduction*

This paper seeks to critically assess the relevance of quantum physics for contemporary Christian theology in an ecumenical context. My method here will be to use philosophy as a bridge between physics and theology, in particular focusing on a philosophy of nature informed by quantum physics and addressing questions both to metaphorical and systematic theology. Since, as I take it, the task of theologians is to rethink their heritage of Scripture, creed and tradition in terms of contemporary culture, it is particularly relevant that theologians now engage with scientists and philosophers of science in understanding the radical changes occurring in contemporary natural science and discover the effects these changes can have on our own theological agenda.

Quantum physics, as it developed from 1900 to the late 1920's, has become a primary a source of deep change within the physical sciences. It is now an irreducible part of fundamental physics.[1] In combination with special relativity it is essential to the whole range of research at the frontiers of physics today, from high energy physics to superconductivity to astrophysics and cosmology. Hence the task of this paper is to investigate ways in which the philosophical implications of (pre-relativistic) quantum physics might be relevant to the work of contemporary theology, both as a heuristic source of theological metaphor and as a systematic factor in constructive theology.[2]

2. *Philosophical Issues In Quantum Physics*

2.1 *A Short Tour of Quantum Physics*

The over-riding impression one gets about the data from scintillation counters, bubble chambers, and Geiger counters is that of unremitting chance. As ordered, regular and constant as are the rocks, tables and flowers of our ordinary world of experience, at close range matter seems ruled by a chaos of unpredictable change. For example, though all the atoms in a kilogram of uranium are identical, and though on the average a predictable number of them will decay in a specified amount of time, no one knows how to predict in advance which atom actually will decay, or why it did and its neighbors didn't. After nearly a century of study, this process remains an anomaly to the causal explanations of the past, though a number of ways have been found to address this challenge at the epistemological and ontological levels.

Without digressing too far in this short paper we can disclose a bit more of the quantum paradox by considering what is for many physicists the paradigmatic example of the quantum world: the "2-slit experiment." Indeed most of the profound philosophical issues raised by quantum physics can be extracted from this simple example. Imagine a beam of electrons aimed at a metal plate with two narrow slits. Beyond the plate, a flourescent screen registers the impact of individual electrons as bright dots. The fact that dots are produced by individual electrons suggests the particle-like nature of the electrons. However the pattern these dots produce on the screen suggests that the electron is a wave. For example, particles should be distributed in two clumps, one below each slit. Waves, however, would go through both slits at once and produce a ripple pattern on the screen (technically a diffraction pattern with interference fringes). In our experiment the particle-like dots are actually distributed in a wave-like ripple pattern, not as two clumps below the slits; in fact most of the electrons fall at a point directly between the two slits!

The phenomenon is all the more astonishing because the pattern obtained in the 2-slit experiment is independent of the intensity of the electron beam. One could just as well produce one electron per year as millions per second (though it would take a long time to see the pattern form!). Clearly the ripple pattern is *not* due to an interaction between incoming electrons. How then can matter display both wave and particle attributes? What can one say about the structure of matter at the atomic and sub-atomic level such that these contradictory properties result from the same kind of matter? Finally, the experiment can be repeated for photons, or for any other kind of matter one likes, and the same phenomena occurs. Viewed from the perspective of quantum physics, the world is strikingly counter-intuitive!

Adding to the conceptual challenge of quantum physics is the fact that quantum chance come in two distinct varieties, *both* of which are radically different from classical statistics. Many phenomena, such as the pressure of a gas at room temperature, the laws of gambling, and the genetic variation and mutation in evolutionary biology, obey classical (Maxwell-Boltzmann) statistics with its familiar "bell-curve" shape. Classical statistics, in turn, can be seen as the limiting case for two distinct types of quantum statistics.[3] "Bose-Einstein" statistics describes the behavior of particles such as photons, gravitons, and gluons, which mediate the four fundamental interactions: electromagnetic, gravitational, strong and weak. These kinds of particles are generically called "bosons". Particles which act as sources of the fundamental interactions, and hence as the "constituents" of matter, such as electrons, protons, and, at a more elementary level, quarks, obey what is called "Fermi-Dirac" statistics and are called "fermions".[4]

The mathematical difference between these two types of chance leads to striking differences in the effects they describe. Bosons tend to act more like waves than particles, occupying the same place at the same time and adding together like waves to produce a single effect. For example, photons from many sources — the sun, light bulbs, candles, reflections from walls — will superpose at each point in a room. Vision occurs when our eye samples the resulting electromagnetic field at a particular point and our brain unravels this complex signal, interpreting it as a spatial display of objects. Moreover, though bosons tend to clump together, this does *not* occur through an attractive force, and it admits of no classical explanation. The superfluidity of liquid helium (called Bose condensation) and the coherence of laser light are further examples of this clumping effect.

Fermions act in an entirely different manner from bosons. The Pauli exclusion principle is a direct result of Fermi statistics, according to which no two electrons can occupy the same (quantum) state. This principle describes the fact that electrons in atoms cannot spiral down into the nucleus. Instead they form a series of shells which surround the nucleus, giving to atoms their structure and accounting for the impenetrability of matter. Moreover the number of electrons per shell produces chemical valence, which in turn leads to the bonding of atoms into molecules and compounds, chemical reactions, and so on. Fermi statistics accounts for electrical resistance, the heat conductivity and heat capacity of metals, the operation of semiconductor devices such as transitors and computer chips, and such striking low temperature phenomena as superconductivity. Again, as in the Bose case, the Pauli exclusion principle does *not* arise from a force between fermions; it is a uniquely quantum mechanical effect.

In sum, Fermi statistics accounts for the impenetrability of matter and provides a basis for our understanding of chemistry. Bose statistics describes the interpenetrability, superposition and cohesiveness of the carriers of the fundamental interactions in nature. Hence quantum chance is intimately linked with the macroscopic world as we know it, both its solidity and transparency, its form and unceasing activity, even its very character as *res extensiva*.

Fermi and Bose statistics also suggest that the behavior of quantum systems cannot be analyzed merely in terms of the behaviour of its parts. Bose condensation and the Pauli exclusion principle, for example, apply to systems of particles and seem to have little meaning for individual particles. The features they describe are not mere extensions of the properties of the individual particles. However one accounts for this (in terms of the several interpretations of quantum physics discussed below), phenomenologically quantum systems display new and irreducible features which are strikingly different from mere composites of those of their components. The laws applicable to quantum systems, such as quantum statistics, are more than mere generalizations of the laws governing its component parts. Quantum statistics thus suggests for quantum systems a wholistic character strikingly different from classical systems.[5]

The mathematical form of quantum statistics suggests another important difference from classical statistics. In the latter case we can assume that what we consider to be a chance event is the juxtaposition of two causally-unrelated trajectories (an example is a car crash or winning a lotto ticket). In the case of quantum physics the statistics suggests that no underlying causal explanation can account for the data with its particular form of randomness.[6]

Lying behind the difference between classical (Boltzmann) and quantum (Bose and Fermi) statistics is the concept of indistinguishability. Classical physics assumes that, regardless of the size or "elementary" status of matter, one can always distinguish one piece from another. Matter can be marked ("painted") and its trajectory tracked continuously. But we now know that this assumption at the heart of classical physics (and hence classical epistemology) is dramatically overturned at the subatomic level. All electrons, for example, are indistinguishable, as are all protons, photons, and so on for each type of elementary particle. The difference between quantum and classical statistics, and in turn the difference in the phenomena they produce, can be seen as depending entirely on this simple but critical fact. Though at the macroscopic level there is countless variety and differentiation to the structures and processes of nature, at the sub-atomic level matter is modular, interchangeable, endlessly repetitive, unremittingly homogeneous. Somehow the macroscopic features are related to and born out of their exact opposite at the sub-atomic level. These striking differences between nature at the sub-atomic and the macroscopic levels must be factored into our philosophical and theological interpretation of the world.

Finally, the counter-intuitive character of quantum chance was brought into sharp relief by the recent discovery by physicist John S. Bell. In 1964, Bell proved a theorem which underscores the dramatic difference between classical and quantum statistics.[7] The theorem can be appreciated through the following thought experiment: An atom in an excited state decays, emitting two electrons in opposite directions. I measure the spin of one of the electrons as it travels through my lab, while my friend measures the spin of the other electron as it travels through her lab. I randomly choose to measure the electron's spin along, x, y, or z axes; my colleague makes a similar, though arbitrary, choice for each measurement. These measurements are taken simultaneously, or at least close enough together in time, so that, according to the special theory of relativity, no physical interaction could be transmitted from one lab to the other to influence the result.

After repeating the experiment many times, the results are as follows: the data I took looks entirely random to me, as does the data my friend took. 50% of the time I got spin up, and 50% spin down. Moreover from the spin taken in the nth measurement I am unable to predict the result of the n+1th measurement, and similarly for my colleague.

However if we later compare the data from the two labs we find the following results: when we both happened to be measuring the spin along the same axis, the data were 100% anti-correlated: if I measured spin up, she got spin down, and vice versa. However, when measuring spin along alternate axes (such as the x-axis for me and the y-axis for her), the anti-correlation was only 25%.

These results now lead to the following paradox: to explain the data taken from the same, though arbitrarily chosen, axes we must assume the particles were produced initially in a perfectly anti-correlated state *along all three axes*. Yet if this were so we should have had 33% anti-correlation in the data taken from alternate axes, in contradiction to the actual result of only 25%. Alternatively, if we try to explain the low anti-correlations from data from alternate axes by assuming that the particles were not correlated when they were initially produced, we cannot account for the perfect anti-correlations in data from the same axes. (We shall hereafter drop the technical point that these are "anti-correlations".)

Perhaps then the correlations arise from the measurement process and not the manner in which the particles were produced. But this conclusion, at least in one sense, can be ruled out immediately. We do not believe that measurements made in lab A influence those made in lab B. The results in lab B remain random whether or not lab A is taking data. Indeed, even the particular type of statistical distribution of the random data in lab A remains the same whether or not lab A is taking data. Furthermore the data in lab B, *taken on its own*, cannot tell us whether data was or was not being concurrently taken in lab A. In this sense lab A appears to have "no effect" on lab B, and we can rule out any possibility of using quantum physics to signal instantaneously between lab A and lab B.

According to the special theory of relativity, no signal or influence can propagate faster than the speed of light; i.e., all causal interactions are restricted to the interior of the light cone geometry of spacetime. Any theory which is consistent with special relativity is said to be a "local" theory in which "superluminal" influences are ruled out. Since the quantum data in either lab, considered on its own, show no evidence of the simultaneous activity of the other lab, they do not explicitly violate special relativity. In this sense quantum theory, which correctly accounts for quantum statistics, is a *local* theory.

On the other hand, if we compare the data from the two labs afterwards, the individual data taken simultaneously along the same axis do match (eg., for electrons the spins are always opposite)! In this sense the *correlations* in the data are *non-local*. So, while we should not say that the measurement process at A influences the measurement process at B, we must admit that the results of the measurements of A and B indicate that the processes producing A and B are to a certain extent inseparable. The image of a hologram, suggested by David Bohm, is a useful heuristic here.

Quantum physics presents us with the predominantly chance character of nature at the atomic and sub-atomic levels. Moreover this statistical character is radically different from the ordinary statistics of classical science in several ways: i) First of all the behavior of elementary particles (and the composite structures they produce) leads to two distinct types of quantum statistics. One type accounts for the impenetrability of matter and for many of its chemical properties, the other describes the interpenetrability of the fields of interaction and their cohesive character; ii) Next, quantum statistics are not the result of our ignorance of the random juxtapositions of causal trajectories (accidents in classical science); iii) All members of a given type of elementary particle are intrinsically identical and hence indistinguishable, a property never fully realized in macroscopic experience, which leads directly to the distinction between Fermi and Bose statistics; iv) Moreover this modular, interchangeable character of elementary particles gives rise to its antithesis, the limitless variation of macroscopic matter; v) Finally, nature reveals a highly non-local and wholistic character at the quantum level which is strikingly different from the separability of nature in our ordinary experience.

2.2 *Survey of Competing Interpretations of Quantum Physics*

Throughout this century, the meaning of quantum physics has been the subject of intense debate, both on the part of such seminal figures as Niels Bohr, Albert Einstein, Louis de Broglie, Max Planck, Max Born and Werner Heisenberg, and of such contemporary exponents as David Bohm, Abner Shimony, John Wheeler and J.S. Bell.[8] Almost all would agree, however, that in discussing quantum physics it is essential to distinguish between physical theory and its interpretation — that while the theory is not directly in question, its interpretation is a continuing subject of discussion. Does the uncertainty principle, for example, imply that natural processes are intrinsically indeterministic? Or is it an epistemological problem arising out of the concurrent use of classical concepts like space, time and causality? Does it represent an epistemological limitation due to the experimental basis of physical theory? Is it a prescription holding for all possible further physical theories, or is our present theory merely incomplete, leaving indeterminacy to be circumvented in some way? Do the terms used in modern physics refer to objective physical reality, or are they merely book-keeping devices for cataloging the results of experiment?

The answer to these and other similar questions have led to fundamentally different interpretations of quantum physics. As physicist and Anglican theologian John Polkinghorne puts it, not only are many of these interpretations counter-intuitive and paradoxical in themselves, "the greatest paradox about quantum theory is that after more than fifty years of successful exploitation of its techniques its interpretation still remains a matter of dispute."[9] How then are we to appropriate the philosophical implications of quantum physics as we undertake the task of Christian theology? For the purposes of this paper I will suggest that we start by grouping these interpretations into three categories, related to the measurement process, the process under observation, and quantum theory.[10]

2.2.1 *Related to the measurement process of quantum properties*

i) In classical physics, dynamic properties such as position and momentum were thought to be inherent attributes (or primary qualities) of the systems being studied. According to Niels Bohr, however, these properties are not inherent in quantum systems. Instead they are relations between the systems and the measuring device; they belong to "the entire measurement situation." (Note, however, the measuring device is treated as a classical system.) In the same vein Bohr frequently stressed the wholistic aspect of quantum physics and the interconnection of observer and observed through the irreducible quantum of action exchanged in any measurement process.

ii) Among Bohr's followers, some argued that, since we choose which properties to measure, the process of measurement in this (weaker) sense creates the properties involved (cf. iv below). Moreover, since the logic of instrumentation precludes simultaneous measurements of certain properties, such as position and momentum, it is meaningless to attribute both simultaneously to the quantum system or to ask whether they exist before measurement.

iii) Werner Heisenberg, at one point, suggested that the properties of quantum systems are real, but only as potentialities, until measurement actualized them.

iv) John von Neumann argued that the measurement problem can only be resolved by appeal to human consciousness. Since even the macroscopic apparatus must in principle be understood quantum mechanically, only the conscious observer can actualize quantum properties and, in this (strong) sense according to von Neumann, consciousness creates the objective world.

2.2.2 *Related to the process under observation*

v) During the early decades of quantum theory, Albert Einstein and others argued against the direction being taken by Bohr and Heisenberg. However, after the debates surrounding the paper by Einstein, Podolsky, and Rosen in 1934, Einstein took a different approach. He agreed that quantum physics was correct as far as it goes, but that it was incomplete: there are causal factors which quantum physics does not include, and these factors are responsible for indeterminacy. Issues raised by the EPR debate remained more or less unsettled until the discovery and testing of Bell's theorem (see below).

In 1951 David Bohm published the first consistent alternative theory to quantum physics based on a "hidden variables" approach, although these factors were not strictly the same as classical variables. Presumably they could lie at the subnuclear level, although their inherently "non-local" (see below) property makes them unusual.[11]

vi) Finally, according to the proposal of Everett, during each quantum process, *all* possible outcomes are realized by a bifurcation of the universe (quantum many worlds theory).

2.2.3 *Related to quantum theory*

vii) Bohr developed what became the standard interpretation of quantum physics in his celebrated principle of complementarity, according to which it is no longer possible to admit the classical assumption that causal explanation is compatible with a spacetime description. He also stressed the need for complementary concepts, such as waves and particles or wave-particle duality, in every complete explanation of quantum data.

viii) Others believe that the indeterminacy in predictions using quantum formalism arises from its basis in classical (two-valued, distributive) logic.

2.2.4 *No difference in predictive power*

To date none of these interpretations offers a direct experimental test which would "prove" it and "disprove" the others. Some interpretations, however, do suggest directions for research into more general theories which could conceivably replace existing quantum theory.

David Bohm, for example, was strongly influenced by Bohr's stress on the wholistic aspects of quantum physics. Although his earlier hidden variables theory produced no decisive experimental prediction Bohm has more recently begun to press for a new ontology underlying quantum processes, radically different from the classical ontology of ordinary objects. In this view, we should abandon such classical concepts as particles and fields and develop a new ontology in keeping with quantum nature as an "undivided whole." Bohm's suggestion is similar to that of Heisenberg, namely that the properties of quantum process are only potentialities until realized by measurement. According to Bohm, the "implicate order" includes a series of potential ontologies which become actual as the "explicate order," the world of ordinary phenomena and quantum data.[12]

Those who adopt von Neumann's approach look to advances in the psychological and neurophysiological sciences which could be fruitful in suggesting how consciousness might be an integral part of the quantum measurement process.

Just as Einstein used non-Euclidean geometry to move beyond Newtonian gravity, some argue that non-classical logic could in principle be used in a more general theory of quantum processes.

Yet to date none of these directions have been predictively advantageous. Quantum physics, as generalized to relativistic quantum field theory, continues to make *extremely* successful predictions even for current measurements at the quark level, several orders of magnitude below the data available two to three decades ago. Theories using alternative (multi-valued or non-distributive) logical systems are still at a speculative stage. New ontologies arising specifically from quantum research have not been developed into broadly coherent philosophical systems comparable to those of classical metaphysics. Alternatively, existing fully-developed systems such as the philosophy of Alfred North Whitehead have, to date, been relatively fruitless in producing a new scientific theory with predictive power to rival quantum physics.[13]

2.2.5 *Philosophical implications of wholism and non-locality*

The wholistic aspect of quantum physics has received considerable attention from physicists and philosophers. Unless one appeals to a hidden-variables explanation of quantum statistics, most of the other interpretations suggest that quantum physics introduces an element of wholism strikingly different from classical physics. It may be due to the measurement process, in which the measurement apparatus and the processes being studied are inter-related through the exchange of (at least) one quantum of action. It might be due to the unitive character of the single wave function which describes a complex quantum system. Again it might be due to the quality of potentiality in nature, that matter takes on states whose properties are only partially actualized at any one given moment. In any case this element of wholeness gives to nature a "social" character even at the atomic level, as Henri Margenau described it.[14] This collective behavior, one must recall, is not the result of forces, interactions, or causal influences in any normal sense of the word, making the phenomenon all that more striking and inexplicable. As Ian Barbour wrote, "the being of any entity is constituted by its relationships and its participation in more inclusive patterns," and thus quantum physics gives, according to Barbour, "a more precise meaning to the statement that 'the whole is more than the sum of its parts.'"[15]

As we have seen, Bell's theorem brought this wholistic dimension to quantum systems into even sharper relief. On the one hand we know that the data in my lab consists of random counts which I could never use to simultaneously "signal" a colleague at a remote location. On the other hand we know that however we refer to the "material world" underlying and producing these data, its structure must be such as to exhibit the kind of correlations which Bell's theorem has underscored, correlations between data which in principle could be taken simultaneously and at cosmic distances.[16] Hence we now know that though quantum phenomena as such are local, the underlying processes must be non-local. Of course I am not suggesting that these underlying processes are any less empirical than ordinary, macroscopic processes, only that the metaphysical categories we may need to explain them may be very different from our traditional categories, such as spatio-temporal location, separability, mass, causal determinism, and so on. As Nick Herbert writes in *Quantum Reality*:

> What Bell's theorem does do for the quantum reality question is to clearly specify one of deep reality's necessary features: whatever reality may be, *it must be non-local.* Since Clauser's experimental verification of Bell's theorem, we know that any correct model of reality has to incorporate explicitly non-local connections. No local reality can explain the type of world we live in.[17]

If one is satisfied with physical theories taken as computational devices, then quantum theory can be considered adequate and further speculation about "the underlying reality" become pointless. On the other hand, if one wants physical theory to yield insights about the ontological character of nature, then the inference to the non-local aspect of these underlying processes may well force one to rethink the metaphysical assumptions (including the atomistic model of matter) on which classical physics is based. Since we can find no consistent way to assign properties to individual electrons that gives a correct account of all the correlations in the data, and if we proceed with some form of a realist philosophy of nature, we may need to abandon the assumption that the electrons in one lab are totally separate phenomena from the electrons in the other lab. Instead the correlations suggest a unitive or non-separable feature about matter at the sub-atomic level which is strikingly contrary to the way matter behaves at the ordinary, macroscopic level. These strong correlations lead to the notion that the electrons still form a single system, though they are being measured in labs which are arbitrarily distant from each other. The result is that local measurements suggest a non-local character to nature.

What does it mean in particular for the interpretations of quantum physics previously advanced?

i) and ii) If Bohr's insight is right, that the "entire measurement situation" is involved in each quantum measurement, then the non-local feature of quantum physics implies that for correlated systems the results of a measurement on one part of the system in my lab could depend on measurements taken simultaneously light-years away on another part of the system. Moreover, if the properties we measure "exist" in part through our choice of instrumentation (that is, through our choice to use, say, a diffraction grating to measure wavelength (and hence momentum) instead of using a phosphor screen to measure position), then they do so as well on the choice of instrumentation of distant colleagues. Hence, as Herbert points out, even if properties are relational rather than intrinsic, they are inter-relational on a *cosmic* scale.

i) and iii) If Bohm's approach is to prove fruitful, the non-local quality of quantum physics must be shown to arise entirely from the non-local properties of the implicate order.

iv) The option to include human consciousness as a determining factor in the measurement process has had few supporters. Still, given Bell's theorem, one might want to suggest that, *if* the minds of the observers in each lab are somehow integrally related to the actualization of the measurement processes in their respective labs, quantum correlations in that data might be a sign of some type of mental correlation between these observers.

Yet the measurement process need *not* involve conscious observers. Indeed, most high energy research is now carried out by fully automated equipment, and the data is only inspected by people well after its initial analysis by computer. Hence there is no substantial basis for an argument involving the human mind in the measurement process itself. It follows then that, contrary to some arguments now being circulated, quantum non-locality, whatever its ultimate implications, does *not* provide a basis for telepathy or any other kind of mental influence at a distance.

Moreover we can clearly rule out using quantum correlations to "signal" or transmit information instantaneously from one lab to the other, since the data in each lab remain entirely random whether or not data are being taken simultaneously at a remote distance. Correlations in remote data are only observable through their *subsequent* comparison, when data from each lab are compared. *No* information can be gleaned from one set of data as to whether or not data are being taken elsewhere at the same time.

v) Perhaps the most serious challenge from Bell's theorem comes to local realist versions of a hidden variables theory. According to local realism, there are minute causal factors, admittedly unknown so far, which actually determine the outcome of my individual, seemingly random measurements. Their, as yet unrestricted and hence unspecified, activity produces the observed statistical scatter in my data. According to Bell's theorem, these factors must be instantaneously connected with those which instantaneously determine the scatter in my colleague's data. Moreover, they must be connected in such a way as to produce the specific type of *strong correlations* found in quantum data produced by systems with a common past.

According to Bell's theorem, these correlations rule out any classical interpretation of the allegedly underlying causal factors. The hidden variables cannot be merely undetected but otherwise ordinary entities with intrinsic classical properties if they are to account satisfactorily for the kind of correlations found in quantum data. Put alternatively, quantum chance cannot be due to the mere juxtaposition of causally unrelated, previously undetected, classical trajectories. No local hidden variables theory can produce the statistical distribution actually found in quantum data.[18]

Hence it seems when working with a quantum perspective on nature one must abandon the *metaphysical* assumptions of local realism, at least in the form adopted by Einstein. As we move into quantum physics, the classical program of assigning attributes to matter seems to break down, and it is extremely hard to see how a realist program can continue unless one is willing to change the classical metaphysical conception of the ontology of matter which has been embodied in modern science. Surprisingly, in a recent book, Henry Folse argues that this was Bohr's real agenda. Folse suggests that Bohr does assume an independent reality but that he rejects the assumption of classical science that it can be described in terms of *traditional ontology*.[19]

What the broader implications are for a realist philosophy of science (even respecting its many types and varieties) is a subject of considerable discussion today. Taken very broadly, I believe one can still work within a realist philosophy as an account of the progress of science: one can still argue that the predictive success and explanatory power of scientific theory is best accounted for by the assumption of reference, that theoretical terms do in fact refer to structures in nature. What seems to be coming under increasing pressure, at least in the realm of quantum physics (and, I would argue, in cosmology as well), is the further and more specific claim of *correspondence*, that the structure of the theoretical concepts corresponds to some extent with the structure of their references in nature. Also being challenged by quantum physics, in my opinion, is the claim of *convergence*, that the sequence of these terms generated by successive theories stand in increasingly more accurate correspondence to these structures. For it is above all true that in quantum physics "picturability" breaks down (though not, I would agree, referentiality in a broader sense). We surely believe there is something "out there" and that we are in some sense gaining a more complete understanding of it as we gain more increasing predictive accuracy through successive theories. However our ability to think up an *ontology* for what's "out there" is now under serious and sustained attack. At the limits of our experience, the phenomenal world being discovered through increasingly complex instrumentation seems to be increasingly alien and uninterpretable in terms of our ordinary human experience.

The *philosophical* challenge of quantum physics makes it particularly hard to give quantum physics a fair *theological* appropriation. Most authors in theology and science work from a critical realist perspective, and hence the value of their theological work in some measure depends on the strength of this philosophical position. Critical realism may work quite well for classical science: thermodynamics, classical mechanics, biochemistry, evolution, and so on. Hence, as we shall see, the insight urged by Arthur Peacocke and others that classical chance does not undermine a theist argument is extremely helpful and to be celebrated. The issue here is whether this philosophical bridge, critical realism, will continue to bear the weight of traffic between quantum physics (and cosmology) and Christian theology.

Peacocke is not alone in urging a realist view both in science and as a bridge between science and religion. For example, Stanley Jaki is particularly critical of the Copenhagen interpretation of quantum physics. He argues against what he takes to be Bohr's denial of causality, his rejection of ontology and with it objective reality, through the undue restriction placed on epistemology by the principle of complementarity. In essence Jaki sees Bohr's move as undercutting metaphysics and hence the foundation for his own work in natural theology.[20]

Support for critical realism has also come from John Polkinghorne, who writes that the intelligibility of quantum physics warrants a realist interpretation of its implications.[21] Like Heisenberg, he too adopts a somewhat Aristotelian perspective in which the measurement process actualizes the potential properties of microscopic matter.[22] Similarly, Ian Barbour, Ernan McMullin, Janet Soskice, Bill Stoeger, and to some extent Sallie McFague, argue in this volume for a form of realism.[23] On the other hand, Chris Isham points out in passing that, unlike classical statistics, quantum physics poses a serious challenge to classical realism.[24] It should be noted that some forms of the "many-worlds" argument circumvent the challenge to realism posed by Bell's theorem.[25] Clearly then the challenge posed by quantum physics to realism continues to be a highly controversial and stimulating area for research.

3. *Quantum Nature in Theological Perspective*

I would now like to suggest several options for relating quantum physics to theology. They are meant to reflect the working presuppositions of contemporary theologians and represent possible directions for future research. I would first like to suggest ways in which quantum physics can play a heuristic role in metaphorical theology. Later I will look at the constructive role of quantum physics in systematic theology.

3.1 *Heuristic Role in Metaphorical Theology*

In religious language, metaphors are often more than illustrative figures of speech. According to Sallie McFague, metaphors like "God is love" or "the Lord is my shepherd" assume a central role in theology.[26] Metaphors "fund" theology, providing the language and images out of which theological concepts grow; they describe the unknown in terms of the known. They do so by asserting both a simile and a "dissimile" and the tension between them, between "is" and "is not," is essential to their power to communicate the ineffable.

As McFague warns, when metaphors lose their original meaning and fruitfulness, the theology built upon them must be reconstructed, drawing upon new metaphors appropriate for a new age. Moreover, not only do they serve a cognitive role, metaphors also illumine our spiritual life of prayer and devotion. It seems reasonable that physics, as well as biology and the other sciences which infuse our culture, can be a source of religious metaphors. I would like to propose two new types drawn from quantum physics.

3.1.1 *Indeterminacy: nature as surprise and as hidden nature*

The overwhelming impression one gets from quantum physics is of the irreducibly statistical character of experience. I am reminded here of the New Testament parables of the Kingdom of God, where divine providence is at work bringing about the redemption of the world even in the face of evil and injustice. The sower of seeds does not stop to direct each seed to its target; indeed, many fall on rocky soil or among the weeds. What is guaranteed is that some seeds will fall on good soil and there take root and grow to maturity. Quantum chance suggests that the structures of the Kingdom are constructed out of the random flow of ordinary processes, and that a hidden pattern seems to correlate, if not direct, all that happens.

Quantum statistics can also provide a new metaphor for surprise: nature is full of the unpredictable, and "expect the unexpected" is the norm. One prepares a sample containing trillions of "identical" atoms and simply waits. Suddenly atoms literally at random begin to decay. Each event is, as far as we can tell, without cause. Quantum chance is not just accident, the unforeseen (but, in principle, predictable) intersection of two causal streams. Quantum events behave as though they are uncaused; their surprise is of a different order than we experience in our daily lives.

Moreover, these surprise events radically change the history of the system involved. Atoms decay; they do not "reassemble" on their own. When nuclei fuse and emit light, they become an entirely different kind of nucleus. Particles annihilate and pair produce. Particles don't just change their properties, they are transformed: the old perishes, the new is born, and the event of transformation is a surprise. Quantum physics reveals that nature is full of surprise: deterministic causal explanation falls short of the reality being revealed, and the world is radically changed and transformed at each quantum event.

This leads to further metaphors of the Kingdom of God. According to the eminent Biblical scholar John Dominic Crossan, the parables of Jesus are structured around three terms or categories which reflect our fundamental experience of the Kingdom of God. These categories are: advent, reversal and action.[27] According to Crossan, the Kingdom comes like an advent, when we least expect it, opening up a world of possibilities which were previously unforeseen. We respond to this advent by reversing our entire past and acting in a radically new way. I suggest that the unpredictability of a quantum event is analogous to the surprise of advent and that the transformation of matter seems like the transformation of the person as we reverse our life's journey and act anew in the Spirit of God. Quantum chance seems to capture the non-cognitive aspect of advent as well, the feeling of joy, fear and astonishment we experience when the *totally* unexpected truly occurs.

Quantum physics also teaches us that the ordinary experiences of everyday living — seeing, tasting, touching, hearing, smelling — and the ordinary realm of classical science — measuring, weighing, locating, comparing, moving — all have a hidden dimension. Through the metaphor of nature as hidden, quantum physics illumines the existence of the mysterious within the mundane; nature discloses a mysterious quality and an extraordinary reality otherwise enclosed within everyday attire. In a similar vein, the parables of the Kingdom underscore the extraordinary within the ordinary, the mystery of the divine working everywhere in the world we know. Jesus often used parables of nature to depict the Kingdom, comparing it with a pearl of great price, a thief in the night, leaven in a loaf, a grain of mustard seed, bread, water, new wine. Now quantum physics depicts nature as filled with the extraordinary. Moreover this extraordinary quality produces the mundane world we know. Viewed from this perspective nature anticipates and provides the seeds of human history in which the Kingdom fully appears as a promise for hidden surprise and transformative power. And so even at the microscopic level, the metaphors of quantum chance convey something of the joy we experience at the discovery of the truly hidden, at the revelation of ultimacy and authentic existence we find when the Kingdom, invisible yet most real of all, is at work in our midst.

3.1.2 *Non-local correlations: nature as gossamer*

However we are to think about sub-atomic processes, one thing seems increasingly clear: when studying distant processes with a common past origin, the matter involved, from a quantum perspective, seems gossamer-like, somehow globally co-present.[28] Less than a mechanical connection (which, like vibrations of a rigid rod, transfer energy and information), more than a numerical coincidence or an optical illusion, quantum systems which were once united remain strongly correlated no matter how distant.

What metaphors can be found to introduce this complex non-local feature of quantum physics into the theological arena? Studying the role of chance and law in the creation of ordered structures, Arthur Peacocke has suggested the metaphor of God the creator as a composer "...who, beginning with an arrangement of notes in an apparently simple tune, elaborates and expands it into a fugue by a variety of devices."[29] Peacocke draws his arguments from thermodynamics and evolutionary biology, where chance has a strictly classical interpretation, as reflected in his metaphor of the composer.

Though I find this helpful for relating *classical* science to theology, its limitations might be clearer if we alter the metaphor a bit in light of *quantum* physics. Suppose we imagine that we are each a single voice in God's universal orchestration, and that the melody we each sing is a sequence of notes generated by quantum processes in our vicinity. If an audience were to listen to all these melodies, individually recorded and then played back simultaneously, what would they hear? The lesson of quantum physics would lead us to expect that the composite symphony might or might not display an internal pattern. If the individual melodies were generated by quantum processes of previously unrelated systems, their composite would be as featureless as the individual melodies. However, if the particular notes sung by each voice were produced by quantum processes which were themselves the decay products of previous single quantum systems, then, though the melodies sung by each voice would still be intrinsically random, and though the composition would show no structure in time (from one beat to the next), our listener would detect a pattern at each beat of the symphony. Hence if God the creator is like a composer, rather than producing the focused creation of a Bach masterpiece, God's symphony is indistinguishable from raw static both from the perspective of each voice and from one beat to the next even for the audience. Only with each beat would "harmony" appear and disappear.

Another prominent metaphor in the literature on theology and science is that of "body". Both Arthur Peacocke and Sallie McFague have suggested this metaphor in dealing with God's relation to the world. For example McFague's paper in this volume, as well as her recent book, *Models of God*, depend critically on two models: the world as God's body and God as mother, lover and friend of the world, and she draws on the biological sciences to support her metaphor.[30] Similarly Peacocke has developed a model of God's relation to the world in terms of mind and body and he explicitly uses feminine models of God in describing divine creation as *creatio continua*.[31]

The theological arguments which McFague and Peacocke are attempting to make can contribute a critically important balance to the male-dominated language prevalent in traditional Christianity. Nevertheless, their case is somewhat weakened by its dependence on the analogy of "body". I doubt whether this analogy is applicable to the universe either from the perspective of quantum physics, or from its originating biological context.[32] The universe may be more than "matter in motion" but from a quantum perspective it is less cohesive than a biological organism. Though helpful in their own domain, neither mechanical nor organic metaphors for nature drawn from classical physics, biology or evolution are adequate to fully conceptualize the philosophy of nature implied by quantum statistics.

Where then can we look for a new source of appropriate metaphors? Interestingly, a fruitful path to explore actually takes us into the rich variety of metaphors of coherence and co-presence found in our scriptural, spiritual, liturgical and philosophical literature! Here we find the metaphor of "body" again, but within a different underlying context more consonant with quantum physics. One example comes from the Pauline epistles where we are called to become members of one body, the Body of Christ. By this metaphor St. Paul seems to have meant a body not in the ordinary sense but rather a spiritual body understood through the context of the Resurrection. We are born into this new unity by the Divine Spirit and through this body we form a communion of saints stretching across time and space, back to the inception of the Church at Pentecost, forward towards the horizon of the eschaton. Jesus, the Divine Word, is with us always, not only as that which binds us together but more deeply as that in which we are bound together. The theme, "where Christ is, there is the Church," speaks of the universal presence of the Logos through which, as Paul writes in Colossians, all things are created and are bound together. Hence we find a remarkable resonance between the fragile, wholistic character of quantum correlations and the transcendent, invisible, and unitive character of the Body of Christ.

Another area to which the gossamer-like quality of quantum correlations might be relevant is inter-religious unity. As highlighted in the Papal message, this is a century of extraordinary movement towards the goal of inter-religious unity. Today many pray for Christian unity in liturgy, creed and mission, and we hunger for greater understanding between Christians, Jews and Muslims, and among all world religions. It is extraordinary that, according to quantum physics, even at its most elementary levels nature shows a novel kind of "unity amidst diversity." Although rudimentary and unconscious, nature at the sub-atomic level is like a fragile web extended through space and time. Of course the world from an ordinary perspective often looks like a disjointed array of "matter-in-motion," a world we often experience as a veil of sorrows. Yet if we look more closely, a hidden unity underlies the universe.

In this way quantum correlations offer stimulating metaphors for our unity in Christ and our search for wider ecumenical unity in the global religious perspective. Moreover the insights from quantum physics can be extended as well to the constructive theological agenda.

3.2 *Constructive Role in Systematic Theology*

I now turn to constructive questions in Christian systematics. For the purposes of this paper I will focus in some detail on two of the competing interpretations of quantum physics presented above in survey form which have received particular attention from the theological community. I will discuss some work presently underway within these particular interpretations and suggest additional questions within these approaches.

3.2.1 *Complementarity in physics and theology*

In 1927 in an historic address to the International Congress of Physics at Lake Como, Niels Bohr proposed the principle of complementarity, which was soon widely accepted among physicists and philosophers. Bohr began by characterizing classical physics as resting on the assumption that forces could be thought of as acting on material bodies in space. This in turn rested on the assumption that the physical state of a system, i.e., its position and momentum at a moment in time, can be observed without disturbing these variables. In quantum experiments, however, measurement entails irreducible interference with the system being studied. Bohr therefore concluded that the "claim of causality" and its "space-time coordination" must be reinterpreted as "...complementary but exclusive features of the description, symbolizing the idealization of observation and definition respectively."[33]

Quantum physics thus represents a radical shift in epistemology, if one follows Bohr's interpretation. Not surprisingly, his views have been incorporated into a diverse spectrum of philosophical positions. Subjectivists stress the role of choice in measurement and proclaim the demise of classical objectivity. Neo-Kantians build on Bohr's criticism of the classical epistemological assumption that one can simultaneously employ spatio-temporal and causal language. Positivists insist on abandoning all metaphysical concerns and instead stick close to the empirical content of physics. For our purposes, I will focus on yet another view, one that is frequently advocated: that quantum complementarity is based on the inevitable occurrence of contradictory models in physical theory. Quantum data display wavelike and particlelike features. Though these features are manifestly different and even contradictory as such, the principle of complementarity requires that both be included in a complete description of the data.

But does such "wave-particle duality" represent a limiting condition on classical epistemology, one which must hold for any future physical theory and which reflects the irreducible role of ordinary language in physical theory, as Bohr argued? Or does it actually arise from an ontological feature of complementarity in nature? On the whole, since Bohr believed that we cannot talk meaningfully about nature independent of measurement, we cannot attribute indeterminacy to the ontology of nature *per se*. Complementarity would thus be an epistemological limitation, and one which cannot be overcome by any conceivable future physical theory. Still I would argue that it carries at least one ontological implication, namely that we can no longer specify completely some of the key influences that determine reality, for if Bohr is right, they are forever beyond our reach.

How might this epistemological shift be relevant to theological epistemology? Clearly the warrant for its relevance would ultimately have to come from theology proper, with complementarity in quantum physics serving primarily as a heuristic device.[34] There are, however, several areas where theological doctrines *per se* do in fact seem to invoke complementary language. A striking example is Karl Barth's analysis of the perfections of God. Another is Dietrich Bonhoeffer' s insistence on the interwoven roles of belief and obedience in Christian discipleship. Patristic Christologies, especially the Chalcedonian formulation with its duality of human and divine in hypostatic union, are particularly suggestive of epistemological complementarity.[35]

Similarly, Hans Küng argues that the Resurrection must be understood as "intangible" and "unimaginable". To amplify his point, he turns specifically to physics: "Certainly we can attempt to convey this intangible and unimaginable life, not only graphically but also intellectually (as for instance physics attempts to convey by formulas the nature of light, which in the atomic field is both wave and corpuscle, and as such, intangible and unimaginable)." It is noteworthy that, in a style strongly reminiscent of Bohr, Küng attributes this situation to the "limitations of language," leaving us only one alternative: "to speak in paradoxes: to link together for this wholly different life concepts which in the present life are mutually exclusive."[36]

Clearly other examples of complementarity in religious doctrine include the relation of nature and grace, justification and sanctification, flesh and spirit, and so on. Barbour gives particular attention to Paul Tillich's use of personal and impersonal models of God as involving a possible form of theological complementarity.[37] According to Barbour, each of Tillich's polarities of existence — individualization and participation, dynamics and form, and freedom and destiny — reflect the complementarity of the personal/numinous and the impersonal/mystical elements of our experience of God. Similarly he suggests that Tillich's description of Christ as personal Word and as impersonal Logos may require a form of theological complementarity. Religious experience, too, provides striking examples of complementarity, such as the duality of numinous encounter and mystical union.

Some argue that theology and science are complementary fields of inquiry. It is, however, far from clear that these fields share a common referent and common rules for the construction and testing of their theories. Actually such arguments tend to use the term "complementarity" in the traditional sense of two alternative and supporting views; as such they have little in common with the meaning of complementarity in the context of quantum physics.[38]

In my opinion two key issues emerge out of this discussion. The first one is this: is theological complementarity an intrinsic limitation on any possible reformulation of church doctrine, as advocates of Bohr's interpretation of quantum complementarity would insist about their field? The answer to this question depends in large measure on the status of theological doctrine and the warrant for its reformulation. It might be possible to argue that the theological concepts themselves, or the kinds of religious experience they represent, enforce a kind of complementarity on theology. Alternatively for those sectors of the Church which take confessions, creeds and councils as normative formulations binding on faith, one could argue that complementarity in such documents is irreducible. Yet for many Christians who stress the relativity of doctrine to culture, language and history, the authoritative role of these documents is more restricted, their hypothetical aspect more accentuated. From this point of view, the occurrence of theological complementarity in historical documents or texts would be less binding on new explications of Christian faith.

Secondly, does the occurrence of theological complementarity carry a religious ontological implication? If we followed Bohr, the answer would probably be mixed: negative, in the sense of a direct ontological statement about "objective reality," yet open-ended in so far as the real challenge of complementarity in physics is directed against classical realist metaphysics, not against inference in general if understood in terms of a new ontology. Theologians, too, might give a similar argument. Clearly the God who is disclosed in religious experience and Scripture is always the hidden God, the wholly other, the ultimately unknowable. Yet what is disclosed is about this same God and no other! This problem comes to the forefront, for example, in the relation of the economic and immanent Trinity. To the extent that these are identified, the complementarity of Trinitarian language reveals something essential about the Godhead. To the extent that they are distinguished, it remains beyond our capacity to describe the divine mystery. Clearly then if the ontological implications of complementarity are related to the metaphysical elements of classical philosophical theology, then as these elements are recast a new vision of God and creation would emerge.

3.2.2 *Indeterminacy in physics and theology*

Werner Heisenberg defended a different approach to quantum statistics in some of his earlier writings. He argued that the statistical character of quantum theory implies that nature is inherently indeterministic, that chance is an actual feature of atomic processes. Heisenberg spoke in terms of Aristotelian categories of potentiality when describing his views on quantum indeterminacy. Suppose we too go further than the Copenhagen interpretation described above would seem to allow and suggest that nature is inherently statistical, that chance and law as dialectic elements in our theory reflect an irreducible element of chance as well as regularity in nature herself. How would this effect Christian theology? And what further effects would we find if we then introduced the important distinctions between classical and quantum chance?

Very broadly, we can anticipate several theological areas in which quantum chance has a bearing. Regarding the doctrine of creation, quantum physics is relevant to both *creatio ex nihilo* and *creatio continua*. For example, we may view God as creating *ex nihilo* through both law and chance, for as transcendent,[39] God is the author of the laws of nature, including those of quantum physics, as well as the statistical processes which they describe. Moreover, the transcendence of God is one in which God is present to all of creation as the power of immanent, redemptive love, as the tradition of continuous creation emphasizes.

Quantum physics underscores the irreducibly random character of these ongoing processes, and yet the hidden patterns and surprises within them.[40] Since, quantum chance is involved in the production of order and life, this suggests that even the random character of elementary processes contributes something essential to the greater panorama out of which emerges the conditions for genuine alternatives, and eventually the reality of free will and authentic relationship characterized by love. Hence quantum chance bears on the problem of evil and theodicy, as well as on theological anthropology. Free will may have its initial previsioning in the indeterminacy of quantum processes, although it can only emerge fully in the complexities of human biochemistry and neurophysiology. Moreover, as William Pollard has argued extensively, we understand something about the divine activity in the world if we view God as influencing our choices without violating the lawful processes which govern them.[41] Moreover, from the quantum perspective, providence is not so much a time-independent or fixed *telos*, as a constant persuasive re-directing of our choices through the creative divine immanence in all processes and events.

Quantum physics can illuminate the contents of Christology, particularly when we start with the Easter event as the normative disclosure of God's work in Jesus and on this basis develop the doctrine of Incarnation and the efficacy of Jesus' earthly ministry. To avoid a docetic Christology we must wrestle with the irreducible role of chance in the life of Jesus as fully human and hence as open to all the uncertainties of our own human lives. The theological task will be to understand the significance quantum physics gives to the role of chance in Christology, in terms of non-locality, coherence and the analysis of chaos and structure. In a similar way, the meaning of eschatology will be influenced by the wholistic character of nature at the quantum level, as well as by the distinction between Bose and Fermi statistics. However, an extended discussion would take us into the area of quantum cosmology, a tantalizing topic but one lying beyond the limits of this paper.

Hence I will focus here primarily on the first topic, law and chance in the doctrine of creation, leaving for another time these other areas. My goal will be to suggest that, far from thwarting the meaning of this doctrine, the irreducible role of chance, as well as law, in nature augments the meaning and subtlety of creation theology. Yet as we distinguish more carefully between classical and quantum chance, whole new areas open up which have not yet been explored.

Many contributions have been made to the discussion of the relevance of science to the doctrine of creation. It is here that the writings of Arthur Peacocke have, in my opinion, been particularly fruitful.[42] Rejecting both reductionist materialism and dualistic vitalism, Peacocke views nature in terms of emergent continuity among the levels of nature understood in terms of fields as diverse as physics, chemistry, biology, psychology, and theology. God is involved with the evolution of the universe, creating new and emergent levels of organization through the open, statistical processes of this world, including quantum indeterminacy, irreversible thermodynamics and biological evolution. Against Jacques Monod and others who see chance in nature as either antithetical to divine purpose or as a reason for denying divine existence, Peacocke takes both chance and law as instruments of God's creative will.

Recognizing that this view of God could seem deistic (although the emphasis is on dice-playing rather than clock-making), Peacocke stresses the immanence of God in the processes of the world as well as the utter difference between God and the world through the divine transcendence. Though not a process theologian, this does make his theology a form of *pan*en*theism* to use Hartshorne's term, since it combines immanence and transcendence in a single unifying perspective. He draws on models from physics, biology, Scripture, feminist theology, and numerous other sources to describe God as immanent in the world, "'exploring' and 'composing' through a continuous, open-ended process of emergence" (or *creatio continua*), ringing the changes and creating novelty in the as-yet undetermined future. At the same time God is also regarded as transcendent creator *ex nihilo* of the world.[43]

Peacocke's claim that the presence of disorder in nature need not be a compelling reason against Christian theism seems sound. The new scientific picture, with its intimate relation of chance and law in thermodynamics and biology, can provide us with a theological model for the way God both participates in, and yet transcends, the whole of creation.

Yet as we have seen, quantum statistics is of a radically different sort from that of classical mechanics, thermodynamics and biology. Fermi statistics involve the impenetrability of matter and the wholistic character of complex systems obeying the Pauli exclusion principle, while Bose statistics describe the coherence and superposition of the particles of nature's interactions. Both types of quantum statistics arise from the indistinguishability of elementary particles, a characteristic totally foreign to the world at a macroscopic level. Quantum chance cannot be explained away as ignorance of local causal streams, or mere accident in the classical sense. Most importantly, nature at the quantum level displays a highly non-local and wholistic character.

Hence the difference between quantum and classical chance of the sort employed in mechanics, thermodynamics and biology can no longer be overlooked either in terms of its philosophical implications or its theological significance. It is undoubtedly helpful to view God as working through both law and chance, as Peacocke and others urge. Given the predominantly statistical character of nature as revealed by the natural sciences, this notion of God's manner of acting is certainly more adequate than the traditional one of God as somehow working against, rather than through, chaos to produce order, the order being thought of as a pre-conceived goal or *telos*.

However, the question to be asked now is how to take into account the difference between classical and quantum chance. In essence, because of quantum physics it is the kind of chance God the Creator works through which poses both a radical challenge and unique opportunity to theologians. I would argue that the shift from "law versus chance" to "law and chance" which has already been recommended by others is rather modest compared with the shift which we must now undertake from "law and classical chance" to "law and non-local, Bose/Fermi quantum chance."

In this light I would make the following suggestions for our theological consideration of quantum chance. First of all, we know that Bose and Fermi statistics, respectively, give rise to many of nature's cohesive features on the one hand, and its extended, structural, ordered features on the other. Hence when we affirm that God creates order out of chaos, we can begin to see that chance in the quantum domain plays a double role in the characterization of that order. We may take this one step further. From the point of view of Fermi statistics, chance is, in a sense, embodied in order; we might almost say that chance gives order its structure. The entities of our world — the nuclei, atoms, molecules, rocks, organisms, buildings, planets and stars — depend on the Pauli exclusion principle for their being a structure, for their very existence as structure. Moreover, from the point of view of Bose statistics, chance adds to the forces of interaction in nature a cohesive quality that allows many different forms of interaction to superpose, occupying the same state at the same time. This type of chance adds a unitive factor to the structures that make up our world.

Now according to Ilya Prigogine, Manfred Eigen and colleagues, the evolutionary processes which generate increasing levels of organization are not only consistent with thermodynamics and the law of increasing entropy, they occur precisely because of the entropy-generating processes in dissipative structures characterized by non-linear, non-equilibrium thermodynamics.[44] But we can augment this view by bringing onto center stage the creative role played by quantum chance in the formation of ordered systems. Here our view of chance in the quantum context helps us understand the basis for the geometry of organization. The Pauli exclusion principle, in particular, underlies the extendedness and impenetrability of matter. What is astonishing is that the geometries arising through quantum processes are directly related to the statistical distribution of these processes, and hence the domain of chance, of chaos, of randomness, contains within it the seeds of structure and order. From a theological perspective we can add to the view that God creates the universe through chance and law the claim that the order God is creating is in some sense the order of quantum chaos. Rather than saying that God creates order in place of (i.e., out of) chaos, from a quantum perspective we could say that one way God creates order is through the properties of chaos.

My second suggestion involves the traditional conception of the immanence of God in the world and the model of non-locality offered by quantum physics. Here the divine immanence was taken to imply that events which are physically separate are somehow co-present or coherent *to God.* In classical theism the contingent unity and intelligibility of the world was understood in terms of the divine reality which is its ground and source. Quantum physics supports this notion by suggesting that this contingent unity is a complex combination of local randomness and global correlation. And so the God who is immanent "in, through and under" even the basic physical processes of nature might now, through the lessons of quantum physics, be more fully understood as offering the particular kind of ground of being in which differences need not be contradictions, distinctions need not be isolations, separate identities need not produce alienation, and in which, even at the elementary physical level, distant and simultaneous events need not be ultimately unrelated. One might even suggest (though I would not want to press this too far) that such an ultimate coinherence of all events within the divine reality may leave a trace of that same coinherence in nature, not only in the human community where it reaches its fullest expression, but even in the complex of events at the quantum level.

A final theological suggestion might be to take a methodological lesson from the development of quantum physics in the first decades of our century. Moving from classical to quantum modes of thought proved extremely arduous; a really satisfying approach only came about when physicists produced a self-consistent quantum theory and then showed how classical physics could emerge under the proper limits. Although it may be unusual in the history of science, in this case at least starting afresh with a new conception of the whole and then breaking it down to study its parts was more fruitful than starting from the parts and trying to construct a new science out of them. Curiously this seems a bit like explaining the known in terms of the unknown, instead of the usual converse approach!

There may be a lesson in this for theology. The relation of God to creation might be understood more fruitfully by endeavoring to grasp creation first in its essential reality and from this working out a theological interpretation of God's relation to the world in what Tillich would call its existential ambiguity, rather than starting with the world as we know it and then asking how it relates to it as its Creator. Both in the history of its construction and in the insights it offers about the underlying unity of nature even in the face of chance, quantum physics suggests the value of such an approach. It would also lead naturally into the problem of theodicy and the meaning of grace and redemption.[45]

4. *Summary and Closing Comments*

This paper has undertaken an assessment of the relevance of quantum physics for contemporary Christian theology in an ecumenical context. Philosophy can offer a fruitful bridge between science and theology. Hence the approach here was to suggest key elements of a new philosophy of nature informed by quantum physics. As developed from 1900 to the late

1920's, quantum physics is now an irreducible part of fundamental physics. The task of this paper was to investigate ways in which the philosophical implications of quantum physics might be relevant to the work of contemporary theology, both as a heuristic source of theological metaphor and as a systematic factor in constructive theology.

Quantum physics is the continuing subject of conflicting interpretations. Nevertheless, it already presents us certain key insights into the character of nature both at the sub-atomic level and, by implication, at the macroscopic level as well. As in evolutionary biology, thermodynamics, and statistical mechanics, quantum physics depicts nature as predominantly statistical. However, wherever quantum theory applies, this statistical character is radically different from classical statistics in several ways: i) The behavior of elementary particles (and the composite structures they produce) leads to two distinct types of quantum statistics. One type accounts for the impenetrability of matter and for many of its chemical properties; the other describes the interpenetrability of the fields of interaction and their cohesive character; ii) Quantum statistics are not the result of our ignorance of otherwise random juxtapositions of causal trajectories (i.e., of accidents as understood by classical science); iii) All members of a given type of elementary particle are intrinsically identical and hence indistinguishable, a property never fully realized in macroscopic experience which leads directly to the distinction between Fermi and Bose statistics; iv) Moreover this modular, interchangeable character of elementary particles gives rise to its antithesis, the limitless variation of macroscopic matter; v) Finally, nature reveals a highly non-local and wholistic character at the quantum level, which is strikingly different from the separability of nature in our ordinary experience.

Quantum physics poses a special challenge to realism. Taken very broadly, one can still work within a realist philosophy as an account of the progress of science. What seems to be coming under increasing pressure are the further, more specific claims of epistemic correspondence and convergence. Given Bell's theorem and the emphasis it has brought on the non-local character of nature at the quantum level, it is becoming increasingly hard to construct a realist ontology in light of quantum physics.

The philosophical challenge of quantum physics makes its theological appropriation particularly complex. The task is further complicated by the fact that most researchers in theology and science work out of a critical realist perspective, and hence the value of their theological work in some measure depends on the strength of this philosophical position. Critical realism may work quite well for the interpretation of thermodynamics, classical mechanics, biochemistry, evolution, and so on. The issue here is whether critical realism will bear the weight of relating quantum physics and Christian theology.

Turning to the theological agenda, quantum physics provides metaphors which illuminate central religious concepts. As a new source of theological metaphors, quantum chance suggests that the structures of the Kingdom are constructed out of the random flow of ordinary processes, and that a hidden pattern seems to correlate, if not direct, all that happens.

Quantum events behave as though they are uncaused; their surprise is of a different order than we experience in our daily lives. The world is radically changed and transformed at each quantum event. Nature anticipates and provides the seeds of human history in which the Kingdom fully appears as a promise of hidden surprise and transformative power.

Quantum correlations provide us with rich metaphors for the mysterious and transcendent unity for believers in Christ, and even for our search for wider ecumenical unity in the global religious perspective. For quantum correlations point to an underlying unity in nature, a nature which essentially includes human nature and thus the entire theological agenda. This unity far transcends the traditional dialectic of "many and one"; if it leads to a new ontology, the reverberations will be felt throughout the whole realm of constructive theology.

Theological complementarity, as an *epistemological* parallel to complementarity in physics, may illuminate many of the apparently contradictory issues in theology. Whether it is a limitation on all future theological doctrine, as it would be by analogy in physics according to the Copenhagen school, and whether it points to an underlying ontological duality in nature, in contrast to the Copenhagen opinion, is a key research topic today.

Alternatively, if quantum physics carries an *ontological* lesson, the difference between quantum and classical chance ought to no longer be overlooked. As we affirm that God as Creator can act through the means of chance, the real issues become the kinds of chance through which God works. This change in our concept of chance, from ignorance of causal trajectories to non-local Bose-Fermi chance, poses a radical challenge to theology.

One direction being explored is to see chance as contributing to the structure and cohesiveness of nature. From quantum physics we now know that structure occurs, not in spite of but rather through chaos; indeed, chaos plays a constitutive role in the structure of matter. Hence from a theological perspective we can understand God not only creating the universe through the mixture of chance and law but creating order as embodied chaos. Rather than saying that God creates order in place of chaos, from a quantum perspective we could say that God creates order by displaying the properties of chaos. In addition we can now think of God the Creator as immanent in the world such that events which are physically separate are somehow co-present or coherent to God.

This leaves us with several intriguing questions. If, for example, any one interpretation of quantum physics proves clearly advantageous to the theological task, would this suggest that it too might prove more fruitful for the development of physics? Here it would seem again that the channel of influence is philosophy, both explicitly as a field of inquiry which can act as a bridge between theology and science, and implicitly in so far as philosophical assumptions and concepts infuse theology and science *per se*. In this task the analysis given in this volume by Bill Stoeger of the philosophical dimensions of science is particularly helpful.

If there are reasons to rethink the metaphysical presuppositions of either quantum physics or Christian doctrine, would it be advantageous to attempt to formulate a single new metaphysics which would be adequate and applicable to both fields jointly? This may be an incredibly complex task, but one whose realization could prove extremely valuable.

Finally, given the complexities of relating quantum physics and contemporary theology, it may be that what is needed here is something much more radical than a redesigning of the traditional fabric out of which our theological garments have been sewn. There are signs that much of contemporary theology is moving into a new period of challenge and growth following the extensive debates between such theological schools as neo-Orthodoxy, existentialism, linguistic analysis, and neo-Thomism which were dominant during most of this century. Although we are only now catching a glimpse of the road ahead, the natural course for theology seems to be one of increasing interaction with the sciences and technologies which shape so much of contemporary culture. A theology for *our* time will be increasingly articulated in the context of these sciences, for they disclose many of the mysteries of the universe which have made and are making us. We also have before us, for all who take the Biblical faith seriously, an ecumenical and even inter-religious task, for affirming that "God so loved the world" means accepting the challenge that this world, in its cultural pluralism *and* its empirical complexity, is in fact the world "God so loved/s."

ACKNOWLEDGEMENTS

I want to give special thanks to Ernan McMullin and Bill Stoeger who read a preliminary version of this manuscript and offered many very helpful suggestions.

NOTES

[1] Alternative formulations of quantum physics were developed early in this century by Werner Heisenberg (in terms of matrices), Erwin Schroedinger (via the wave function formalism) and Paul Dirac (using Hilbert space), but these were shown to be formally equivalent by Dirac. More recently, Richard Feynman has given another, equivalent formulation (the path-integral approach).

[2] This paper will focus primarily on ordinary or non-relativistic quantum physics. The philosophical and theological implications of relativistic quantum physics, quantum field theory, quantum gravity, and other topics will be undertaken in future research.

Two initial challenges to the agenda of this paper should be faced at the outset. One could well ask why theologians should bother about the issues raised by quantum physics, since, presumably, they only apply to the microscopic realm. For example in his essay in this volume, John Polkinghorne writes: "...quantum theory in general only manifests its idiosyncratic character in processes of a smaller scale than normally concerns us in the great part of what is going on."

I find this view curious. Certainly the minute value of Planck's constant limits the effects of many quantum phenomena. Yet, though the domain of quantum physics may be the realm of the atom, the repercussions of quantum physics are felt throughout the levels of nature from the atom to the cosmos. For example, quantum physics underlies human vision and the sensation of taste, the expansion of water as it freezes, the color of the sky, the light from our Sun and the glow of embers in a cooling fire. It underlies all of astrophysics, chemistry, and molecular biology. Without quantum physics we would not have such technologies as electrical lighting and power, computers, nuclear fission, solar power or communication satellites. On a cosmic scale, without quantum physics we would not have life in this universe, or possibly even the universe itself (if one believes in the Anthropic Principle at all)!

It is particularly interesting that visual perception, without which classical science would not function and on which classical epistemology depends to large measure, actually arises through a quantum process: individual photons are produced by a quantum transition in atoms in the Sun or an ordinary light bulb, and a quantum process is involved in the firing of a receptor in the retina, leading to the experience of vision. One could even say that vision itself, so integral to the classical philosophy of nature, actually raises the whole measurement problem to be discussed below, leading us to a quantum philosophy of nature!

Hence, though I would *not* want to defend a reductionist view of quantum physics, since I firmly believe that no one field, including physics, can provide a sufficient explanation of the world or a privileged access to the way the world "really" is, I would defend a form of *inter-disciplinary epistemology* (for which I prefer a network over the ever-popular hierarchical model). In this light I must then consider what quantum physics offers when attempting to construct a world-view which is consistent with even ordinary empirical experience. To put it glibly, atoms may be small but they're everywhere — they can't be ignored! More seriously, it is hard to think of examples of phenomena studied by physical or biological scientists today where quantum physics does not play at least an indirect role.

A more serious objection frequently raised is that, until disagreements abate over the philosophical implications of quantum physics, we should abstain from giving serious consideration to its potential theological implications. From my point of view, however, the presence of these competing philosophical interpretations does not mean we cannot draw some general, highly probable conclusions about quantum theory.

As I shall attempt to argue, we already do know that nature displays an irreducibly statistical character in the data arising from the atomic level; that quantum statistics are radically different from classical statistics, leading away from classical notions of chance as the juxtaposition of unrelated causal streams, and underlying the kind of order and structures of nature at the macroscopic level; that there is a deep relationship between micro-nature as modular and macro-nature as heterogeneous; that correlations between distant and simultaneous data from previously coupled systems challenge the explanatory power of classical realism at either the epistemic or ontological levels; and that, unless a highly exotic explanation be found true (such as many-worlds or the role of consciousness), quantum systems are somehow inherently non-local (see below). We thus have reason to believe that either classical epistemology or classical metaphysics is inapplicable to quantum problems. And we know that these quantum problems are directly or indirectly a part of all supposedly classical phenomena, and hence of ordinary experience.

In active areas of scientific research, there are always numerous competing theories as well as competing interpretations of theories. If our strategy is to wait for agreement, I fear we will be limited to historical studies. Moreover, agreement is seldom univocal: when is it really reached? what about reversals after a theory was considered settled?

Finally I believe that, to varying extents, both science and religion have been involved in the formation of major theories, and in the process of theory choice, on each side. Science arose in a Western culture heavily influenced by the Biblical concept of God, expressed in particular through the doctrine of creation (see the paper by McMullin in this volume). Even in contemporary science, criteria of theory choice include aesthetic and religious values, while the passion for and dedication to science can be seen as a genuinely religious experience. The scientific imagination often bears a strong resemblence to religious insight. Theology too has been shaped by the philosophy and science of its time, as even a cursory reading of Protestantism since the Newtonian virtuosi shows only too well. Therefore, to assume that as theologians we can chose not to engage with scientists until their issues are settled is, in my opinion, inconsistent: to a surprising extent we already have so engaged!

Rather than see theology and science as two entirely separate fields with clear and unambiguous lines of demarcation, fields which then might interact in one way or another, I would prefer to view theology and science as the designations for two fields which, to some limited but irreducible degree, already include something of the discoveries, histories, visions, and commitments of one another, both intentionally and inadvertently.

[3] A more complete theoretical explanation of quantum statistics would involve a discussion of *relativistic* quantum physics and quantum field theory, and in turn would include other such phenomena as matter/anti-matter, CPT invariance, vacuum fluctuations, and so on. Though the theoretical questions involved here lie outside the scope of this short paper, I have included an elementary discussion of the Bose/Fermi distinction because it is central to an appreciation of quantum statistics, and because, as far as I know, it has not previously been introduced in the theological discussions of quantum physics.

[4] Bosons have zero or integral spin, whereas fermions have odd half-integral spin. The spin-statistics relationship, although unexplained by non-relativistic quantum physics, is a fundamental theorem of relativistic quantum physics.

[5] One should note, however, that there is now growing evidence of phenomena in the domain of classical physics which display collective properties. Still,

simple classical systems have routinely been taken as paradigmatic of reductionist analysis and epistemology, leading to ontological reductionism.

[6] This line of argument was given additional support by Bell's theorem, although a modified form of hidden variables theories is still possible. See below.

[7] John S. Bell, "On the Einstein-Podolsky-Rosen Paradox," *Physics I* (1964) 196. See also John S. Bell, "On the Problem of Hidden Variables in Quantum Mechanics," *Reviews of Modern Physics* 38 (1966) 447.

[8] These issues were the subject of a conference in October, 1987, at the University of Notre Dame, the proceedings of which will soon be published (edited by Ernan McMullin).

[9] John Polkinghorne, *One World* (London: SPCK, 1986) 47.

[10] For further reading at an introductory level, see Heinz Pagels, *The Quantum Code* (Toronto: Bantam Books, 1983) and Nick Herbert, *Quantum Reality* (New York: Anchor Press, 1985). The organization of my presentation was suggested in part by Herbert's approach. A particularly helpful discussion of the relation between quantum physics and its broader philosophical and theological issues can be found in Chapter 10 of the outstanding textbook by Ian G. Barbour, *Issues in Science and Religion* (New York: Harper and Row, 1966). At a more technical philosophical and mathematical level see Max Jammer, *The Philosophy of Quantum Mechanics* (New York: John Wiley and Sons, 1974). John Polkinghorne gives an especially inviting introduction to more advanced topics, such as quantum field theory, in his *The Particle Play* (San Francisco: W. H. Freeman and Company, 1979).

[11] David Bohm, "A Suggested Interpretation of the Quantum Theory in Terms of 'Hidden' Variables. I.," *Physical Review* **85** (January 15, 1952) 166-79, and II., *Physical Review* **85** (January 15, 1952) 180-193.

[12] David Bohm, *Wholeness and the Implicate Order* (London: Routledge and Kegan Paul, 1980). In my opinion, Bohm's views are often underestimated, both from a scientific and a philosophical perspective. Polkinghorne, for example, is incorrect when he claims that Bohm's "approach" has been unable to incorporate special relativity since Bohm's theory is, in fact, a non-local realist theory in compliance with relativity! (See Polkinghorne's paper this volume.) Whether the approach of Bohm is "contrived" is, in my view, a judgment call for which Polkinghorne gives little warrant. For a detailed discussion of possible criteria of assessment, see Robert John Russell, "The Physics of David Bohm and its Relevance to Philosophy and Theology" in *Zygon: Journal of Religion and Science* **20** (June, 1985) 135-158.

[13] A possible exception to this is the very recent work by Geoffrey Chew and Henry Stapp in topological bootstrap theory. Stapp in particular is highly influenced by the philosophy of Alfred North Whitehead.

[14] Henri Margenau, *The Nature of Physical Reality* (New York: McGraw-Hill Book Company) 442.

[15] Ian Barbour, *Issues*, 297. This quotation is from a very useful section on wholes and parts in Chapter 10 of Barbour's book. Not only does this chapter still provide the best introduction to the significance of quantum physics; though written over two decades ago, this book as a whole remains for me one of the most important general works in the field.

[16] Again, though we knew this from the outset of quantum physics, Bell's theorem has forced the issue in a truly startling way.

[17] Herbert, *op. cit.*, 245.

[18] For a recent review of the experimental tests of Bell's theorem, see John F. Clauser and Abner Shimony, "Bell's theorem: Experimental Tests and Implications," *Reports on Progress in Physics* **41** (1978) 1881.

[19] Henry J. Folse, *The Philosophy of Niels Bohr: The Framework of Complementarity* (Amsterdam: North Holland Personal Library, 1985).

[20] Stanley L. Jaki, *The Road of Science and the Ways to God* (Chicago: The University of Chicago Press). See especially Ch. 13. See also his recent book, *Chance or Reality and Other Essays* (Lanham: University Press of America, 1986). Clearly Folse's defense of Bohr (cf. footnote 19), if sustained, would seriously disarm Jaki's often vitriolic attack against Bohr.

[21] See Polkinghorne, this volume, and his recent book, *One World*, 22, 44.

[22] See Polkinghorne, this volume: "...quantum uncertainty arises (from) the intrinsically indeterminate character of the quantum world." One also finds this in *One World*, where Polkinghorne supports Peacocke's argument, that, contrary to Jacques Monod, one can view God as working through the balance of chance and necessity. Here chance serves "in the realization of potentiality" (*One World*, 68-69; also 45, 47). It should be noted, however, that in the discussion which follows he rather brusquely disavowes panentheism (p. 73), the position Peacocke defends articulately and extensively throughout his writings.

[23] See the papers by Barbour, Soskice, Stoeger and McFague in this volume; see Ernan, McMullin, "A Case for Scientific Realism," in *Scientific Realism*, ed. by Jarrett Leplin (Berkeley: University of California Press, 1984).

[24] See the paper by Isham in this volume.

[25] I am indebted to Frank Tipler for pointing this out in private discussions at Castel Gandolfo. Essentially put, if the universe is understood to bifurcate in the decay process leading to measurements at distant labs A and B, *both* kinds of correlations (A up, B down and A down, B up) result (though only one in each universe), reflecting the superposition of both kinds of correlations in the original excited state. Hence rather than focus on the end results as "non-local" we can redirect attention to the original excited state which causally produces both results. I am still not convinced that this approach can solve the assignment problem of consistently attributing quantum properties like spin up to the initial state in a realist fashion.

[26] Sallie McFague, *Metaphorical Theology: Models of God in Religious Language* (Philadelphia: Fortress Press, 1982) and *Models of God: Theology for an Ecological, Nuclear Age* (Philadelphia: Fortress Press, 1987); see also this volume, 249-271.

[27] John Dominic Crossan, *In Parables: The Challenge of the Historical Jesus* (New York: Harper and Row, 1973) 33 ff.

[28] In referring to this feature, Polkinghorne writes that "...even at the level of fundamental constituents, the world does not fall apart but it exhibits a degree of mutual cohesion (See his paper this volume, 338).

[29] Arthur Peacocke, *Intimations of Reality* (Notre Dame: University of Notre Dame Press, 1984) 72.

[30] See McFague, this volume, 249-271; also *Models of God, op. cit.*

[31] Arthur Peacocke, *Creation and the World of Science* (Oxford: Clarendon Press, 1979); *Intimations, op. cit.*, 64.

[32] Though a detailed response would take us too far afield in this paper, it should be noted that several authors have challenged the "body" analogy, and their objections have not been satisfactorily met by its proponents. For a useful analysis of the arguments as well as his own critique, see Ian Barbour's *Myths, Models and Paradigms* (New York: Harper and Row, 1974) 160-161. See also Barbour's article in *Religion and Intellectual Life*, March 1988, 59-63.

I would add to this that, even from the perspective of quantum physics, the universe does not seem to have the cohesiveness of a body. What non-local characteristics it may display via quantum correlations are more like synchronizations than information-bearing transmissions.

However, if we were to extend the topic to (relativistic) quantum field theory (QFT), there might be more grounds for arguing for the "bodily" character of the universe. QFT combines particle and wave images in a unitive model in which the quantum field is coterminus with the universe and its excitations are what we think of as the universe's discrete material structures. (Of course here I am intentionally using realist language about what is, quite arguably, a non-referring theoretical term, the quantum field.)

If we include quantum gravity, the universe itself as a spacetime continuum can be seen as a quantum field and the "Big Bang" as a quantum fluctuation, with interesting implications for *creatio ex nihilo*. Yet there are other ways to interpret the creation of the universe in terms of quantum gravity which do *not* involve an initial singularity. For a very interesting analysis see the paper by Chris Isham in this volume.

[33] Niels Bohr, "The Quantum Postulate and the Recent Development of Atomic Theory," *Atti del Congresso Internazionale dei Fisici, Como, 11-20 Settembre 1927* (Bologna: Zanichelli, 1928) **2**, 565-588.

[34] Ian G., Barbour, *Issues*, 1966, 292-294; Ian G., Barbour, *Myths*, 1974, 77-78.

[35] See Christopher B. Kaiser, "Christology and Complementarity," *Rel. Stud.* **22**, 37-48. Kaiser (private conversations) does not assume a realist interpretation of his argument in this paper. It should be noted that Barbour agrees in large measure, as do I but with more reservations, with William Austin's criticism of this type of argument (see Ian Barbour, *Myths, Models and Paradigms*, 152 ff.)

[36] Hans Küng, *On Being a Christian*, translated by Edward Quinn (Garden City: Image Books, Doubleday and Company, 1984) 350.

[37] Ian G. Barbour, *Myths, Models and Paradigms*, Ch. 5.

[38] For an example of this sort of claim about complementarity, see A. R. Peacocke, *Science and the Christian Experiment* (London: Oxford University Press, 1971) 128. Barbour gives an important critique of such usage in *Myths, Models and Paradigms*, 77-78. I am grateful to Ernan McMullin for pointing out the traditional sense in which "complementarity" is being used here.

[39] Here as elsewhere I want to underscore the problematic character of any language about God, and the metaphorical character of theological language. The traditional distinction between transcendence and immanence, though itself a complex and controversial distinction, can provide a framework for making explicit the particular tension in our language about God. We can take transcendence to stand for the irreducible insight that God is wholly-other, mysterious, hidden, and unknowable, so that our constructive language about God is always qualified by a denial of its adequacy. By immanence, then, we mean that which we can say about God's will and activity through analogies and metaphors drawn from experience, though always delimited by the governing themes of divine mystery and the finitude of human epistemic categories.

[40] A fuller treatment of the bearing of quantum physics on the doctrine of creation would lead to the role of quantum physics in cosmology, a topic to be pursued in a further study. For example, in some grand unified theories, the "origin" of the universe is viewed as a fluctuation in a quantum field of which the vacuum is the quiescent state. How might this notion of the vacuum relate to the concept of non-being in the classical formulation of the doctrine of creation? Should we say that the quantum field is more like *me on* than *ouk on* (to borrow a distinction Paul Tillich emphasized) and hence that it too is subject to creation *ex nihilo*? The papers in this volume by Ted Peters and Chris Isham develop similar issues. Isham's paper, for example, makes some very helpful proposals about these questions in terms of the Hartle-Hawking theory. See also Wim B. Drees, *CTNS*

Bulletin **8.1** (January 1988) 1-15. Similarly a proper discussion of quantum physics and eschatology (see below) would involve physical cosmology.

[41] William G. Pollard, *Chance and Providence* (New York: Charles Scribner's Sons, 1958). For a challenging response, see Ian Barbour, *Issues*, 428-430.

[42] A. R. Peacocke, *Creation*, and, *God and the New Biology* (San Francisco: Harper and Row, 1986).

[43] See in particular A. R. Peacocke, *Creation*, 209.

[44] For a very useful introduction to their work, see A. R. Peacocke, *An Introduction to the Physical Chemistry of Biological Organization* (Oxford: Clarendon Press, 1983).

[45] I am not in the least suggesting that quantum physics is somehow inherently more related to essential reality and classical physics to distorted existence. My point is that a methodological lesson for theologians might be learned from the history of quantum physics.

CREATION OF THE UNIVERSE AS A QUANTUM PROCESS

C. J. ISHAM, The Blackett Laboratory, Imperial College, London

1. *Introduction*

The doctrine of *creatio ex nihilo* enshrines the Christian belief in the creation of the universe by God, and in God's continuous and sustaining relationship with all things. As such, it is one of the central tenets of the Faith. But, during the last ten years, theoretical physicists have been developing their own ideas on the creation of the universe as a quantum fluctuation "from nothing." Do these quite different attempts to come to grips with the fundamental "question of Being" have anything to say to each other? If so, what are the implications for modern theologians in their efforts to formulate the Christian message within the conceptual and sociological framework of life in the 20th century?

This recent attempt of science to add to the tally of creation myths is rooted in the steady accumulation of astrophysical evidence suggesting that the universe has been expanding for the last 10-20,000 million years, starting from an initial "point" of extreme compactness and density.[1] The theoretical underpinning for such a picture lies in Einstein's general theory of relativity and the existence of solutions to his field equations that exhibit precisely such a singular behaviour.

Of course, the idea that there *was* such a Big Bang has been around for some time, but it is important to appreciate that work in this area has, for the most part, aspired only to demonstrate the broad consistency of this notion with the large-scale features of the observable physical universe. There exist many different solutions to Einstein's equations, corresponding to a vast number of different possible histories for the universe, and the theory cannot of itself select one rather than another. The actual solution must simply be taken as a contingent, "given" fact of the physical world; in this sense, the theory is intrinsically incapable of throwing any light on the idea of the creation itself.

From a mathematical perspective, this limitation stems from the particular way in which the concepts of "time" and "causality" are built into general relativity. But, in general terms, it may not seem surprising that theoretical physics cannot describe the actual creation of the universe. Indeed, to do so would seem perilously close to providing a proof of the *a priori* necessity of the existence of the world, an epistemological pit into which Western philosophers have been tumbling for over two thousand years!

Nevertheless, the new, quantum theory-augmented accounts of the Big Bang do in fact claim: (i) to yield predictions about the actual state of the universe; and (ii) to give a precise mathematical meaning to the concept of the evolution of this universe from "nothing." In particular, the mathematical theory allows a specific answer to be given to the old philosophical problem of what is meant by the "beginning of time." The introduction of quantum theory does, however, introduce profound new conceptual problems of its own, especially in relation to the meaning that can be ascribed to a quantum state of the entire universe.

The main purpose of this paper is to give a reasonably comprehensive, but non-technical, introduction to this rather difficult branch of modern theoretical physics. A crucial question in any scientific account of creation will inevitably be the role played by the linked concepts of "causality" and "time," and Section 2 of the paper is devoted to a rather general discussion of the issues involved. This is followed in Section 3 by a very brief account of some basic ideas in general relativity. Of particular importance is the fluid way in which "time" is interpreted in that theory. Section 4 contains a short introduction to some of the conceptual problems posed by quantum theory and the subtleties that arise when attempts are made to view general relativity from within a quantum framework. Once again, the role played by "time" is of great significance. The heart of the paper is the account in Section 5 of the Hartle-Hawking approach to constructing a unique quantum state for the universe and the associated picture of its creation "from nothing." A striking feature of this theory is the introduction of an "imaginary" (in the sense of complex numbers) time. This is an essential ingredient in the quantum theory, but it entails a radical reappraisal of the conventional concept of "real" cosmological time, which, rather than being an *a priori* property of the background structure, now becomes a phenomenological construct in terms of the gravitational and/or material content of the universe.

I cannot pretend that these scientific accounts of the creation make the easiest reading and, by way of motivation, it might therefore be sensible to conclude this introduction by considering briefly some of the reactions which a theist might have to these highly abstract (and more than a little speculative!) ideas.

Much has been said in general about the relation between the scientific and theological world-views (see Peacocke [2] and Polkinghorne [3]), but there are still many unresolved and difficult issues. In the context of creation, it is frequently emphasised that the Christian doctrine does not necessarily entail a beginning of the world in time; it is sufficient that at every point of time the world is totally dependent on God for its being. From this perspective, the recent scientific developments could be argued to be of little relevance to the theological concept of the "ground of Being." However, like much creedal material, the doctrine of *creatio ex nihilo*, is aimed partly at affirming the Old Testament/Judaic view of the world and partly at negating certain Greek ideas that were current at the time of the Church Fathers. The idea that the world is "sustained by God" is mainly of the first sort and, as such, is somewhat decoupled from modern scientific thought. However, the same cannot be said for the second, anti-Greek components of the Christian doctrine, if for no other reason than that of the crucial role played in the scientific account by the idea of causality — a concept that caused no little problem for the Church Fathers and whose roots lie deep in the history of Greek philosophy.

These days, most people interpret the Genesis story of creation as an existential statement of the relationship between God and Man rather than as a literal account of a time-bound cosmic process. Indeed, throughout the Old Testament "causality" is understood mainly in terms of "who is responsible" rather than of the "efficient cause" that informs both later Greek thought and modern science.[4] Thus Yahweh is the ultimate ground of all things, and His covenant with Israel and associated intervention in her destiny involves a dynamic sense of purpose and historicity.

This personalized perspective is central to Christian theology, but its conceptual framework is far removed from that of present day physical science. On the other hand, in proclaiming a belief in "....one God, maker of Heaven and Earth and of all that is seen and unseen...." Christianity is not only reaffirming the Judaic view but is also denying the existence of both pre-existent matter (or Babylonian primordial chaos) and pre- existent Platonic form: two concepts that have authentic parallels in the modern scientific accounts of creation.

The statement that matter does not exist independently of God but was brought into being by a free act of God's will, replaces the Greek idea of the creation by a demiurge whose power is fatally limited by the prior existence of the recalcitrant *materia prima* from which he shapes the world. The opposition of Christian doctrine to the Platonic split of form and matter stems from the inevitable tendency of all such dualisms to emphasise the spiritual at the expense of the material. Such a denigration of the physical world played a key part in much Gnostic speculation on the source and dynamics of the force of evil, but it is manifestly incompatible with the central Christian themes of Incarnation and Resurrection. Similarly, Christianity stands opposed to the emanationism of Neo-Platonic thought in which the physical world is perceived to lie at the bottom of a hierachical ladder of being and meaning.

An important role is played in Christian theology by this understanding of Creation as just one part of an essentially inseparable triad of Creation-Incarnation-Resurrection.[5] It strengthens the Hebraic concept of a purposeful and "linear" time via its temporal centering and ordering of BC-AD. This eschatological view of time replaces the string of events that make up the non-historical (and often cyclic) concept of time as portrayed in much Greek thought. Indeed, it has often been claimed that our present day, quasi-scientific concept of time owes much to the intellectual labours of the early Christian thinkers.

Some of the ideas expressed here are genuinely archetypal and re-emerge with starting force in the current scientific attempts to explain the creation of the world from "nothing". But let us ask once more: what are the implications for theism? Three views on the significance of the scientific picture seem to have emerged so far: (i) it provides active support for the Christian view; (ii) it provides yet another nail in the coffin of the theistic account of the world. Thus the "quantum creation from nothing" negates the *creatio ex nihilo*; (iii) the scientific and Christian conceptions of "creation" are epistemologically incompatible and should be left to their own devices. The fact that the universe apparently "began" a finite time ago is certainly prone to raise ideas of a necessary creator. This view was even endorsed by Pope Pius XII in a famous address in 1951 to the Pontifical Academy of Sciences. But it has been attacked on a variety of grounds,[6] not least because of its obvious susceptibility to the "God of the gaps" syndrome, in which God is relegated to filling in the blanks in an otherwise complete scientific theory. Perhaps the best argument in favour of the thesis that the Big Bang supports theism is the obvious unease with which it is greeted by some atheist physicists. At times this has led to scientific ideas, such as continuous creation or an oscillating universe, being advanced with a tenacity which so exceeds their intrinsic worth that one can only suspect the operation of psychological forces lying very much deeper than the usual academic desire of a theorist to support his/her theory. Thinkers of this type will inevitably welcome the scientific attempts to "explain" the creation and will cite them as further evidence in support of the atheist view.

On the other hand, the "God of the gaps" danger is sufficiently pronounced to encourage many theologians to adopt the third approach and deny any meaningful connection between the scientific and religious accounts of creation. This is understandable, but the fact that the new scientific ideas *have* been used to justify atheism means there is certainly some work which needs to be done by Christian apologists, at least to the extent of ensuring that parties on both sides understand the statement of these ideas. This is not a minor matter, since there is a regrettable, but recurrent, tendency for the results of science to be mis-stated and mis-used in the propagation of world views that are not in themselves scientific.

The issue under debate is really a specific example of the more general question of the potential relations between science and theology. A cautious approach will see the dialogue as being mainly concerned with the different epistemological frameworks of the two disciplines. From this perspective, the theological relevance (if any) of a new scientific doctrine is not primarily the facts which it proclaims but rather the extent to which it may reflect a shift in the general conceptual framework with which science views the world. Thus the meeting ground of science and theology is philosophy, not the fine details of contingent existence.

This conservative approach is based on an awareness of the fluid nature of the scientific *Weltanshauung* and the tendency for the "facts" to change as a theory is developed. This avoids the more obvious "God of the gaps" pitfalls but, in a certain sense, at the expense of restricting the partners in the dialogue to be science and theology rather than science and religion. The account which follows of the scientific ideas on creation should be read from within this "safe" framework of an analysis of concepts; I shall relegate to a few remarks in the final Section the more "dangerous" (but exciting?) possibility of taking seriously the complete scientific picture.

2. *Scientific Perspectives on Creation*

In this section I would like to consider, in rather general terms, the types of scientific statement that might meaningfully be made concerning the creation of the universe. In coming to grips with this problem it is helpful to keep in mind that, pragmatically speaking, the aspirations of theoretical physicists fall into two classes. The first is concerned with the study of specific properties of an object at a fixed time; an example is the use of a theory of quarks and gluons to predict the masses of the elementary particles. The second class of activity is aimed at discussing the way a physical system evolves in time. This might involve a theory which predicts how a physical object will move through space, for example the trajectory of a stone that has been thrown, or the paths followed by elementary particles as they collide in an accelerator. Or the time-dependent property of the system might be an internal one; an example would be a theory that involves a variation in time of the mass of an elementary particle.

This division is not meant to be dogmatic, but it has some heuristic value in indicating how theoreticians do in fact see their subject. However, reflection on what might be desired from a theory of creation shows that both classes of prediction are of interest. Thus a good creation theory would (i) predict many properties of the observed universe, (ii) say how they are changing in time, and (iii) give some account of how they came into being, preferably in terms of a well-defined concept of the beginning of time and, if appropriate, an associated evolution from "nothing". One might suspect that these features will be interlocked and that a prediction of the properties of the universe cannot be entirely divorced from discussions of its "coming into being," even if for some purposes its subsequent evolution in time can be treated as a separate problem.

Clearly there are a number of fairly general questions that could be asked of almost any scientific attempt to deal with the creation of the universe. However, the interpretation of such questions is usually a highly non-trivial matter and, from a scientific perspective, can only be done with any safety from within the epistemological and/or mathematical framework of a specific piece of theoretical physics. Indeed, it is always important to remember that the conceptual language employed in physics is closely bound to the mathematical and internal structure of the particular theory under consideration.

With these thoughts in mind, let us consider the following examples of general questions that could usefully be asked of any scientific theory that purports to deal with the creation of the universe:

(i) What does *creation* mean in general within the context of physics? And what meaning can be ascribed to this concept when it is applied to the particular example of the universe in its totality?

(ii) *What* is created? What are the types of entity whose creation is predicted by the theory?

(iii) Is it meaningful to ask *from what* the created entities were created? In particular, what is creation from *nothing?*[7]

(iv) *When* was it created? Or, more precisely, what is the role of *time* in discussions of creation?

A fifth, and rather critical question concerns the precise set of *a priori* ingredients that are assumed in formulating the theory. As always in theoretical physics, this set is required to be minimal in a sense which, while difficult to quantify, plays a crucial role in the acceptance of new ideas by the scientific community. This question is of particular importance in discussions of the philosophical and/or theological implications of creation theories, since it is not unusual for physicists who work in these areas to be unaware that there is even a meaningful question to be asked, let alone to know how to answer it.

Note that the one question which is *not* asked is "*Why* was the universe created?" or, perhaps equivalently, "*Why* is there anything at all?" Notwithstanding the numerous affirmations by philosophy of the meaninglessness of such questions, they continue to be asked, and with a seriousness and intensity that reflects the depths of the existential and ontological insecurity from which they arise. However, even the most irreverent physicist is usually still content to defer to other disciplines the task of formulating an acceptable response, unless of course the recent expositions of the so-called "anthropic principle" are seen as the first steps in a scientific takeover! [8]

It might appear that the logical way to tackle the four questions above is to study them in turn but, as we shall soon see, they are a closely coupled set and really need to be discussed as such. Let us start by considering the first question: "what is creation in the context of theoretical physics?" An (essentially tautologous) answer might be "bringing into being that which was not." But this illustrates at once the strong interdependence of our questions. For example:

(a) "... that ..." raises question (ii) concerning the "what" of creation;

(b) "... was not" leads compellingly to the rider "from that which was not" and hence to question (iii) in the guise of the old Greek problem of the use of the negative in the existential verb "to be;"

(c) The use of the past tense, "... was not", also raises the question (iv) of the "when" of creation and of the meaning of time in this situation.

Evidently some care is needed when using the word "creation" in a physical context. One familiar example is the creation of elementary particles in an accelerator. However, what occurs in this situation is the conversion of one type of matter into another, with the total amount of energy being preserved in the process. This sounds more like demiurgic creation than the *ex nihilo* which we are seeking (but see Sec. 2.3).

2.1 *Causality and State Space*

To explore this matter further we need to consider rather carefully the role of causality in theoretical physics, especially in relation to the time evolution of the system. The notorious philosophical difficulties surrounding the concept of causality can be sidestepped to some extent by giving it a precise definition in terms of certain properties of the mathematical equations that determine the development of the system in time. The central ingredient in such a theory is an abstract mathematical space **S** whose points represent the possible states of the system at any fixed time. Since it is intended that the "state" should encode the maximum amount of information that relates to the system, the theory must also possess a mathematical algorithm, whereby a specification of the state can be converted into a set of predictions of the results of any measurements made on the system. (Those not liking the positivist flavour of this sentence may wish to change the end to read "... to determine the value of any numerically representable attribute of the system.")[9]

As the system evolves in time, the state will change, and to each possible history of the system there corresponds a curve in **S** parametrized by time. Each such curve is associated with a specific solution of the dynamical equations of motion, and the system is said to be *causal* if there is just one curve passing through each point in the state space. Thus specifying the state B at any particular time t_B uniquely determines the state C into which the system will have evolved by a later time t_C; it also determines uniquely the state A from which it must have evolved at an earlier time t_A.[10, 11] See Figure 1.

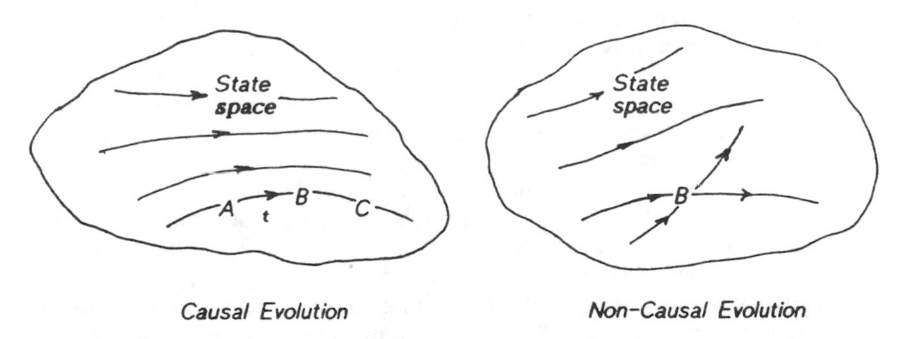

FIG. 1. State Space in causal and non-causal evolution.

The concept of "state" plays a central role in both classical and quantum physics. We will discuss the quantum case in Section 4, but for the moment let us restrict our attention to the classical situation and consider how one sets about constructing the state space for such a physical system. This is often done via the intermediate concept of the *configuration space*. As the name suggests, this is a mathematical space whose points represent the different configurations that the system can have — an idea that is best illustrated with the aid of a few specific examples:

(i) The configuration space **Q** for a point particle moving in three-dimensional (Newtonian) space is the set of all triples (x, y, z) of real numbers representing the three coordinates of the position of the particle. Another example is a particle constrained to lie on a circle: **Q** is then the one-dimensional space of all angles lying between 0 and 2π which parametrize the points on the circle.

(ii) The configuration space for a solid object moving in three dimensions is as for case (i) but augmented with the extra angular variables needed to represent the orientation of the body in space.

(iii) In the theory of magnetism, a configuration of the system specifies the three components of the magnetic field. This must be done at all points in space, and hence the configuration space is infinite-dimensional.

To illustrate how the state space is derived from the configuration space, consider the example of a point particle moving in three dimensions. The possible trajectories are determined by the solutions of Newton's equations of motion. But specifying the position of the particle at some initial time is *not* sufficient to determine the actual trajectory — we also need to choose the initial *velocity*. Thus the state space of this particular system is the set of all possible values for the position and velocity. This mathematical space has six dimensions, three for position and three for velocity.

This particular example has wide-ranging generalizations and, for most physical systems, a unique solution to the equations of motion can be determined by specifying a point in the configuration space at some initial time plus a value for the rate of change of the configuration at that time (for example, velocity is the rate of change of position). Thus the state space of a classical system with configuration space **Q** is the set of all pairs (q, p), where q is a point in **Q** and p represents the "direction in **Q**" (and the "rate") along which the trajectory will emerge from q.[12]

It is important to note that, although fixing a single point in **Q** does not lead to a unique trajectory, a unique curve can be obtained if *two* points are specified. More precisely, given a pair of points q_1 and q_2 in **Q** and a pair of times t_1 and t_2, there will in general be a unique solution to the equation of motion such that the configurations at t_1 and t_2 are q_1 and q_2 respectively. This will play a significant role in our discussion of quantization in Section 4.

It is clear that one of the major tasks for a theoretical physicist charged with "explaining" the time evolution of a physical system (be it classical or quantum) is to find: (i) a mathematical model for the states of the system; (ii) a causal dynamical law describing the evolution through state space. This way of looking at time evolution emphasises the fundamental distinction traditionally made by physics between the dynamical laws, which determine the possible motions of a system, and the initial conditions/boundary conditions, which determine *which* actual motion is realised. The former are "laws of nature" whereas the latter are adjustable by the experimentalist as s/he seeks to test a postulated law.

Of course, there might be no such causal model, in which case the search would be a hopeless task. However, the desire for an ordered and orderly world inspires many scientists to strive to impose causality on a physical system even if, as in the case of quantum theory, the concept undergoes a fairly radical revision in the process. Such neo-Kantian operations were cast into psychological terms by Jung with his profound analysis of the archetypal modes in which reality is experienced and comprehended.[13] In the present context, there is a particularly apposite quote from one of his earlier works: "Another inexhaustible source of happiness can be the gratification of the causal instinct."[14]

While on psychological matters, it should be emphasised that of course physics knows nothing of the "passing of time"; indeed, in many respects the fundamental theoretical entities are the complete *histories* of the system in the state space **S**, rather than the points of **S** as such. This is true of both general relativity and quantum theory, and it is a particularly significant factor in the quantum accounts of creation. From this perspective, the use of phrases like "evolution in time," "beginning", "end" should be understood as a psychological hangover from the peculiarly human experience of time, not as positing any sense in which a point actually "moves" along the path in **S**. This "anti-personalistic" account of time, with its denial of the human "now", is part of the general reductionist tendency of science and, as such, poses obvious difficulties for certain parts of the theological enterprise.

Note that a discussion of time evolution does not necessarily have to take place within a theoretical framework which can also predict the "static" properties of the system. For example, many attempts have been made to construct a theory of the scattering of elementary particles in which the masses of the particles have of necessity to be inserted as free parameters. Such a deliberate limitation of the potential domain of applicability of a theory is best discussed as part of the general philosophical question concerning the role of mathematical structures in physics and the sense in which they "model" features of the world. See, for example, the work of Ian Barbour.[15]

2.2 *The Acausality of Classical Creation*

Any account of the creation of the universe formulated within the framework sketched above must confront the problems of (i) the applicability of the concept of "state" to the entire universe, and (ii) the ontological status of the space of all such states. Classical physics appears to admit such a construction, with the state being defined as the positions and momenta of all the particles in the universe (plus the values, and their first derivatives with respect to time, of any fields that are present). However, the path of the actual universe involves a unique sequence of time-ordered states, and hence a certain amount of creative activity is required by the theoretician in deciding in *what* space **S** this curve should lie. In a normal physical system, the choice of **S** is determined by the different states which the system *could* have and in which, if desired, it could be placed by an ubiquitous external observer/experimenter. I have deliberately emphasised this "operational" nature of the concept of state in order to contrast the situation with that pertaining to the universe as a whole where there is a single, once-and-for-all history, and no external observer to set initial conditions at his/her whim.

With this caveat, the history of the universe, as represented by its time-ordered path in **S**, could perhaps be viewed as a continuous process in which the state at a given time is "annihilated" and the state at the next instant of time in "created". This is vaguely suggestive of Whitehead's philosophy and its application to process theology, although it is not easy to reconcile such a picture with the absence of any objective "now" in the scientific view of time. In any event, for the purposes of describing the creation of the universe, we need to consider the possibility of a situation in which the state C "evolves" from a state X with the property that there is *no* state lying prior to X on the path in **S**:[16]

$$X \longrightarrow C \longrightarrow$$

Some insight into what is involved can be gained by considering the time-reversed situation in which the universe is "annihilated" via a process in which a state C is followed by a state X which lies at the ("future") end of the path in state space. This could happen if, for example, the concept of time breaks down at X:

$$\longrightarrow C \longrightarrow X$$

Such a "big crunch" is indeed one possible fate of the universe. (Another is the "cold death" associated with a never-ending expansion.) However, the crucial point for us is that mathematical theories possessing solutions of this type are likely to violate causality in the sense of admitting many trajectories in state space ending at the *same* final state X:

$$\begin{array}{c} \longrightarrow C \longrightarrow X \text{ ——— } C' \text{ ——— } \\ \nearrow \\ C'' \\ / \end{array}$$

This is precisely what happens in the phenomenon of gravitational collapse (see Section 3) and it suggests that, in general, the "creation" process in such a theory will also be acausal with many different trajectories in state space (i.e., possible "histories of the universe") emerging from the same initial state X:

$$\begin{array}{c} \nearrow \\ C'' \\ / \\ \longleftarrow C \longleftarrow X \longrightarrow C' \longrightarrow \end{array}$$

If such a system is known to be at a point C at a particular time, then the causal structure of the theory will enable us to compute which trajectory the system is on, and hence to predict the state at any other time. However, if we start at the point X, there is *no* way from within the theory of predicting which trajectory in state space the system will follow.

But then we seem to be driven back inevitably to a view of creation in which the most to which science can aspire is the use of present day observational material to compute the current state and hence, via a knowledge of the dynamical laws, the particular trajectory being followed by the deterministic universe. In such a situation, we can (at least, in principle) say what the state must have been at any earlier time (indeed, this is what is meant by the claim that there *was* a Big Bang in the past), but there is no way in which we can *predict* that this trajectory, rather than any other, is the history of our actual universe. In this sense, nothing can be said about creation itself.

2.3 *Creation Within the Framework of a Pre-existing Spacetime*

This singular failure of classical physics stems directly from the basic dualism between the dynamical laws and the boundary conditions needed to fix a specific solution to these equations of motion. It is clear why it might be attractive to the religious mind to invoke a "deistic" creator who sets the initial conditions and thereafter leaves the universe to evolve according to a precise set of causal laws. But this is a dangerous line of thought which opens up the "God of the gaps syndrome" with a vengeance.

The possibility of using the recent creation theories to plug this gap lies in quantum theory and in the subtle relation of its state space to that of an underlying classical system. But let us turn aside for a while from the problem of the meaning of creation and consider the question of *what* it is that is created. We want the answer to be, in some way, "the universe," but this is a difficult concept, and it is useful to consider first what type of entity might in principle be subject to creation within the framework of a mathematical theory.

The critical question is how the division is made between those features of the observed universe which are fed into the creation theory, and those whose appearance is predicted by it. The simplest possibility to envisage is perhaps the creation of matter in a fixed spacetime, and with the physical laws specified in advance. For the evolution of the universe to be uniquely determined thereafter we also require that the theory should predict a precise initial state for this matter, i.e., what is created is a specific state of the entire material content of the universe. In a structure of this type the "nothing" from which the universe is created is "no thing," but within the framework of a pre-existent space and time.

A more radical theory would be one in which the laws obeyed by the created matter were also an outcome of the theory. A special case might be the prediction of the numerical values of a set of fundamental constants appearing in an otherwise specified set of physical laws. Such a theory would of necessity involve some type of meta-law, and there is an obvious danger of generating an infinite regression of theoretical structures. In practice, we do build, *a priori*, a number of features of the universe into the formulation of our theories; even the currently fashionable "grand unified theories of everything" are subject to this limitation!

It is not surprising that creation within a pre-existent spacetime was the first possibility to be considered seriously in a scientific context:[17] implicitly or explicitly, most theoretical physicists still have a mental picture of space as a large box in which is contained the material stuff of the universe. The postulated process is quantum mechanical, but the basic idea is simple enough and rests on the fact that the gravitational energy of a pair of particles is negative.[18] Thus the possibility arises of using the gravitational energy of a set of particles to cancel out their $E = mc^2$ self-energies, giving a total energy of 0. Clearly the energy of "no thing" is 0 too, and hence the creation process takes place with a "demiurgic" conservation of energy of the form:

$$0 = E_{Nothing} = E_{Something}.$$

In this simple version, the argument is somewhat dubious, but the basic idea is workable, and it can be developed in a more rigorous way within the framework of relativity theory.

The physical picture is one in which a small seed of matter nucleates in a pre-existent, but empty, spacetime. The quantum fluctuations of this matter produce a varying gravitational field whose energy gets transmuted into elementary particles, whose quantum fluctuations produce more varying gravitational fields, whose energy, etc., etc. (See Figure 2). This process is claimed to produce a fire-ball explosion away from the seed-point, and this is our Big Bang! Since nothing can move faster than the speed of light, the spacetime picture is of a material universe contained within a cone whose apex lies at the seed-point and whose surface corresponds to the possible paths of light emitted at this point.

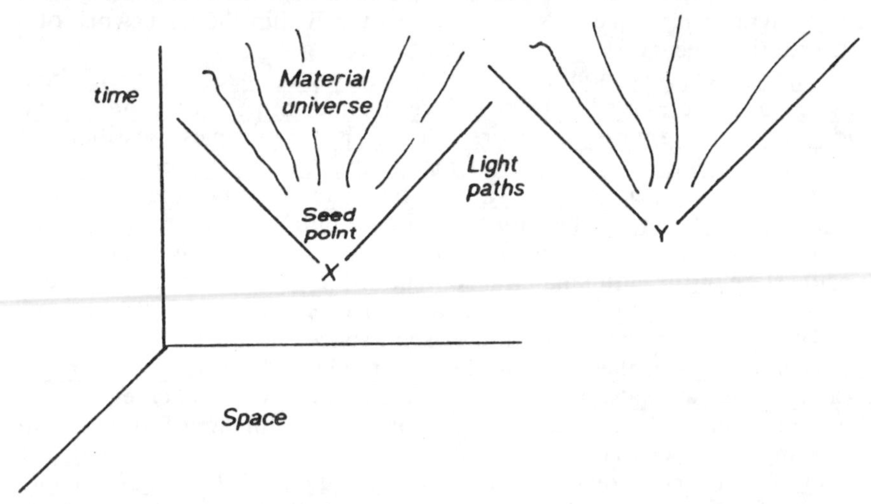

FIG. 2. The material universe from the quantum fluctuations of a nucleating seed-point.

Theories of this type have not found wide acceptance, and their interest for us lies mainly in some of the rather general problems that arise in attempting to implement them successfully. In particular, any such theory must inevitably encounter the extremely awkward question of how the precise time at which creation occurs is to be determined. This problem arises because, within an infinite, pre-existent and homogeneous timeline, there is simply no way of distinguishing any particular instant of time.

As an almost inevitable concomitant, these theories are prone to predict, not a single creation/seed-point, but rather an infinite number of them, with a given probability of one occurring in any particular period of time. The existence within a single spacetime of infinitely many "conefuls" of matter might be theoretically acceptable if they did not interfere with each other. But this is far from being the case. For example, in Figure 2, the matter emitted from a seed-point *Y* will eventually interact with that emerging from the point *X*. This is rather a peculiar picture, and not one that seems particularly consistent with large-scale astronomical observations![19]

2.4 *Creation of Time*

At this point it is conventional to observe that St. Augustine pre-empted this particular problem many centuries ago with his profound reflections on how best to reply to an obstinate interlocutor who insists on asking, "What was God doing before He made heaven and earth?"[20] Augustine's refutation of this demiurgic concept of God lay in his argument that, like matter, time was God-created, and that before heaven and earth were made there was no time. Hence the question, "What was God doing then," is without meaning since, if there was not any time, there was not any "then".

It is singularly striking that, sixteen centuries later, theoretical physicists have considered precisely the same subterfuge as a means of avoiding the question of the "when" of "before" of creation. I shall give some details shortly, but for the moment it suffices to note that any suggestion that space and/or time are themselves subject to creation raises a number of general questions concerning the ontological status of these entities. Once again, inspiration can be drawn from a Master of the past; the relevant source in this case being Philo of Alexandria who, in "On The Creation," was bothered, like Augustine, by the fact that:

> There are some people who, having the world in admiration rather than the Maker of the world, pronounce it to be without beginning and everlasting, while with impious falsehood they postulate in God a vast inactivity;...[21]

Philo's response is most instructive. He affirms that:

> Time began either simultaneously with the world or after it. For since time is a measured space determined by the world's movement, and since movement could not be prior to the object moving, but must of necessity

> arise either after it or simultaneously with it, it follows of necessity that time also is either coeval with or later born than the world.[22]

The basic idea is that, rather than speaking of things moving *in* time, we should view time as determined *by* the motion of things. This gives to matter an ontological status that is prior to that of time — a reversal of the container image that would doubtless have appealed to Leibniz and which has a precise technical analogue in the current scientific theories with which we are concerned. Indeed, a critical ingredient in these theories is the replacement of what we normally call "time" with a phenomenological construct in terms of the constituents (fields or particles) of the universe. This allows the possibility that, as we approach the "point" of creation (going backward in time), the phenomenological time will look less and less like the usual one, and in such a way that the problem of the "beginning" of time is resolved with the realization that conventional "time" is an inapplicable concept within the highly quantum mechanical, very early universe. But these are difficult matters, and to make further progress we must take a careful look at some aspects of the role played by time in the theory of general relativity.

3. *General Relativity*

The theory of general relativity is a sophisticated and highly geometrical description of the force of gravity. One of its most significant deviations from the older Newtonian theory is the prediction that, not only mass, but also energy and pressure can generate a gravitational field. This leads to the remarkable phenomenon whereby the gravitational forces in a piece of matter whose density surpasses a certain critical value will overcome all repulsive forces and cause the matter to collapse to a point of zero size and infinite density. This final state is a singular point in the theory and is essentially independent of the property of the matter which caused it. Since the theory cannot predict what happens beyond this point, we have a striking example of the annihilation process $\rightarrow C \rightarrow X$ discussed in Section 2.2. The existence of such a process immediately raises the question of whether the solutions to the equations of general relativity also include examples of time reversed "creation" processes $X \rightarrow C \rightarrow$? In an appropriate sense, the answer is "yes", but to understand this properly we need to be more precise about what it is that is created at the Big Bang and, in particular, how this relates to Augustine's conception of the creation of time.

The key point is the radical reappraisal in general relativity of the concepts of space and time. The old Newtonian picture was of a fixed "transcendent" spacetime which is ontologically prior to the material contents of the universe (be it in the form of particles or fields) and which decomposes into a fixed family of three-dimensional spatial planes parametrized by the values of a universal, one-dimensional time. See Fig. 3.

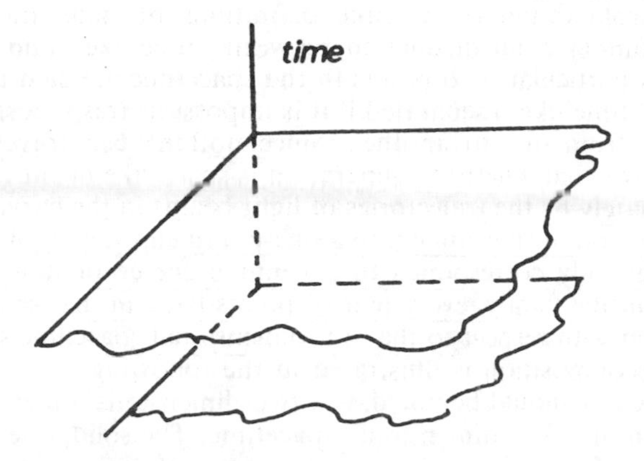

FIG. 3. Surfaces of equal time in Newtonian spacetime.

This is the appropriate time parameter to use if the discussion in Section 2 of paths in state space is applied to the causal evolution of a Newtonian system. However, this sharp distinction between three-dimensional space and one-dimensional time is incompatible with electromagnetic theory and was removed by Einstein in his special theory of relativity. The new picture is of a four-dimensional spacetime which admits a variety of choices of time, each associated with a different inertial frame of reference. This introduces a subtlety into the discussion of causality since the space of states (and the time variable parametrizing the paths), now depends on the choice of reference frame. A relativistic theory is deemed to be causal only if the causal structures associated with all of these different choices are mutually consistent.

In spite of their differences, the spacetimes of Newtonian and relativistic physics share the property of being fixed, background structures within which the material content of the universe has its being. However, in general relativity, this rigid image is replaced by one in which spacetime is a curved space whose curvature is identified physically with the gravitational field and depends (via the Einstein field equations) on the distribution of mass, energy and pressure of whatever matter (fields or particles) is present in the system. The spacetime of special relativity is then simply a special, flat solution to these equations corresponding to the total absence of all matter and gravitational effects. It can be shown that, for any choice of inertial frame, the three-dimensional subspaces of equal time are also flat.

However, a crucial feature of general relativity is the absence of any special choice, or class of choices, of time. The primary concept is the complete four-dimensional spacetime in its entirety, and there are many ways in which this can be decomposed into a one-parameter family of curved, three-dimensional spaces. In each such decomposition, the parameter represents a particular choice of "time." Note however that, even with this proliferation of possible definitions of time, the theory still imposes a fundamental distinction between "time-like" and "space-like" intervals. In particular, two points in the spacetime are said to be "space-like" (resp. "time-like") separated if it is impossible (resp. possible) to send information from one to another. Since nothing can travel faster than light, it follows that whether a separation is space-like or time-like is determined ultimately by the trajectories of light beams in the curved spacetime. A decomposition of spacetime into a one-parameter family of three-dimensional spaces only corresponds to a genuine choice of time if, for every three-space in the family, every pair of points lying in that space are space-like separated with respect to the four-dimensional spacetime structure.

This decomposition is illustrated in the following "cylinder" picture, (Fig. 4) which, it should be noted, is a two-dimensional diagram representing a portion of a four-dimensional spacetime. The solid, one-dimensional, lines labelled Σ_1 to Σ_5 represent a sequence of five, three-dimensional, spaces corresponding to five values of a particular choice of time. The dotted lines Σ'_1 to Σ'_5 represent another way of decomposing the spacetime, and hence a different choice of time. Note that, in this particular example, the subspaces of equal time are all closed on themselves. Many discussions of cosmology assume a large-scale structure of this type.

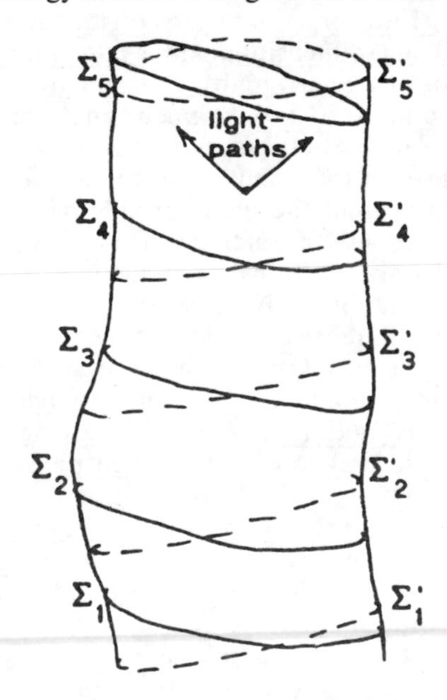

FIG. 4. Two different ways of decomposing a four-dimensional spacetime.

If desired, the structure of general relativity can be recast in the language of "dynamical change, evolution," etc. by studying the way in which the curvature of the three-dimensional subspaces changes with respect to different values of some particular choice of time. Such a view leads to the identification of the configuration space of general relativity with the set of all possible curved three-dimensional spaces. A four-dimensional spacetime can then be related to a path in this set parametrized by the time associated with some particular decomposition. Note however that a single four-dimensional spacetime can be associated with many different paths in the space of curved three-spaces, corresponding to the many different ways in which the decomposition can be performed. This important, but complicating, feature of the theory explains why, for many purposes, it is natural to view the fundamental entity in general relativity as a single four-dimensional spacetime rather than as a time-parametrized family of three-dimensional spaces. These two views are blended together in a subtle and important way in the quantum framework we will be discussing later.

Note that, if any matter is present, its possible configurations will be included in the configuration space for the full theory. This applies, for example, to the Big Bang, which is described in general relativity with the aid of an approximate model in which matter is regarded as being distributed uniformly throughout the universe. The associated "expanding universe" solutions to Einstein's equations admit a slicing of spacetime in which each three-space can be viewed as the three-dimensional spherical boundary of a four-dimensional ball. The time variable associated with this decomposition is the radius of the sphere, and the spacetime itself can be pictured as somewhat resembling an ice cream cone as in Fig. 5, which is a degenerate version of Fig. 4.

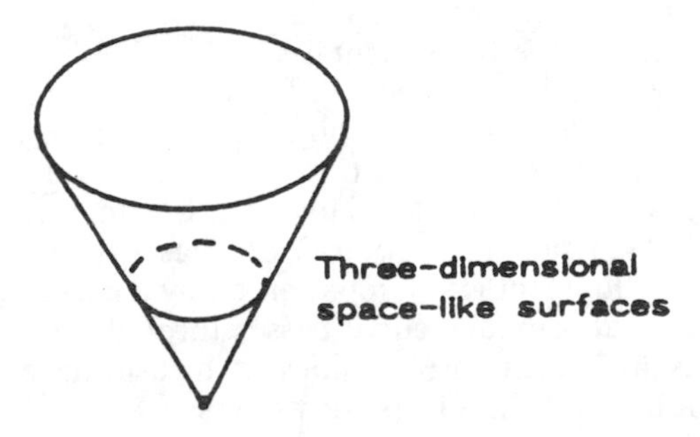

FIG. 5. Spacetime picture of the expanding universe following the Big Bang.

An absolutely crucial idea here is that "time" can be defined internally in terms of a particular property (i.e., the radius) of the curvature of the three-dimensional space, and hence in terms of the gravitational field which it represents. Such an approach is almost inevitable if we wish to deal with generic curved spacetimes in which, *a priori*, there will be no preferred way of performing a decomposition. This possibility of formulating a "phenomenological" definition of time in terms of the contents of the universe (in this case, just the gravitational field) plays a central role in quantum cosmology and is one of the most fundamental features of general relativity.

Note that at the tip of the cone (the Big Bang), the three-space has zero radius and the theory cannot be extrapolated back beyond this point. This is essentially the sense in which space and time can be said to "come into being" at the point of creation. However, we are still locked into the old dualism of dynamical law *versus* boundary conditions, since there are many such singular solutions to Einstein's equations, each corresponding to a different distribution of gravitational fields and matter, and there is no way within the theory of selecting any particular one. Thus the most that can be achieved with classical general relativity is to list the vast number of possible universes (i.e., possible distributions of matter and gravitational field). There is no *a priori* way of giving weight to one over another and hence no way to "explain" the creation itself.

4. *Quantum Theory and Quantum Gravity*

The crucial step in escaping from this circle is the introduction of quantum theory. This leads ultimately to the rather mystical (and potentially misleading) statement that the universe was created as a "quantum fluctuation in the vacuum."

Quantum theory was invented in the 1920's in response to a number of striking failures of classical physics in the realm of atomic phenomenona. It was highly successful in this area, albeit at the expense of employing a mathematical structure that was considerably more abstract than any seen theretofore in theoretical physics. However, the most revolutionary feature of quantum theory is its insistence on the existence of an intrinsically probabilistic component in the fundamental laws of physics. Thus, in a quantum theory, the specification of a state of the system will not in general yield a prediction of the *results* that will be obtained if a measurement is made; it can predict only the *probabilities* that certain results will be found. In this sense, the theory is not deterministic. However, it is important to appreciate that the general discussion in Section 2 of causality still applies and, in particular, time-development is still represented by a curve in a state space **S** with the property that a unique curve passes through each point in **S**. But what evolve causally are the probabilities of measuring certain values for observable quantities, not the values themselves.

4.1 *Some Conceptual Issues in Quantum Theory*

The philosophical problems associated with quantum theory are grounded in the claim that the nature of the probabilistic predictions is such as to render meaningless any statements to the effect that an observable quantity actually *possessed* the value obtained in a measurement before the measurement was made. The only exception is a prediction that the probability of getting a certain value is one: in this case the observable *is* said to "possess" this value.

This negation of the "realist" interpretation of classical theory leads to a radical revision of the conceptual framework employed in discussions of the physical world. It should be emphasised, therefore, that the use of probabilities does not in itself rule out an interpretation based on philosophical realism. For example, classical statistical physics deals with large complex systems (e.g., a box full of gas) within a theoretical framework in which a limited knowledge of a state yields a limited prediction of the results of measurements. But this is perfectly consistent with the assumption that the observables *have* certain values, even if we do not know what they are. Indeed, it is assumed that this uncertainty can be made arbitrarily small by a sufficient increase in the accuracy of our knowledge of the full state of the system. Thus, in this "epistemological" interpretation of probability theory, probabilistic statements refer to our knowledge, or ignorance, of the actual state of affairs.

Quantum theory is profoundly different with its insistence that the probabilities are *irreducible* in the sense that there are no "hidden variables" whose values would enable us to refine our knowledge of the state and hence to improve the accuracy of our predictions. A particularly significant feature in this respect is the occurrence of pairs of so-called "complementary" variables. These have the peculiar property that the more accurate is our prediction of the value of one element of the pair, the less accurate will be our prediction of the other. This is a difficult phenomenon to incorporate within the framework of classical realism.

There is still much disagreement about the correct conceptual framework for quantum theory. Most pragmatically minded physicists settle for an essentially instrumentalist account within a frequentist interpretation of probability. Thus a state is assumed to give only statistical information about what happens if a large number of measurements of the same observable are performed on an ensemble of identically prepared systems. In particular, nothing is asserted about the properties of an individual system until *after* a measurement has been made, at which point it *is* deemed meaningful to say that, for that particular system, the attribute possesses the value measured. Note, however, that, if a measurement is made on a single system yielding a specific result, conceptual consistency then requires that an immediate repetition of the measurement on the same single system will of necessity reproduce the result obtained by the first measurement (assuming that the measurement is an "ideal" type that does not do anything drastic like destroying the system in the process!). Thus, after making a measurement, the state of the sub-ensemble of those systems for which a particular result was obtained will not in general be the same as the state of the entire ensemble prior to the measurement.

This so-called "collapse of the state vector" seems quite natural within the context of a strictly statistical interpretation and the selection of a sub-ensemble. However, in interpretations which attempt to give meaning to the idea of the quantum state of a *single* system, it generates a picture in which the act of making a measurement causes a "jump" in the system itself. This issue has been much discussed in the literature, particularly since approaches of the statistical type incorporate a strict object-subject dualism between an observer (who makes the measurements) and the quantum system (on which the measurements are made). Such distinctions have been criticised frequently as being artificial or, even worse, meaningless. Similarly, the idea that a state evolves causally between measurements, but undergoes a jump when an observer intervenes, has been attacked for its extreme instrumentalism and negation of the unity of the laws of the physical world.

These difficult and contentious issues become acute when applied to cosmology, where it is hard to ascribe *any* meaning to an absolute distinction between observer and system. Many of those working in this area agree that any discussion of quantum states of the entire universe must be within an interpretative scheme in which: (i) there are no references to measurements of the entire system being made by an external observer; (ii) there is no concept of collapse of the state vector induced by such measurements; (iii) an interpretation of probability is used that avoids the notion of "ensembles" of universes; (iv) it is possible to reproduce the usual statistical results when applied to a sufficiently small sub-system of the physical universe.

4.2 *Quantum States and Transition Probabilities*

I will return later to the difficult question of the existence of such an interpretative framework, but to proceed further we must turn to some of the more technical aspects of quantum theory. We will start with a brief discussion of the way in which the concept of probability is conventionally coded into the mathematical structure of the theory.

The main ideas can best be illustrated by considering the "quantization" of a classical system with configuration space **Q**. The state space of the quantum theory is totally different from that of the classical theory and is defined to be the set of all complex-valued *functions* Ψ defined on **Q**. Thus each state Ψ is a function which associates a complex number $\Psi(q)$ with each point q in **Q** (to emphasise this feature I shall frequently refer to a state as a "state-function"). The central probabilistic axiom is that if the configuration of the system is measured, the probability that it will be found to lie in a small region around q is proportional to the real number $|\Psi(q)|^2$. (More precisely, the probability of finding the configuration in any subset B of **Q** is proportional to the integral of $|\Psi(q)|^2$ over B.)

A simple example is the quantum theory of a single particle moving in three-dimensional, Newtonian space. The configuration space **Q** is the set of triples of real numbers (x,y,z) and the states of the quantum system are functions $\Psi(x,y,z)$. A more sophisticated example is the quantization of the theory of electromagnetism. The classical configuration space is the set of all possible values $(B_x(x,y,z),\ B_y(x,y,z),\ B_z(x,y,z))$ of the three components B_x, B_y and B_z of the magnetic field evaluated at all points (x,y,z) in space. In the quantization of such a system (an example of a "quantum field theory") the state is a function of these three functions and is interpreted as giving the probability of finding various values of the magnetic field if all three components are measured simultaneously at every point in space.

As time "passes", the state function Ψ will change causally (as the solution of an appropriate first-order differential equation) and, therefore, it may be difficult to see why quantum theory should be any more successful than classical physics at producing the type of "acausal" creation process discussed in Section 2. The reason is rather subtle and depends on the phenomenon of "quantum tunnelling," which, although causal, produces a behaviour for the system that would be quite impossible within the framework of the classical theory. The quantum creation of the universe can be understood as an example of this effect, albeit with a significant change in the concept of time so as to render the causal nature of tunnelling consistent with the idea of a "beginning" of time.

It will evidently be useful to discuss the time development of a state Ψ in a way which makes this tunnelling effect reasonably clear and, to this end, I shall use the so-called "path-integral method." For each point q_1 in **Q** this yields an expression for the state at a time t_2, conditional on the state at an earlier time t_1 being such that, with probability one, a measurement of the configuration of the system would have yielded the result q_1. It is convenient to write the argument of this state-function as q_2 and to denote its modulus squared as $K(q_2, t_2;\ q_1, t_1)$. This is the "transition" probability that, at time t_2, the system will be found to lie in a small region around q_2 given that, at time t_1, it definitely had the configuration q_1. Thus this method gives directly the causal time evolution of that special class of states in which, with probability one, the configuration has a particular value at some initial time. It is easy to show that the evolution of an arbitrary state-function can be computed from this information.

The heart of the method is an expression for the state-function as a sum of terms, one coming from each possible path in **Q** joining together the two points q_1 and q_2 at times t_1 and t_2, respectively. Like the state-function itself, these terms are complex numbers and, consequently, it is possible for their individual phases to augment, or to cancel, each other. In particular, it can be shown that if there exists a solution to the classical equations of motion passing through q_1 and q_2 at the times t_1 and t_2, respectively (cf. the discussion in Section 2.1), then the contributions of all the other paths almost completely cancel. Thus the transition probability is mainly determined by the classical solution to the equations of motion, with the other paths combining together to give a small quantum correction.

4.3 *Quantum Gravity*

We must now consider the important question of what happens when the quantization scheme described above is applied to general relativity. For the sake of simplicity we shall assume from now on that all matter in the universe can be described in terms of various quantized fields. Where appropriate, elementary particles can be viewed as quanta of such fields. Such a particle-based interpretation of quantum field theory is well-defined in a weak gravitational background but breaks down in the presence of strong fields, and especially in the very early universe. For this reason I will emphasise the "field" rather than the "particle" picture of quantum field theory.

As explained in Section 3, the configuration space of general relativity is the set of all curved, three-dimensional spaces plus whatever is needed to specify the configuration of any matter that is present. This suggests that the states in a quantum theory of general relativity will be functions of the form $\Psi(c,f)$ where c and f are, respectively, a curved three-space and a point in the configuration space of the matter fields. But note that: (i) we still have to decide how such states are to be interpreted. In particular, we do not wish to invoke an observer who "measures" the configuration of the entire universe; (ii) it has often been suggested that a quantum theory of the universe will involve a selection of a *single* quantum state from the infinite-dimensional state-space. This is so in the proposal we will discuss in Section 5, but it makes the problem of interpretation even harder.

We might assume that the main task in a creation theory is to determine how a/the state function $\Psi(c,f)$ evolves in time. But, as we have argued already, "time" is a fluid concept in general relativity that can best be understood as a phenomenological construct in terms of the gravitational fields (or matter) present in the system. Implementing this idea involves splitting the variables (c,f) into two types: those that contain information on the intrinsic choice of time, and the rest (which we shall denote $(c,f)_{Phys}$). It can be shown that these remaining variables are sufficient to specify uniquely the physical configurations of the gravitational fields and the matter. Thus the original set (c,f) contains "redundant" information (viz., the value of the internal time) in addition to these "physical" degrees of freedom.

Since the variables (c,f) include an internal definition of time, it would be incorrect in a quantum context to add an external time label to the state-function. Thus the entire history of the quantum gravity system is coded into the *single* function Ψ, and we no longer talk about paths in the state space. Instead (modulo the general issue of interpretation), for each pair (c,f), $\Psi(c,f)$ should be understood as giving the probability of finding a particular physical distribution $(c,f)_{Phys}$ of curvature/gravity and matter at the internal time determined by the values of c and f.

This is a subtle concept and one might wonder what has happened to the dynamical equation that arises in ordinary quantum theory and whose solutions yield the possible paths of the system in the state space? In fact, there *is* still such an equation (known as the "Wheeler-DeWitt equation"), but its role now is to describe how the dependence of Ψ on the "physical" parts of c and f is related to its dependence on the internal time. An important feature of the theory is that this equation does not of itself select any particular definition of intrinsic time but rather gives the appropriate results for any choice. This is the sense in which the quantum theory respects the fundamental independence of classical general relativity on the choice of time.

However, it is not true that every solution to the Wheeler-DeWitt equation admits an unequivocal interpretation in terms of an evolution of the (probabilistic) properties of $(c,f)_{Phys}$ with respect to an internal time. In classical general relativity, "internal" is understood in relation to a decomposition of a four-dimensional spacetime. But, in the quantum theory, the state-function Ψ is a function of the *three*-dimensional spaces and, in general, it will not be true that the set of curved three-spaces c on which $\Psi(c,f)$ is non-zero (which implies the probability is non-zero) could all arise as space-like sections of a single four-dimensional spacetime. Certain states *will* admit such an interpretation, namely those that essentially reproduce the predictions of the classical theory. But the more "quantum mechanical" is the state, the harder it becomes to sustain an interpretation in terms of anything "evolving" in time. In effect, the concept "spacetime" only has an unambiguous meaning within the framework of non-quantum physics, whereas the idea of three-dimensional "space" can be applied to both the quantum and the classical theories.[23]

Modulo these subtleties, the ideas on quantization discussed above are still applicable. Thus we can find an expression for both the state function and the transition probability $K(c_2, f_2; c_1, f_1)$ as a sum of terms associated with all possible "paths" joining (c_2, f_2) to (c_1, f_1). However, these quantities have to be reinterpreted in the light of the remarks above. Thus: (i) $K(c_2, f_2; c_1, f_1)$ refers to the probability of finding a result $(c_2, f_2)_{Phys}$ for the physical variables (at the internal time determined by the pair (c_2, f_2)) given that, at the internal time determined by the pair (c_1, f_1), the physical variables had the configuration $(c_1, f_1)_{Phys}$. There are no explicit t_1 or t_2 labels; (ii) A "path" between (c_1, f_1) and (c_2, f_2) means: (1) a curved four-dimensional spacetime with two three-dimensional ends (as in the "cylinder" picture in Fig. 4) such that the curvatures of the two ends are c_1 and c_2 respectively, and (2) a field defined on this spacetime whose values on the two ends are the fields f_1 and f_2 respectively.

The main contribution to $K(c_2, f_2; c_1, f_1)$ will come from a classical solution to the gravity-matter Einstein field equations and will be augmented with the small quantum fluctuations associated with gravity-matter fields that do not correspond to these classical solutions. This quantity $K(c_2, f_2; c_1, f_1)$ tells us the probability of getting from one curved three-space to another and, duly interpreted, is a *bona fide* prediction of the theory. However, this still does not give any information on the actual creation of the universe. To proceed any further we must delve deeper into the peculiar nature of time in general relativity and its implications for the quantum theory.

5. *Quantum Creation of the Universe*

We must now pull together the strands from the previous sections and discuss some of the deeper implications of trying to apply quantum theory to the universe in its entirety. This subject of "quantum cosmology" began with seminal work by DeWitt and Misner and was much in vogue in the early 1970s. The difficulties encountered in constructing a consistent theory of quantized general relativity led eventually to a drop in activity in this area, but enthusiasm was rekindled about five years ago under the influence of several rapidly expanding research programmes into the role of elementary particle physics in the early universe. The current theories of "creation from nothing" were tentatively advanced a few years ago and have been the subject of much interest since.

5.1 *The Hartle-Hawking Proposal*

In considering how the creation process might be approached, we start by recalling that the "paths" employed in the calculation of the transition amplitude $K(c_2, f_2; c_1, f_1)$ in Section 4 involved four-dimensional spacetimes whose boundary consisted of the pair of three-dimensional spaces with curvature c_2 and c_1 respectively. Thus the appropriate diagram is the "cylinder" picture in Fig. 4 in which the two three-spaces at the "ends" of the cylinder are this pair of spaces. This suggests that one way of describing the creation process would be to choose a spacetime "path" whose boundary consisted of a curved three-space with curvature c_2 and a single initial *point* (rather than another three-space). In this case, the appropriate diagram is the conical spacetime in Fig. 5. But a singular point is not a smooth three-space, and the technique for computing $K(c_2, f_2; c_1, f_1)$ breaks down, not least because the classical solution to Einstein's equations represented by Fig. 5 is itself singular and ill-defined at this point.

Had this procedure worked it would have described the creation of the universe from an initial "point". However, we are interested in creation from "nothing", which suggests that a more appropriate "path" would be a spacetime of the type shown below whose boundary is just a *single* three-dimensional space. See Figure 6.

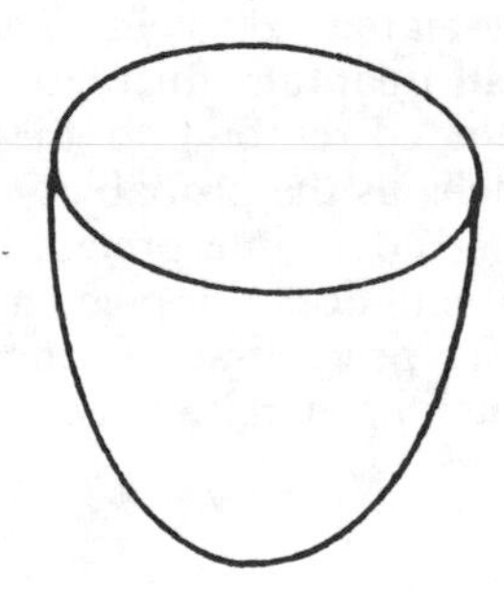

FIG. 6. A four-dimensional space with a single three-dimensional boundary.

At a first glance this seems even worse than the conical spacetime of Fig. 5, and in fact curved spacetimes of this type invariably involve pathological features, like the existence of closed time-like loops. Nevertheless, it is clear that, in some sense, Fig. 6 *is* what we need to describe creation "from nothing," but the way in which this can be realised is subtle and, from the perspective of the classical theory, rather unexpected.

The crucial step is to invoke the idea of "imaginary" (in the sense of complex numbers) time. This idea is common enough in normal (i.e., non general-relativistic) quantum field theory and is little more than a convenient mathematical trick. However, the situation in general relativity is more obscure. For example, which of the possible choices for time is to be singled out for multiplication by $\sqrt{-1} = i$? The answer is "none" or, at least, none of the internal times about which we have been speaking. What is done can be motivated by noting that the difference between space (x, y, z) and time coordinates t in special relativity is reflected in the difference of sign in the wave equation describing the propagation of a light wave:

$$-\frac{\partial^2 B}{\partial t^2} + \left(\frac{\partial^2 B}{\partial x^2} + \frac{\partial^2 B}{\partial y^2} + \frac{\partial^2 B}{\partial z^2}\right) = 0$$

where, for example, B is one of the components of the magnetic field. But note that changing t to it has the effect of introducing an extra minus sign in front of the first term, and hence the space and time variables now enter in the *same* way.

The crucial assumption is that this should also apply to general relativity, viz., the analogue of employing an imaginary time variable is to change the fundamental mathematical entity used to describe the curvature of the four-dimensional spacetime so that there is no longer any distinction between space-like and time-like directions; i.e., we work with a four-dimensional space in which all directions have equal status. This sounds highly non-physical, but we push on regardless and remark that there exist plenty of non-singular solutions to Einstein's field equations for such spaces. In particular, this includes cases which have a single three-dimensional boundary of the type depicted in Fig. 6. This picture should be compared with the conical Big Bang of Fig. 5. The boundary of the latter consists of the three-dimensional space *plus* the singular point of "creation" at the apex of the cone. But in Fig. 6 there is only the three-space; the singular point has been avoided by the trick of placing all four dimensions on an equal footing.

Now comes the fundamental ansatz postulated by Hartle and Hawking[24] for constructing the "state-vector for the universe":[25] The quantum theoretical properties of the universe are represented by a unique state-function $\Psi(c,f)$. This is obtained from the path integral method but with the "paths" involved being restricted to: (i) All four-dimensional curved spaces with time and space on an equal footing and with a single three-dimensional boundary as in Fig. 6. The curvatures of these spaces must be such as to induce the given three-curvature c on this boundary; (ii) All fields on these four-spaces such that the value on the three- boundary is the given field f. Note that: (a) The "transition" probability associated with this state-function is $K(c,f) = |\Psi(c,f)|^2$. Hence, unlike the more conventional object $K(c_2,f_2; c_1,f_1)$. $K(c,f)$ is a function of just a *single* configuration point (c,f): there is no (c_1,f_1) corresponding to an earlier configuration and time from which the system has "evolved". This is the precise sense in which the theory is said to predict the probability that the universe is created in various configurations "from nothing"; (b) The space illustrated in Fig. 6 has no conical singularity and, correspondingly, there no actual "point" of creation; the four-dimensional space simply "is". This is a direct consequence of the assignment of equal status to the space and time directions; (c) Notwithstanding this use of "imaginary" time, the state-function $\Psi(c,f)$ thus defined satisfies the Wheeler-DeWitt equation. Thus it still describes the evolution of the physical parts $(c,f)_{Phys}$ of the system with respect to an internal/phenomenological time. This is one of the most attractive features of the Hartle-Hawking scheme; (d) When referred to such an internal time, the quantum and classical pictures of the Big Bang give very similar results far from the conical singularity. However, the behaviour of $\Psi(c,f)$ is such that, as the three-boundary in Fig. 6 gets smaller (the analogue of getting near the "creation point"), it becomes harder to sustain an interpretation of an evolution with respect to a genuine time variable. Basically, as the underlying equality of space and time directions starts to assert itself, the phenomenological time begins to pick up something like an imaginary part with its associated non-physical features. By this means, the problem of the "beginning of time" is adroitly averted.

The reason why this is sometimes regarded as a tunnelling process is very briefly as follows. We concluded the discussion on conventional quantum theory in Section 4 by remarking that the dominant contribution to the transition probability $K(q_2, t_2; q_1, t_1)$ would come from a solution to the classical equations of motion which passed through the configuration points q_2 and q_1 at times t_2 and t_1, respectively. If there is no such classical path, the transition from q_2 to q_1 can only arise at the quantum level, and this is essentially what is referred to as "quantum tunnelling." The dominant contribution to this process can be shown to come from a solution to the classical equations in which the time variable t is replaced everywhere[26] by it.

But now we recall that the "paths" employed in the calculation of the creation probability are four-dimensional spaces in which space and time are treated on an equal footing, and this is the general relativistic equivalent of replacing Newtonian time t with it. For sufficiently small three-boundaries, the dominant contribution to the path-integral will come from a classical solution to the Einstein equations of this special type. It is in this rather analogical sense that the creation of the universe is said to be a process of quantum tunnelling.

Let us now summarize what has been discussed from the point of view of theoretical physics.

5.2 *Achievements*

(i) The problem of the "beginning of time" has been tackled via the appreciation that "time" is a phenomenological construct in general relativity. In the quantum context this means that the exact interpretation of part of the (c,f) variables as intrinsic time depends on the state $\Psi(c,f)$ itself, and that in the extreme quantum region the Wheeler-Dewitt equation can no longer be interpreted in terms of an evolution of the physical parts of (c,f) with respect to an internal time. Thus "time" (i.e., our familiar time) comes slowly into focus as the size of the three-surface gets larger.

(ii) The use of "imaginary-time" paths in the computation of $\Psi(c,f)$ has enabled us to use solutions to the classical field equations that are free of the conical singularity in the traditional picture of the Big Bang spacetime. The initial space from which the universe "emerged" can be defined to be that part of the boundary of the four-dimensional space which is *not* part of the (later) three-surface. But this is the empty set, which gives a precise mathematical definition of the concept of "nothing"!

(iii) The theoretician's traditional demarcation between equations of motion and boundary conditions is broken by the prediction of a single, unique state-function which determines the entire (probabilistic) history of the quantum universe.

(iv) The creation from nothing is precisely that. In particular, there is no sense of a bouncing or oscillating universe.

(v) The history of quantum gravity research has been marked by many deeply-felt disagreements about whether the primary object of study should be three-dimensional space or four-dimensional spacetime. The use of four-dimensional spacetime "paths" to construct the state-function on three-spaces constitutes an elegant resolution of this dichotomy.

(vi) The theory is capable (at least in principle) of making predictions that could be positively refuted by experiment. For example, model calculations suggest that, with probability one, the universe will emerge from the Big Bang in an "inflationary" phase, a prediction that appears to agree well with current cosmological data.[27]

5.3 *Assumptions*

It is important to be clear about what has been fed into these theories, especially since they have been hilled by anti-theists as further evidence of science's ability to explain "everything". The greatest supposition is perhaps that shared by all cosmologists, namely that (i) the material universe exists, and (ii) it is susceptible in its entirety to mathematical analysis. More specific assumptions are:

(i) The ideas of quantum theory can be extended from their home in the atomic world to the universe at large. Many theoretical physicists have strong reservations about both the technical and the philosophical validity of such a vast extrapolation from microcosm to macrocosm. The entire procedure is particularly inimical to those whose view of quantum theory is primarily instrumental.

(ii) The representation of space and/or spacetime by a mathematical continuum is correct, even at the minute distances where quantum effects would be expected to dominate. This is a non-trivial matter. Many workers in the area (including myself) expect that a complete theory of quantum gravity will require a radical revision of the simple ideas of continuum spaces.

(iii) Einstein's field equations correctly describe the large-scale, classical gravitational properties of the universe.

(iv) The material content in the universe can be satisfactorily described within the language of local, interacting quantum fields (or superstrings, or whatever else is currently fashionable in the world of elementary particle physics). This is a non-trivial supposition and is a (usually unacknowledged) ingredient in all the so-called "theories of everything."

These are strong assumptions, and they will of necessity be satisfied in any "universe" predicted by the theory. Like most things in life, theoretical physics does not yield something for nothing, and what you get out is what you put in.

5.4 *Problems*

(i) It is possible to contemplate the creation of more than one universe in the sense that, for example, we could use "paths" involving two disjoint spaces of the type in Fig. 6. Then the two three-dimensional boundaries correspond to the equal-time spaces of "two universes." But note that if we live in one of them there is no way of communicating with the other. So as far as we are concerned, we can forget about it! (This is in sharp distinction to the creation within a pre-existent spacetime discussed briefly in Section 2.3. There, the matter produced at different seed-points will ultimately come into causal contact.) However, the possibility of "real" universes other than our own is not without its implications for philosophical and theological speculation.

(ii) It is most unlikely that the theory is mathematically consistent in the form presented above. Quantum theories of gravity tend to be plagued with ill-defined expressions which are singularly difficult to remove. The current favourite for combining general relativity and quantum mechanics is superstring theory, although it is not yet clear if this really works. This uncertainty enhances my general feeling that the broad conceptual ideas of quantum cosmology are more relevant for the science-religion dialogue than the technical details of the theories themselves.

(iii) Similarly, the Hartle-Hawking ansatz is not the only way of constructing a "unique" quantum state or of obtaining a picture of creation from nothing. Other methods have been suggested which would yield different physical predictions,[28] although they are not so well-developed and, to my mind, neither do they fit in as well with a general approach to the problem of quantum gravity. A particularly striking feature of the scheme we have discussed is the way in which "imaginary" time is introduced. This is a commonplace in ordinary (i.e., special relativistic) quantum field theory and is really just a simple mathematical trick. But the situation in general relativity is otherwise, and there are radical differences between spaces with real or imaginary time. For some time (real), Hawking has been vigourously pursuing a quantum gravity programme based on the latter, but the issue is still a topic of much debate amongst those who work in these areas.

(iv) What does the theory mean? Or, more precisely, what is the interpretation [29] of the (unique) state-function $\Psi(c,f)$?

This final problem is extremely non-trivial. Even setting aside the general problem of probabilities for the universe, it seems unlikely that $|\Psi(c,f)|^2$ can be regarded as a probabilistic distribution for (c,f), since these variables include an intrinsic time which should presumably be removed first. It seems more appropriate to interpret $|\Psi(c,f)|^2$ as a probability distribution for the physical modes $(c,f)_{Phys}$ at the internal time specified by the pair (c,f). However, as we have remarked already, the latter can only be interpreted as a genuine time variable well away from the creation region. This suggests that, like the notion of classical time, probabilities may "emerge" from the formalism and that, near the creation region, $\Psi(c,f)$ may not have any direct physical interpretation at all! I rather like this view.

However, even if the physical modes can be isolated from the internal time, we are still faced with two major problems: (i) How should we interpret probabilities as applied to the physical modes of the entire universe? (ii) We have claimed that, far from the creation region, the quantum state is "almost classical." However, model calculations suggest that this state vector will not correspond to just *one* solution to Einstein's equations but will rather be a linear superposition of many such vectors. This is the problem of Schrödinger's cat writ large! It raises in an extreme form the general question of the ontological status of the entities described by quantum theory and the duality between realism and idealism that lies at the heart of much of the debate on the interpretation of the mathematical symbols.

In the context of quantum cosmology, frequentist interpretations of probability seem ruled out,[30] as do any references to "measurements" of "observables", and the like. Much has been written about the type of conceptual scheme that might be applicable in the context of quantum cosmology, but this is a highly contentious area and there is no general agreement about what is correct. My own views on this problem are rather fluid, but at the moment I am inclined to support the school of thought which maintains that the only meaningful probabilistic statements are those affirming something with probability one.[31] The utility of such a cautious position relies on the adoption of something like Everett's ideas of "relative" states [32] and the implication that the state-function of quantum cosmology can be used only to predict guaranteed *correlations* between the values of observables associated with small subsystems of the universe. This is similar in many respects to the "many-worlds" interpretation but without the excess metaphysical assumptions that sometimes become attached to the latter.

This approach (which allows a modicum of philosophical realism) has the advantage of being geared to reproduce the conventional statistical interpretation of normal, small-scale quantum theory but without making any fundamental distinction between an observer and a system (they are both quantum subsystems of the universe). The main idea is to use the theory to study the quantum state of an actual ensemble of sub-systems, and then to show that, with probability one, an appropriate observable coupled to this ensemble will reproduce the usual results.[33] Whether or not such a scheme can really be applied to the entire universe is still a matter for considerable debate. But note that, in any event, it will only work if the universe is sufficiently complex to contain real ensembles of physical systems; the interpretation of probability as applied to a *single* electron remains as mysterious as ever.

6. *Implications for Theology*

In his contribution to this volume Nicholas Lash has remarked rather laconically that the "dialogue" between science and theology was prone to be rather one-way, and with the former doing most of the talking![34] If the present paper (and especially the title of this final section) appears to substantiate this view, it is not because I have succumbed to the hubris of maintaining that the epistemology and methodology of science are normative for every area of human experience. My reasons for presenting the ideas underlying a modern scientific theory stem rather from a belief that philosophy and theology are indeed the "queen of sciences" and, as such, are charged with the awe-inspiring task of overseeing *all* modes of enquiry and of cohering them in a unity of vision that is both emotionally and intellectually satisfying.

What then can these new scientific ideas on creation contribute to the theological archive of metaphysical wisdom? I suspect that the honest answer is "not very much," although some attention should surely be paid to the shifting forms in which the archetypes of space and time are impinging on the scientific world. The role played in these theories by the concept of internal/phenomenological time emphasises strongly the absence in general relativity of any notion of "absolute time," or of the "now" of human experience. Indeed, all "times" are co-present and have an equal ontological status. The metaphysical implications of this move away from the conceptual structure of Newtonian physics would certainly seem worthy of consideration by any "process" theologian who desires to ascribe a time-dependence to the relation between God and the physical world. Similarly, and notwithstanding the peculiarities introduced by their quantum content, these new theories are still deterministic at heart and, as such, are difficult to reconcile with any genuine sense of an "openness to the future."

At a more adventurous level, one might consider the possible implications of the actual "content" of the theories and, in particular, the eradication of the conical singularity in the conventional Big Bang picture of Fig. 5. There is no doubt that, psychologically speaking, the existence of this initial singular point is prone to generate the idea of a Creator who sets the whole show rolling. The new theories would appear to plug this gap rather neatly, although in saying this we must keep in mind the many *a priori* assumptions made in the formulation of these theories, and the existence of several different schemes for constructing a "unique" state-function of the universe. There is clearly a danger that the original problem of the multiplicity of possible initial states will be replaced by an equivalent multiplicity of theories claiming to determine it uniquely!

From an aesthetic point of view, there is something rather attractive about the completeness of spacetime as represented in the Hartle-Hawking proposal; one can almost imagine the universe represented in Fig. 6 being held in the cup of God's hand. This picture of God "sustaining" the world in all its manifold "times" has been developed in some detail by W. Drees in his recent study of the possible theological implications of the work of Hartle and Hawking.[35]

However, the God of Christianity is not only "the ground of Being." He is also Incarnate, and the absence in the scientific account of any sense of the "passing of time" opens up a significant gulf with the eschatological experience of personal religious life. This is exemplified by Torrance's discussions of these matters and his emphasis on the specifically Christian intertwining of Creation, Incarnation, and Resurrection.[36] Thus he writes of our deep sense of pathos at the passing of time and the loss of the past, and their concomitants of decay, finitude, and death. This is counter-balanced by the vision of the Resurrection as the "new creation out of the old order" and thus to the almost Gnostic, but profound, notion of the "redemption of time" though the life and death of Christ. I think it will be rather a long time before theoretical physics has anything useful to add to that.

Acknowledgements

I am most grateful to Noah Linden and Chris Marooney, who read through an early draft of this paper and made a number of helpful suggestions for improving the clarity of the exposition. I would also like to express my considerable gratitude to Wim Drees for his detailed and comprehensive set of comments on the draft version; these were a most valuable aid in the preparation of the final form of the paper. Finally, I would likc to thank all the participants at the Castel Gandolfo meeting for their friendly, but informed and insightful, remarks and I can only express my deepest thanks to the organisers of the conference for producing such a stimulating and memorable occasion.

NOTES

[1] More precisely, if the picture of expansion given by the current experimental data is extrapolated backwards, it appears "as if" the universe began at a specific time in the past. However, the data can be rendered compatible with a range of disparate conjectures concerning the very early stages of the universe, and it follows that any theory of "creation" (including the one discussed in this paper!) must be regarded as being highly speculative.

[2] Arthur R. Peacocke, *Creation and the World of Science* (Oxford: Oxford University Press, 1979).

[3] John C. Polkinghorne, *One World: The Interaction of Science and Theology* (London: SPCK, 1986).

[4] G.C. Henry, *Logos: Mathematics and Christian Theology* (New Jersey: Associated University Presses, 1976).

[5] T. Torrance, *Space, Time, and Incarnation* (Oxford: Oxford University Press, 1968) and *Space, Time, and Resurrection* (Edinburgh: The Handsel Press, 1976) has developed in a powerful way the relation of the modern views of space and time to the Christian triad of Creation-Incarnation-Resurrection.

[6] S. L. Jaki, *Cosmos and Creation* (Edinburgh: Scottish Academic Press, 1980) gives a lively account.

[7] P. Edwards, ed., *Encyclopedia of Philosophy* (New York: MacMillan, 1967) **6,** 524-525, gives an illuminating account of why the concept of "nothing" is nothing to worry about.

[8] J.D. Barrow and F.J. Tipler, *The Anthropic Cosmological Principle* (Oxford: Clarendon Press, 1986).

[9] When applied to quantum theory, the significance of this operationalist-realist distinction is the subject of much argument and debate.

[10] The causal nature of the time evolution is typically associated with a first-order differential equation (in time) for the numerical parameters (i.e., the "coordinates") specifying the location of a point in the state space **S**. Generically, a solution to such an equation is uniquely determined by the point in **S** through which it passes at some reference time.

[11] The physical relevance of such "causality" depends to some extent on the stability of the solutions under small variations in the point B through which the curve passes at time t_B. A phenomenon much discussed in recent years is "chaotic" motion, in which the solution changes wildly under such variations. For all practical purposes, systems of this type cannot be regarded as deterministic.

[12] The crucial point is that the equations of motion for a typical classical system are second-order differential equations in the coordinates of **Q.** Thus it suffices to specify the initial point q in **Q** and the tangent vector to the trajectory at that point. In particular, this means that the dimension of the state space is twice the dimension of the configuration space.

[13] Jung's collected works should be compulsory reading for all theoretical physicists! They are published (currently twenty-one volumes) by Routledge and Kegan Paul in London, and by Princeton University Press in the USA.

[14] It seems that Jung was preempted somewhat by Aristotle. In the introduction to *On the Parts of Animals, Book 1,* he speaks of the "immense pleasure" felt by "all those who can trace the links of causation." As usual, there is nothing new under the sun! I am grateful to Ernan McMullin for drawing my attention to this reference.

[15] Ian G. Barbour, *Myths, Models and Paradigms* (London: SCM Press, 1974).

[16] It should be noted that, in a mathematical sense, the singular point X will typically lie on the *boundary* of the space **S** of physical states. Thus, strictly speaking, X is not actually in **S** itself but rather on its "edge."

[17] E.P. Tryon, "Is the Universe a Vacuum Fluctuation?" *Nature* **246** (1973) 396-397 and R. Brout, F. Englert, and E. Gunzig, "The Creation of the Universe as a Quantum Phenomenon," *Annals of Physics* **115** (1978) 78-106.

[18] If the particles have masses M and m and are distance r apart, the potential energy of their gravitational interaction is $-GMm/r$, where G is Newton's constant.

[19] This particular problem could possibly be avoided by postulating that the creation of matter takes place in an expanding De Sitter background spacetime.

[20] The relevant reference is Book 11 of *The Confessions of St Augustine*. A number of editions of this work are available although, unfortunately, some of them end with Book 10!

[21] Philo, *On the Account of the World's Creation Given by Moses,* trans. F.H. Colson and G.H. Whitaker (London: William Heineman, Loeb Classical Library, 1981).

[22] *Ibid.*, Sec. 26.

[23] Another way of stating this is to observe that, in this particular approach to quantum gravity, the "time" variables are operators and can therefore be regarded in some respects as if they were observable quantities. This is in sharp contrast to the normal quantum mechanical interpretation of "time" as a fixed, external parameter.

[24] S.W. Hawking, "The Boundary Conditions of the Universe," in *Astrophysical Cosmology*, eds. H.A. Bruck, G.V. Coyne, and M.S. Longair (Vatican City: Pontifical Academy of Sciences, 1982) 563-572; and J.B. Hartle and S.W. Hawking, "Wave Function of the Universe," *Physical Review* **D28** (1983) 2960-2975.

[25] It should be noted that other ways of finding a state function have also been proposed recently, although, in this paper I have discussed only the Hartle-Hawking approach. This is partly because it is the easiest to visualize in a geometrical way, and partly because it fits into a general quantum gravity framework and has an aesthetic appeal which, for me, is lacking in any of the competing suggestions. See, for instance, A. Vilenkin, "Boundary Conditions in Quantum Cosmology," *Physical Review* **D33** (1982) 3560-3569.

[26] In the context of a quantized Newtonian system with a potential energy V, one way of understanding this effect is to note that a path which is forbidden by the equations of motion may be perfectly feasible if V is replaced by $-V$. Rather remarkably, it can be shown that the *quantum* tunnelling probability for the original system is determined mainly by the contribution to the path integral given by this *classical* solution to these new equations. However a transformation of V to $-V$ in Newton's second law of motion can also be obtained by leaving V alone but replacing t with it. This is basically the origin of the invocation of an imaginary time parameter.

[27] The idea that the universe underwent an inflationary expansion shortly after its creation has been discussed extensively in recent years. "Inflationary" refers to an expansion that is much faster than any that can be obtained from conventional matter within the framework of classical general relativity. This rapid expansion is driven by certain peculiar quantum mechanical properties of matter fields coupled to gravity, but it is not directly relevant to the idea of quantum creation itself.

[28] See especially the work by Vilenkin; see Note 25.

[29] It should be remarked that most of the conceptual problems are common to *any* quantum theory of gravity that involves state-functions of the type $\Psi(c,f)$. Creation theories "merely" add the extra difficulty of understanding the implications of a structure that predicts a unique such state.

[30] Unless it can be argued that the ensemble is in some way "physical". For example, one might try and apply the statistical predictions to the different cycles of an oscillating universe, or perhaps to the causally disconnected regions in a single universe arising from a spontaneous symmetry breakdown (and associated production of domain walls) occurring near the Big Bang.

[31] R. Geroch, "The Everett Interpretation," *Nous* **18** (1984) 617-633.

[32] See the many articles reprinted in B. S. De Witt and N. Graham, eds. *The Many Worlds Interpretation of Quantum Mechanics* (Princeton: Princeton University Press, 1973).

[33] See J. B. Hartle, "Quantum Mechanics of Individual Systems," *American Journal of Physics* **36** (1968) 704-712.

[34] See Lash in this volume.

[35] W. B. Drees, "Beyond the Limitations of the Big Bang Theory: Cosmology and Theological Reflection," *Bulletin of the Center for Theology and the Natural Sciences* (Berkeley) **8**, no. 1 (1988),

[36] See Note 5.

INDEX OF NAMES

Aaron, 156
Abelard, Peter, 81
Abraham, 164,166,180
Abram, 164
Adam, Charles, 102n
Adam, 57,164
Albēri, Eugenio, 100n
Alexander, S., 314
Alter, R., 169n
Anderson, Bernhard W., 75n,169n
Andison, Mabelle J., 102n
Anselm, St., 91,180
Anshar, 153
Anu, 152-153
Apel, Karl-Otto, 215n
Apsu, 152-153
Aquinas, St. Thomas, 25,28,36-38, 60-63,67,72,76n-77n,82,108-110, 113-114,116-118,121,122n,174, 182,215n,245n,271n,288-292, 294n-296n,317-319,323
Arbib, Michael A., 201n-202n,267n
Archimedes, 126,143
Aristotle, 25,37,53-55,58-63,67, 76n,84,92,100n,110,114,126,143-144,252,254,259,268n,271n,288-289,294n,317-318,334,361,406n
Armstong, A.H., 75n
Arnaud, Daniel, 168n
Asshur, 152-153
Aten, 158
Augustine, St., 25,56-59,70,72-73,75n-76n,82,91,268n,270,271n, 279,281,283,294n,319,329n,387, 388,407n
Austin, William, 47n,373n
Averroes, 76n
Avicenna, 76n
Ayala, Francisco J., 315,329n

Baal, 154
Bach, Johann Sebastian, 357
Bacon, Roger, 60,200n,214n
Barbour, Ian G., 47n,247n,267n, 270n,274,276,286-291,293n,295n-296n,329n-331n,351,354,360, 371n-374n,383,407n
Barden, Garrett, 215n
Barr, James, 168n
Barrett, David B., 102n
Barrow, John D., 48n,78n,86,184n, 195,201n-202n,314,325-326,329n, 341,342n,406n
Barth, Karl, 28,32,46n,264n,360
Baynes, K., 122n
Behemoth, 163
Bell, John S., 346,348-349,351-354,366,371n
Bennet, D.C., 329n
Bentley, Richard, 65,86,89-90, 101n
Bergant, Dianne, 75n
Bergier, Nicolas-Sylvain, 98,102n
Bergson, Henri, 72
Berkeley, Bishop, 64,335
Bernstein, Richard J., 215n,268n
Bertotti, B., 246n
Betty, L.S., 122n
Bildad, 161
Bingham, Alfred J., 102n
Bion of Borysthenes, 91
Birch, L. Charles, 44,48n
Birtel, Frank, 328
Black, Max, 265n, 267n
Blacker, C., 168n
Boeder, H., 122n
Bohm, David, 336,338-340,342n, 347-350,352,371n
Bohman, James, 100n,122n
Bohr, Niels, 335-338,342n,348-350,352-354,359-361,372n-373n
Boltzmann, Ludwig, 344,346
Bonaparte, Napoleon, 94,181
Bonaventure, St., 289
Bonhoeffer, Dietrich, 360
Born, Max, 348
Bose, S.N., 344-346,348,362-364, 366-367,370n
Boslough, J., 201n
Bottero, J., 168n
Boyd, Richard,178-179,184n
Boyle, Robert, 64-67,77n,86,89
Bradley, W., 122n
Bramah, Ernest, 302

Brück, H.A., 245n,407n
Brewster, D., 140n
Brout, R., 407n
Brown, Delwin, 78n
Bruno, Giordano, 91,96-97
Buber, Martin, 211
Buccellati, G., 169n
Buckley, Michael J., 100n-102n, 214n,231,242,247n
Bultmann, Rudolf, 29,46n,264n
Burian, Richard, 46n
Burns, J.O., 245n
Burrell, David B., 79n,215n
Butterfield, Jeremy, 202n
Butts, Robert, 77n
Byers, David M., 47n,75n
Bynum, Caroline, 269n

Cain, 166
Cajori, Florian, 100n-101n
Calvin, John, 25,34,60,264n
Caplan, Arthur, 46n
Capra, Fritjof, 38,48n,78n,340, 342n
Caputo, John, 122n
Cardano, Geronimo, 91
Carr, Bernard J., 70,201n
Carter, Brandon, 78n,201n,307
Cassirer, Ernst, 96, 102n
Charron, Pierre, 91
Chaudon, Louis-Mayeul, 98,102n
Chew, Geoffrey, 371n
Chopp, Rebecca, 268n
Christ, see Jesus of Nazareth,
Cicero, 76n,91-92
Clark, Elizabeth, 268n
Clarke, Samuel, 89-90,95,97-98, 101n
Clarke, W. Norris, 122n
Clauser, John F., 351,371n
Clement VI, Pope, 61
Clifford, Richard J., 168n-169n
Cobb, Jr., John B., 43-44,48n, 78n,286,295n
Cohen, I. Bernard, 77n, 100n
Cohenand, R.S., 246n
Collins, Anthony, 97
Collins, Harry M., 200n
Collins, James, 92,100n,102n
Colson, F.H., 407n
Columbus, Christopher, 177,181
Copernicus, Nicolas, 200n
Copleston, Frederick, 113
Cordell, B., 122n
Cotes, Roger, 90
Cousins, Ewart H., 270n
Coyne, George V., 101n,245n,311n, 407n
Craig, Edward, 214n
Crew, Henry, 100n
Crick, Francis, 23,46n
Crossan, John Dominic, 356,372n
Crosson, Frederick, 76n
Cumming, John, 215n
Cupitt, Don, 184n
Czerny, Robert, 267n

Daecke, Sigurd, 264n
Darwin, Charles, 26,58,66-69,73, 105,134,151,168n,307,309,326-327
David, 50
Davidson, Donald, 184n
Davies, Paul C.W., 86,196,201n, 273,293n,296n,301,342n
de Alanus, Insulis, 81
de Broglie, Louis, 348
de Felice, F., 246n
de Lamennais, Felicite, 98
de Rūjula, A., 245n
de Salvio, Alfonso, 100n
De Vaucoleurs, G., 246n
De Witt, Bryce S., 195,200n-201n, 396-397,400-401,408n
Delson, Eric, 329n
Democritus, 285
Dennett, Daniel C., 329n,331n
Derham, William, 65,93
Derrida, Jacques, 106-107,122n, 253,267n
Descartes, Rene, 60,63-66,70,77n, 83-86,88-89,92-97,100n,102n, 107,110-111,198,200n,203,206, 215n,231

Dessain, C.S., 215n
Deutsch, David, 194,201n
Dewey, John, 110,122n
Diagoras of Melos, 91
Diderot, Denis, 93,95-97,102n
Diodati, Elia, 100n
Dirac, P.A.M., 334-335,342n,344, 369n
Dobzhansky, Theodosius, 329
Dodson, Edward, 77n
Drake, Stillman, 100n
Drees, Wim B., 295n,373n,405,408n
Dressler, A., 245n
Dryden, John, 203-204,207,214n-215n
du Nouy, Lecomte, 68
Dyer, Charles C., 245n,247n
Dyson, Freeman J., 78n,103n,122n, 321
D'Arcy, M.C., 269n
D'Espagnat, Bernard, 201n
d'Holbach, Baron Paul, 96-98,102n

Ea, 152-153
Eddington, Arthur S., 32,47n,130, 133,140n
Edwards, P., 406n
Ehlers, Jurgen, 328
Eigen, Manfred, 364
Einstein, Albert, 30,35,47n,109, 141n,148,150n,191,225,274,293n, 323-333,335,337,344,348-350, 353,371n,375,389,391-392,397-400,402-403
Elihu, 161
Ellis, George F.R., 150n,228,244, 245n-246n
Ellis, J., 245n,247n
Englert, F., 407n
Enlil, 153
Epicurus, 90-91
Euclid, 143,227,325,350
Eve, 57
Everett III, Hugh, 193,201n,403, 408n
Ezekiel, 181-182

Farrer, Austin, 79n
Favaro, Antonio, 100n
Ferguson, James P., 101n
Fermi, Enrico, 308,344-346,348, 362-364,366-367,370n
Ferre, Frederick, 47n,267n
Feuerbach, Ludwig, 175,184n,204, 212,254
Feynman, Richard, 369n
Fichte, Johann G., 212
Flew, Antony, 267n
Folse, Henry J., 353,372n
Foucault, Michel, 268n
Fowler, Dean R., 326,331n
Freedman, Daniel Z., 224,245n
Freeman, W.H., 245n
Frei, H., 168n
Freud, Sigmund, 204
Friedman, David, 331n
Friedmann, Alexander A., 230
Frye, Northrop, 199,202n
Frye, Roland M., 46n,75n

Gadamer, Hans-Georg, 111,207-208, 212,215n
Galileo Galilei, 25-26,57,63,82-83,86,92,100n,185-187,200n,231, 309
Gay, Peter, 102n
Gödel, Kurt, 325,330n
Geller, Margaret J., 245n
Geroch, R., 408n
Geyer, H.G., 170n
Gibbons, G.W., 311n
Gilkey, Langdon, 29-30,46n,78n, 275,278,289-291,293n-294n,296n
Gilson, Etienne, 62,76n,99
Gombrich, Ernest, 267n
Gornall, Thomas, 215n
Graham, Neill, 200n-201n,408n
Grant, Edward, 76n
Gratian, 81
Green, M.B., 245n,324
Greenberg, M., 169n
Gregory of Nyssa, 57-58
Griffin, David Ray, 44,48n,286, 295n

Gunzig, E., 407n
Gustafson, James M., 264n
Guth, Alan.H., 78n,245n,300

Habel, Norman, 161,169n
Haber, H.E., 245n
Habermas, Jurgen, 81,200n
Hague, René, 266n
Hahn, Roger, 102n
Haldane, H.S., 102n
Hanson, N.R., 267n
Haran, 164
Harman, Gilbert, 184n
Harré, Rom, 267n
Harrison, Graham, 214n
Hartle, James B., 197,245n,295n, 373n,376,398-400,402,405 407n-408n
Hartshorne, Charles, 43,48n,72, 78n,112,123n,246n,286,363
Hartsoeker, Nicolaas, 93
Hartt, Julian N., 270n
Harvey, William, 85
Hastie, W., 77n
Hawking, Stephen W., 70,123n, 150n,196-197,201n,222,245n, 295n,311n,330n,373n,376,398-400,402-403,405,407n
Hebblethwaite, Brian L., 79n
Hegel, Georg Wilhelm Friedrich, 96,102n,106,204,206,209,212, 214n,271
Heisenberg, Werner, 132-134,140n, 333-336,342n,348-349,350,354, 361,369n
Helck, W., 169n
Heller, Michael, P1,101n,150n, 202n,228,246n-247n,311n,328
Helmholtz, Hermann Ludwig von, 190
Henry, G.C., 406n
Hepburn, R.W., 253
Herbert, Nick, 351-352,371n
Heron of Alexandria, 84,101n
Herschel, William, 94
Hesse, Mary B., 200n-202n,214n-215n,244,247n,253,265n,267n
Hick, John, 266n
Hincmar of Reims, 81
Hobbes, Thomas, 90
Hodgson, Peter C., 184n,214n,270n
Hofstadter, Douglas R., 329n,331n
Holton, Gerald, 247n
Honan, Daniel, 76n
Hook, Sidney, 331n
Hooker, Edward N., 214n
Hooykaas, R., 76n
Horsley, Samuel, 101n
Hosea, 182
Hoyle, Fred, 103,105,281,285-286, 288,290,295n
Hubble, Edwin Powell, 133-134, 140n,220
Hume, David, 175,190,323,329n
Husserl, Edmund, 185-188,192,196-199,200n
Hutten, E.H., 267n
Huygens, Christian, 92

Irenaeus, 270n,286,294n
Isaac, 180
Isham, Christopher J., 122n,197, 200n-202n,214n,228,244,245n, 283,295n,354,372n-373n

Jacob, Edmund, 287
Jacob, 175,180
Jaki, Stanley L., 47n,330n,354, 372n,406n
James, Ralph, 78n
Jammer, Max, 371n
Jantzen, Grace, 271n
Jenson, Robert, 294n
Jeremiah, 50,54-55,70
Jesus of Nazareth, the Christ, 26,28-30,32,36,40,44-45,52,56, 94,99,173,175,183,250,254,258, 266n,268,273,278,292,356,358, 362,367
Job, 52,160-164,169
John of Damascus, 279,294n
John Paul II, Pope, P1,P3,25,273, 275
Johnson, Dr. Samuel, 181,335

Jones, James W., 47n
Joule, James Prescott, 190
Jowett, B., 75n
Julian of Norwich, 261,271n
Jung, Carl Gustav, 247n,383,406n

Kaiser, Christopher B., 373n
Kane, G.L., 245n
Kant, Immanuel, 36,66,77n,103, 105,109-110,122n,168n,175,190, 207,211,215n,227-228,246n,315, 383
Kasper, Walter, 212,215n
Kaufman, Gordon, D., 78n,266n
Kelley, Jerry S., 331n
Kelsey, David, 75n,296n
Kelvin, Lord, 190,
Kenny, Anthony, 76n,329n,
Kermode, Frank, 169n,208,215n
King, Robert H., 184n,270n
Kishar, 153
Kitcher, Philip, 46n
Koch, Klaus, 170
Koelln, Fritz C.A., 102n
Kors, Alan Charles, 102n
Koyre, Alexandre, 100n
Kraus, H.J., 170n
Krauss, L.W., 245n
Kripke, Saul, 177-179,181,184n
Kuhn, Thomas, 38,47n
Küng, Hans, 47n,293n,360,373n

Lafleur, Laurence J., 102n
Lagrange, Joseph, 94-95
Lahamu, 152-153, 168n
Lahmu, 152-153,168n
Lamarck, Jean Baptiste de, 68-69
Lambert, W.G., 168n-169n
Lande, Alfred, 201n
Laplace, Pierre Simon de, 66,94-95,102n
Lash, Nicholas, 150n,202n,215n, 249,264n-266n,268n,404,408n
Leewenhoek, Antoni van, 93
Leibniz, Gottfried Wilhelm, 88-89,94,388
Leplin, Jarrett, 372n
Leslie, John, 78n,86,122n,200n, 202n,328
Lessius, Leonard, 91-92,101n-102n
Leviathan, 158,163
Lewis, C.S., 261
Lewis, John, 46n
Lindbeck, George, 31,47n
Lindberg, David, 75n-76n
Linde, A.D., 245n,301
Linden, Noah, 405
Locke, John, 175,190,200n
Loewe, M., 168n
Lombard, Peter, 81
Lonergan, Bernard J.F., 107,215n, 336,342n
Longair, M.S., 78n,201n,245n,407n
Lovejoy, Arthur O., 327,331n
Lovelock, James E., 41
Lucretius, 90,94,144
Luther, Martin, 25,82,264n
Lyons, John, 184n
Lyotard, J. F., 122n

Maartens, R., 246n
MacCallum, Malcolm A.H., 245n
Mach, Ernst, 94,102n
MacIntyre, Alasdair, 267n
Mackie, J.L., 311n
MacQuarrie, John, 184n
Malebranche, Nicolas, 64
Malpighi, Marcello, 85
Manuel, Frank E., 214n
Marduk, 152-155,157
Margenau, Henri, 350,371n
Maritain, Jacques, 102n
Markus, R.A., 75n
Marsden, G., 168n
Marx, Karl, 96,212
Mary, 174
Mattem, Ruth, 46n
Maupertuis, 190
Maxwell, James Clerk, 189,344
Mayer, Julius Robert, 190
McCarthy, T., 122n
McCool, Gerald, 47n
McCormack, Thomas J., 102n
McFague, Sallie, 37,47n,295n,354-

355,357-358,372n
McGinn, Bernard, 79n
McKeon, Richard P., 100n
McLachlan, H., 101n
McMullin, Ernan, 35,46n,75n-78n, 102,104-105,122n,202n,227-228, 242,245n-247n,264n,286,293n, 295n,328,329n,354,368,370n-373n,406n
Meiklejohn, J.M.D., 77n
Mellor, D.H., 202n
Menu, B., 168n
Mersenne, Père Marin, 83,91-92, 102n
Messenger, E.C., 76n
Migne, J.P., 76n
Millard, A.R., 168n
Miller, James B., 326,331n
Mills, R.L., 324
Milton, John, 130,140n
Misner, Charles W., P3,245n,397
Mittelstaedt, Peter, 246n
Moltmann, Jürgen, 270n,294n,296n
Monk, J. Donald, 330n
Monod, Jacques, 23-24,46n,56,75n, 363
Montaigne, Michel Eyquen de, 84
Montefiore, Hugh, 41,48n
Moran, W.L., 154,168n
Morgan, C. Lloyd, 72
Morooney, Chris, 405
Morris, Henry, 46n
Morris, S.F., 266n
Morris, Thomas V., 98,102n
Moses, 117,156,277,407n
Motte, Andrew, 100n
Mudge, Lewis S., 215n
Mueller-Vollmer, Kurt, 215n
Mummu-Tiamat, 153
Murphy, R.E., 169n
Musschenbroek, Petrus van, 93
Nagel, Ernest, 325,330n
Nahor, 164
Nel, Stanley D., 246n
Neville, Robert Cummings, 281-282,294n
Newman, James R., 325,330n
Newman, John Henry, 207,215n
Newton, Sir Isaac, P1,P3,34,65-67,77n,82-93,95-97,99,100n-101n,104,130,135,141,189,190, 197,203,205,212,214n,231-232, 242,333-336,338,370n,382,388-389,394,400,404,407n
Nichols, Joannes, 101n
Nietzsche, F., 319
Nieuwenhuizen, P., 245n
Nieuwentijt, Bernard, 93
Nineham, Dennis, 251,265n,267n
Nisbet, H.B., 215n
Noah, 165,203-204
Nudimmud, 153
Numbers, Ron, 75n
Nygren, Anders, 268n-269n

Oakeshott, Michael, 207
Ogden, Schubert M., 286,295n
Olsen, R., 122n
Ortony, Andrew, 184n
Outka, Gene, 269n
Outler, Albert, 76n
O'Connell, Matthew J., 215n
O'Donovan, Leo, 47n

Page, Don, 104
Pagels, Heinz, 371n
Paley, W., 104
Palmer, R.P., 102n
Pannenberg, Wolfhart, 35,47n
Pappus of Alexandria, 84,100n
Partridge, R. Bruce, 245n
Pascal, Blaise 180-181
Pascolini, A., 246n
Paul, St., 56,59,164,169n,181, 277,329n,358
Pauli, Wolfgang, 127,133,345,363-364
Peacocke, Arthur R., P1,P3,24,41-42,46n-48n,78n,208,214n-215n, 245n,247n,250,264n-267n,270n-271n,287-288,291,293n,295n-296n,350,354,357-358,362-363, 372n-374n,376,406n
Pedersen, Olaf, 150n, 215n

Peirce, Charles Sanders, 246n
Penrose, Roger, 201n,222,295n, 322,330n
Penzias, Arno A., 285
Peters, Ted, 247n,295n,373n
Pettergrove, James P., 102n
Peukert, Helmut, 81,100n
Philo of Alexandria, 323,387,407n
Pieper, Josef, 268n
Pippard, Sir Brian, 264n
Pittenger, Norman, 269
Pius XII, Pope, 329n,378
Planck, Max, 348,369n
Plato, 53,59,76n,92,122n,126,143, 149,247n,259,270n,277-278,294n, 377
Podolsky, Boris, 191,337,349,371n
Poincare, Henri, 83,334
Polanyi, Michael, 38-39,48n,107-108,111,339,342n
Polkinghorne, John C., 39,48n, 202n,228,342n,348,354,369n, 371n-372n,376,406n
Pollard, William G., 339-340, 342n,362,374n
Popper, Karl, 35,142,150n,192, 201n
Poupard, Cardinal Paul, 46n
Prigogine, Ilya, 201n,284,291, 295n,342n,364
Pritchard, J.B., 168n
Prometheus, 207
Psalmists, 51,56
Ptolemy, 338
Putnam, Hilary, 177-179,183,184n

Quinn, Edward, 373n

Rahab, 161
Rahner, Karl, 35-36,47n,204,206, 209,212,214n-215n
Rainer, E., 154
Ramsey, Ian, 182,184n
Rawleigh, Sir Walter, 101n
Ray, John, 65-66,77n,93
Redi, Francesco, 93
Rees, Martin J., 70, 201n
Reeves, Gene, 78n
Reiss, Hans, 215n
Restivo, Sal P., 78n
Richardson, Herbert, 255,268n
Ricoeur, Paul, 207,210,215n,253, 267n
Robertson, H.P., 202n,224,230
Rogers, D.M., 101n
Rolston, III, Holmes, 39,48n
Rorty, Richard, 214n-215n,268n
Rosen, Nathan, 191,337,349,371n
Ross, Thomas M., 46n
Rowan-Robinson, Michael, 245n
Ruether, Rosemary, 249,265n
Ruse, Michael, 46n
Russell, Bertrand, 83,113,177
Russell, Robert John, P1,48,75n, 123n,228,244,270n,328,371n
Rust, Eric, 78n
Ryan, John K., 76n

Sachs, R.K., 246n
Sagan, Carl, 25,46n
Sagredo, 82-83,100n
Sahlins, Marshall, 46n
Salam, Abdus, 223
Salviati, 82-83,100n
Samuelson, Paul A., 331n
Sarah, 164
Satan, 163
Sauneron, Serge, 168n
Scharlemann, Robert P., 266n
Schelling, Friedrich Wilhelm Joseph von, 314
Schilling, Bernard N., 214n
Schmid, H.H., 169n
Schrödinger, Erwin, 192-194,333, 369n,403
Schwabhauser, W., 325
Schwarz, John H., 324
Scopes, T., 26
Scotus, Duns, 115
Searle, John, 329n,330n
Seux, M.J., 168n
Shamash, 152
Shapiro, Jeremy J., 200n
Shea, William, 77n

Sheldrake, R., 41
Shimony, Abner, 348,371n
Siklos, S.T.C., 311n
Simplicio, 82,100n
Simson, Frances H., 102n
Sitter, Willem de, 407n
Smyth, Kevin, 215n
Socrates, 53
Solomon, 50
Soskice, Janet Martin, 37,47n, 184n,215n,267n,354,372n
Speiser, E.A., 168n
Spinoza, Benedictus, 89-90
Stapp, Henry, 371n
Steinhardt, P.J., 245n,300
Stengers, Isabelle, 201n,295n, 342n
Stevens, Wallace, 266n
Stoeger, William R., P1,245n-247n,328,354,367-368,372n
Strawson, Peter, 189,200n
Streng, Frederick, 47n
Strong, Edward, 86
Stuhlmuller, Carroll, 75n
Swimme, Brian, 123n

Tannery, Paul, 102n
Tarski, Alfred, 325
Taylor, J.H., 75n
Teilhard de Chardin, Pierre, 44, 48n,68-70,77n-78n,266n,270n,314
Temple, William, 78n
Terah, 164,166
Tertullian, 270n
Thales of Miletus, 189
Thaxton, C., 122n
Theophilus of Antioch, 277,286, 294
Thomas, Ivor, 100n
Thomsen, D.E., 122n
Thorne, K.S., 245n
Tiamat, 152-154
Tillich, Paul, 30,265n,294n,326, 331n,360,365,373n
Tipler, Frank J., 48n,78n,86, 122n,184n,195,201n-202n,245n, 329n,341,342n,372n,406n
Toland, John, 96-97
Torrance, Thomas, 30,34,47n,405, 406n
Toulmin, Stephen, 38,47n-48n,264n
Tracy, David, 35-37,47n,202n,249, 264n-265n
Trefil, James S., 295n
Tryon, E.P., 300,407n
Turing, Alan M., 317-318,330n
Turnbal, H.W., 101n
Tyson, J. Anthony, 245n

Unwin, S.D., 301

Van Steenbergen, Fernand, 76n
Van Till, Howard, 79n
Vatier, 102n
Venniere, Paul, 102n
Vilenkin, A., 407n
Viney, Donald, 123n
von Hügel, Friedring, 134-136, 139,140n
von Neumann, John, 321,349-350
von Rad, Gerhard, 169n,277,294n

Walker, J.M., 224,230
Walker, Keith, 214n,224,230
Walsh, Gerald, 76n
Wartofsky, M.W., 246n
Wassermann, Christoph, P3
Watson, Philip S., 268n
Weinberg, Steven, 23,46n,223, 245n,324
Weiss, Paul, 246n
Westermann, Claus, 169n
Westfall, Richard S., 102n
Wheeler, John A., 201n,245n,326, 331n,341,348,396,400-401
Wheelwright, Philip, 253
Whewell, William, 77n
Whitaker, G.H., 407n
White, Lynn, 165
Whitehead, Alfred North, 25,43, 44,47n,72,78n,83,314,339-340, 350,371n,384
Whitehouse, W.A., 46n
Whitman, Anne P., 100n,246n

Wigner, Eugene, 341
Wildiers, N. Max, 78n
Wilken, George Alexander, 134
William of Occam (Ockham), 193
Williams, Rowan J., 169n,210,215n
Wilson, E.O., 24-25,46n
Wilson, Robert W., 285
Winston, Richard and Clara, 268n
Wisan, Winifred, 77n
Wittgenstein, Ludwig, 23,30
Wolters, Clifton, 271n
Woolf, Harry, 102n
Worcester, Bishop of, 200n
Yahweh, 50-52,117,155-159,161,
163,166,377
Yang, Chen Ning, 324
Yoyotte, Jean, 168n

Zukav, Gary, 340,342n
Życiński, J., 101n,247n,311n

TIPOGRAFIA VATICANA